KT-447-232

Animal Biotechnology and Ethics

Animal Biotechnology and Ethics

Edited by

Alan Holland
Department of Philosophy, Lancaster University, UK

and

Andrew Johnson
Environmental Consultant, Cambridge, UK

CHAPMAN & HALL
London · Weinheim · New York · Tokvo · Melbourne · Madras

Published by Chapman & Hall, 2–6 Boundary Row, London SE1 8HN, UK

Chapman & Hall, 2–6 Boundary Row, London SE1 8HN, UK

Chapman & Hall GmbH, Pappelallee 3, 69469 Weinheim, Germany

Chapman & Hall USA, 115 Fifth Avenue, New York NY 10003, USA

Chapman & Hall Japan, ITP-Japan, Kyowa Building, 3F, 2-2-1 Hirakawacho, Chiyoda-ku, Tokyo 102, Japan

Chapman & Hall Australia, 102 Dodds Street, South Melbourne, Victoria 3205, Australia

Chapman & Hall India, R. Seshadri, 32 Second Main Road, CIT East, Madras 600 035, India

First edition 1998

Typeset in 10/12pt Palatino by Acorn Bookwork, Salisbury, Wilts.
Printed in Great Britain by T.J. International Ltd, Padstow, Cornwall

ISBN 0 412 75680 3

A Catalogue record for this book is available from the British Library

Library of Congress Catalog Card Number: 97–69087

Contents

Preface

Advanced biomedical techniques such as genetic engineering are now used extensively in animal-related research and development. To date, practical applications which have reached the market are relatively few: best known are probably the production of clotting factor IX for the treatment of haemophilia from genetically modified sheep, and the cancer-susceptible 'oncomouse'. As the pace of development has quickened, there has been growing public anxiety about the ethical issues involved. Debate among practitioners has been mainly about patenting and 'ownership' of so-called new life-forms, and about the environmental risks associated with release of genetically modified organisms; but more generally, there is considerable unease about the possible harm to the animals involved, the rights and wrongs of interfering with the 'proper nature' of creatures, and the ethics of using animals as sources of pharmaceutical products or spare parts for surgery.

Our purpose in assembling this anthology has been to draw together within a single volume the leading themes and issues which have emerged in the recent debates around biotechnology as applied to animals. Even in such a fast-moving field many of the issues are perennial; but we would argue in any case that the speed of development makes such a reflective exercise more, rather than less, necessary. Given the increasing demand for codes of practice and legislation which relates to animal biotechnology, we hope that a book which brings together the main facts and arguments on the issues involved will also make a particularly timely contribution to the policy-making process. We have deliberately concentrated on animal-related issues: either those which are unique to animal biotechnology or those which though common to plant, animal and human applications, have distinctive features in the animal case. Thus, we have only touched on important ethical questions which are raised equally by animal and plant biotechnology, such as possible threats to biodiversity, impacts on world trade and on agriculture in developing countries, and the general risk of unanticipated environmental consequences. Nor have we dealt at length with such culturally specific matters as conformity with particular religious traditions; for while we recognize the importance of religion in underpinning certain ethical stances, and also such concepts as natural law and the impropriety of human arrogance (*hubris*), we felt that an analysis of the issues in purely secular terms would be accessible to a greater number of likely readers.

The outline plan of the volume is as follows. Following an editorial overview, scientists working in the field explain some of the key points of the new technology, its potential practical applications in medicine and agriculture, and its limitations. Three of these writers are practitioners and advocates of biotechnology, and the fourth describes in some detail its possible implications for farm animal welfare. The next group of contributions contains social comment, including unashamedly pro- and anti-genetic engineering polemics, more neutral analyses, and a description of 'consensus conferences' as one possible means to mediate the debate. In the third section of the book, a group of philosophers who have specialized in animal-related topics tease out what particularly new ethical and conceptual issues are raised by the new procedures in animal biotechnology, compared with the ways of treating animals on the farm or in the laboratory that have gained more acceptance through tradition. This is followed by five chapters on various aspects of policy and legislation, particularly in Europe but also with reference to the USA. Finally, Michael Banner, who chaired a committee reporting to the UK government on the ethical implications of emerging technologies in the breeding of farm animals, brings together a number of the ethical and policy strands from previous chapters.

We are grateful to Alison Johnson for technical assistance and helpful comment.

A. Holland
A. Johnson

List of contributors

Prof. Robin Attfield
School of English Studies, Communication and Philosophy, University of Wales, Cardiff, UK

Dr Michael Balls
ECVAM, JRC Institute, Ispra, Italy

Prof. Michael Banner
King's College, London, UK

Prof. Donald M. Broom
Department of Clinical Veterinary Medicine, University of Cambridge, Cambridge, UK

Prof. Jeffrey Burkhardt
Ethics and Policy Program, Institute of Food and Agricultural Sciences, University of Florida, Gainesville, Florida, USA

Prof. Stephen R.L. Clark
Department of Philosophy, University of Liverpool, Liverpool, UK

Prof. David E. Cooper
Department of Philosophy, University of Durham, Durham, UK

Joyce D'Silva
Compassion in World Farming, Petersfield, Hampshire, UK

Prof. R.G. Frey
Department of Philosophy, Bowling Green State University, Bowling Green, Ohio, USA

Dr Andrew George
Royal Postgraduate Medical School, Hammersmith Hospital, London, UK

Alan Holland
Department of Philosophy, Lancaster University, Lancaster, UK

Dr Andrew Johnson
The White Horse Press, Knapwell, Cambridge, UK

Dr Daniel D. Jones
Cooperative State Research, Education and Extension Service, US Department of Agriculture, Washington DC, USA

Dr Lars Klüver
The Danish Board of Technology, Copenhagen, Denmark

Michelle Paver
Pharmaceutical & Medical Law Group, Simmons and Simmons, London, UK

Dr Caird E. Rexroad
Livestock & Poultry Science Institute, Gene Evaluation and Mapping Laboratory, Beltsville, Maryland, USA

Prof. Bernard E. Rollin
Colorado State University, Fort Collins, Colorado, USA

Prof. George E. Seidel
College of Veterinary Medicine and Biochemical Sciences, Colorado State University, Fort Collins, Colorado, USA

Peter Stevenson
Compassion in World Farming, Petersfield, Hampshire, UK

Dr Donald W. Straughan
Russell & Burch House, Nottingham, UK

Dr Paul Thompson
Center for Biotechnology Policy and Ethics, Texas A & M University, Texas, USA

Dr Ian Wilmut
Division of Development and Reproduction, Roslin Institute, Roslin, Midlothian, UK

1

Introduction

Alan Holland and Andrew Johnson

1.1 HISTORICAL BACKGROUND

It is a commonplace to observe that humans have practised biotechnology – the application of technology to living things – from the earliest times. This is true even if we restrict the scope of the term to the practice, or craft, of genetic modification. Indeed, if we drop the reference to practice or craft, then it could be claimed that every living thing is engaged in the mutual process of genetic modification, since that is the very stuff of natural selection. It could be claimed, furthermore, that natural selection is better understood as a process of mutual modification than as a process of adaptation, which implies too passive a role for the organism.

As an identifiable human practice, genetic modification dates back to the earliest forms of domestication and cultivation. Darwin, famously, distinguishes two forms of this practice: conscious or methodical selection, and unconscious selection (1868, Chapters 20 and 21). He suggests that in early associations between humans and non-humans it would be natural for humans to select strong and healthy plants and animals, and promote their generation. In this way, over time, the race would be subtly modified without any such effect being intended. This was unconscious selection. It may be remarked that this is to describe the association very much from the human point of view. For it is clear that the animals and plants involved were equally exploiting the facility which humans afforded them, and we can see the process continuing to this very day. A great number of flora and fauna are able to flourish in the wake of, and particularly because of, the human presence: dandelions and docks, rats and robins, weevils and woodlice. Accordingly, the process of unconscious selection can plausibly be characterized as a process of mutual exploitation.

Imperceptibly, unconscious selection gave way to more deliberate forms of selection, and the practices of horticulture and agriculture began in earnest whereby, as Darwin puts it '[m]an may select and preserve each successive variation, with the distinct intention of improving and altering a breed, in accordance with a preconceived idea' (1868, p. 4). The conscious methodical process of genetic modification through selective breeding has been practised for several thousand years, though with increased sophistication during the past 200 years. It essentially involves controlling the hereditary characteristics of a plant or animal population, with the intention of making them more serviceable to human needs, by controlling the process of procreation – specifically, who mates with whom. Some would say that by 'improving' breeds in this way, humans can help nature to achieve a more perfect type – the thoroughbred horse, for example. However, there is a widespread feeling that this process embodies a less than equal form of association between human and non-human, connected no doubt with the fact that the good of the plant or animal is regarded as entirely subservient to the good of humans; but also perhaps with the emergence of a kind of human 'authorship' of, and therefore authority over, other forms of life. From a radical animal rights perspective, the whole process of domestication can be seen as oppressive;[1] but most of the more mainstream criticism that has been levelled at farmers and breeders has concentrated on the cruelty of particular methods used, rather than on the enterprise of selective breeding in general.

1.2 THE ETHICAL CHALLENGE

Since the discovery of DNA, and subsequent advances in biological knowledge and techniques, direct manipulation of the genetic constitution of living organisms has become feasible, with the prospect of using genes from other species to achieve more radical modifications than would be possible through traditional selective breeding. The essential difference from traditional breeding seems to lie not in the aim but in the fact that control of the inherited characters is no longer solely achieved through selection of suitable parents. Offspring are no longer merely the offspring of their parents; and parents, therefore, are no longer merely parents.

For a variety of reasons, 'genetic engineering' – as this genetic modification through direct manipulation of DNA is often termed – has become the focus of considerable public interest and concern. This concern is often justified by the claim that the new technology raises

[1] See, for example, Rodman, 1977.

entirely new ethical issues. Contrarily, supporters of genetic engineering tend to emphasize its continuity with established practices, and to deny the existence of any 'further fact' that would justify new checks or controls on its development.

One of the principal challenges facing the authors in this anthology therefore is to address the question of just how new and different are the ethical issues raised by the new animal biotechnology. If they are essentially continuous with those raised by 'traditional' farming or animal research, it may transpire that the only arguments that can be constructed against genetic engineering would equally condemn many established and widely accepted uses to which animals are put. That is not to deny the possible validity of such arguments, but it would mean that biotechnology issues sit firmly within an established tradition of animal welfare and animal rights debate. If, on the other hand, there is some further fact, or some difference in degree so great it should be regarded as tantamount to a difference in kind, that separates animal biotechnology from previous practice, the question of how society should react to the new technology remains much more open.

A number of the ethical issues raised by genetic engineering do appear to be unfamiliar. Indeed, it is fair to say that society at large is somewhat at a loss as to how to address them, and it is no accident that they were an early subject for a consensus conference, as described in this volume by Lars Klüver. Although the structure of DNA has been known for some time, its properties are still being explored. Since DNA is a component part of whole organisms, and organisms are a typical focus of ethical concern, it seems unlikely that the manipulation of DNA can be an ethically neutral matter. On the other hand, it may be observed that organisms are equally composed of chemical constituents; and it has rarely been suggested that their manipulation is, as such, the occasion for ethical concern. However, DNA is clearly a rather special kind of component, in two ways. First, whereas the presence or absence of chemical elements may frustrate or deflect the developmental process, DNA plays a definitive and architectonic role in the development of any organism. Second, whereas chemical elements are common between organisms and between species, DNA, or DNA sequences, are particular to an organism's identity. As discussed by Alan Holland, the logical relation between an organism and its DNA remains unclear, but a clarification of this relation would seem to be of some importance for the articulation of our ethical responsibilities in the matter of DNA manipulation.

1.3 FOLLOWING NATURE

Regarding genetic manipulation, then, some of the familiar handrails which customarily guide us in our ethical decision-making seem to be missing. In this situation, we are inclined to reach out for connections with what is more familiar. One potential source of guidance is nature. On the one hand, genetic engineering is sometimes criticized on the grounds that it is unnatural. Although there are a number of different ways of construing such a criticism, there is a general difficulty in seeing how it can carry any weight. Either the term 'unnatural' is being used in some descriptive sense, as the equivalent of – say – 'artificial', in which case it is hard to see how it carries any pejorative force, or it is already being used in a normative – and presumably pejorative – sense, in which case we still need some account of the basis for the pejorative judgement. On the other hand, it is sometimes suggested that the processes of genetic engineering mimic natural processes. Versions of this claim are made in the chapters by Andrew George and George Seidel. The argumentative force of such a claim appears to be that insofar as the processes which are in fact brought about through genetic engineering might have occurred naturally, then nothing ethically untoward is afoot. We might want to question the positive force of such a claim, because one and the same event which occurs naturally might well be unethical if perpetrated by a human. A further difficulty with this suggestion surfaces with the issue of patenting, discussed in the chapters by Michelle Paver and Peter Stevenson. It would seem difficult to justify patenting 'inventions' in this field if they were simply the kind of thing that might have occurred naturally.

1.4 FOLLOWING TRADITION

Another potential source of ethical guidance is tradition. It can be claimed that genetic engineering is simply a continuation of traditional forms of breeding. The practice remains the same; only the methods and techniques have changed. At a purely conceptual level, however, this defence of genetic engineering is not as straightforward as it might seem. In general, arguments which appeal to the innocence of each step in a series to establish the innocence of the whole are suspect. For instance, a large number of individually insignificant harms might produce a significant cumulative effect; or the conjunction of the difference in aim between unconscious and methodical selection, and the difference in method between traditional and genetic selection, might make a significant difference beyond that made by either factor alone. Moreover, even if no significant change of circumstance has opened up in relation to one's starting point, being further along a given path can change one's view of the wisdom of setting out on the path in the first place. Hence, concerns about the genetic engineering of animals could awaken one to problems with traditional breeding which had so far escaped notice. Finally, there is no guarantee that a series does not contain a 'threshold effect' – a significant discontinuity such as the solidifying of water at zero degrees, or the point at which a given population of animals ceases to be viable.

More specifically, the degree of continuity between 'traditional' breeding and genetic engineering is a matter of dispute. Those who believe there is a substantial difference might contrast the quest of the traditional breeder for a more 'perfect' horse or apple than nature has yet produced, with the aspirations of biotechnologists towards a more radical redesigning of existing species. However, the harmony with nature that is implied by the 'perfectionist' metaphor has scarcely been apparent in the more extreme specimens produced by dog breeders or pigeon fanciers. Nor have the boundaries between species always been treated as sacrosanct: mules were common in ancient times, and the more recent discovery of how to hybridize plants provided the Victorians with 'an endless source of interest and amusement'.[2] Even with hybrids, though, two parents alone are the cause of both the existence and the nature of their offspring – both of the fact that the offspring exists and what it is like. The genetic engineer appropriates part of the parents' originative role, and the new relationships which come into being in the wake of the new technology deserve closer scrutiny than they have so far received. It may be instructive, for example, to compare the role of the genetic engineer with that of the (natural) human parent. Whereas parents are causes of the nature of their offspring by virtue of their own natures, the engineer is a cause of the existence and nature of the animal he or she engineers by virtue of carrying out some plan or design. This comparison may have ethical implications. For human parents are not only responsible for but responsible to their offspring: we can recognize good and bad parenting, and we acknowledge the rights of offspring to a certain standard of care. If the parenting relation carries its own special duties and responsibilities, and there is some analogy between parenting and the role played by the genetic engineer, this suggests that the scientists who engage in this practice may have new kinds of responsibilities in relation to the creatures which they help to (pro)create.

1.5 THE TECHNOLOGICAL METAPHOR

As well as genetic engineering, we have chosen to consider the ethical implications of using animals as 'spare parts' for human transplant surgery. That these issues seem related perhaps offers us a fresh clue about how they both may break new ethical ground. Like genetic engi-

[2] William Herbert, Dean of Manchester and an enthusiastic horticulturalist, quoted by Fisher, 1982, p. 257.

neering, spare parts surgery appears to regard animals as assemblages of components: the 'engineering' metaphor is equally apt in this context, with its implications that animals are mix-and-match structures made out of a kind of super-Meccano, from which the skilled engineer could fabricate something better. By contrast, the traditional practice of animal breeding is far more constrained by its raw material. The analogy there is less with engineering than with a *craft* such as woodwork or sculpture. In such crafts the craftsperson respects and is in sympathy with his or her material, working only with its grain and within its natural limits. Results are contingent on the particular qualities of the base material. Just as sculptors often speak in terms of 'discovering' the finished work within a block of wood or stone rather than 'creating' it, and the diamond cutter 'reveals' the perfect form within, so the horse breeder only hopes to realize in the foal qualities that are already present in its sire and dam. While engineering and technology cannot escape this kind of contingency altogether, it is 'designed out' as much as possible so that provided the components perform within certain limits, results should be precisely defined and replicable. It can be argued that use of the technological metaphor in animal husbandry implies the abandonment of any respect for individuality in favour of a doctrine according to which animals and their component parts are interchangeable cogs or ciphers. The tendency to regard whole animals in this light is already evident in the practice and vocabulary of 'factory farming', but genetic engineering extends this in a way which makes it particularly difficult to maintain a clear notion of the integrity of the individual. Perhaps in recognition of this, the term 'genetic engineering' is now eschewed by many of its practitioners, who are anxious to emphasize its continuity with earlier practice.

1.6 ENDS, MEANS AND MOTIVES

We have indicated some of the issues that seem relevant to general ethical discussions of the pros and cons of the new animal biotechnology. Many of these issues are discussed in more detail in the section on 'Ethical and Conceptual Issues'. Bernard Rollin suggests that the concept of 'human rights', which offers the individual protection against the basically utilitarian government machine, is founded on an Aristotelian idea of the proper end or *telos* of human life; and that the concept of respect for natural *telos* should be extended to other animals. Robin Attfield, however, argues that animal biotechnology is open to criticism on purely consequentialist grounds, without invoking any special constraints such as Rollin proposes. Another consequentialist, Raymond Frey, examines the issue of animals as sources for spare parts surgery, and reaches a surprising and unsettling conclusion.

Consequentialist ethics are often criticized as implying that results are all that matter, regardless of the means by which they are achieved, and some philosophers prefer to consider ethical problems in terms of principles for action. Michael Banner explores this alternative dimension of ethical criticism. Also in this section, David Cooper analyses the emotional dimension of ethical judgements, stressing motives rather than consequences; and Stephen Clark examines the debate in a narrative context, using science fiction examples to illustrate the importance of myths to our understanding of ethical issues.

1.7 RISK

Whether or not it is strictly part of ethics, the issue of risk frequently arises in discussions of animal biotechnology. One of the leading accusations levelled at genetic engineering is that it involves unacceptable levels of risk, or a different order of risk from anything heretofore. Its defenders may point to the irony of this accusation, arguing that one of the main aims of the practical handmaidens of science, such as medicine and agriculture, is precisely to protect the human project and make it adaptable to future contingencies. If technology is supremely devoted to decreasing the risks to human life, it is paradoxical that it should be charged with increasing that risk. However, even if some such general defence of technology were allowed, it would not necessarily assuage sources of concern that are specific to biotechnology. One arises in connection with assurances that are typically offered in response to concerns about risk, to the effect that the technology can be monitored and regulated to reduce potential risks, and that anyway it is futile to imagine that we can live our lives without risk: life is inherently risky. What such a response overlooks is that genetic technology, like nuclear technology, is not 'human-sized': it operates at the microcosmic rather than the mesocosmic level – the realm of medium-sized objects. First, there are concerns over the properties of the specific microorganisms – viral vectors – which are essentially employed in the technology (as described by Ian Wilmut and Andrew George). They are of a kind capable of crossing species, causing disease and transferring antibiotic resistance. But a more general concern with microorganisms must be that 'safety' procedures are exceedingly difficult to operate, and even to envisage. There is a strong suspicion that the language of regulation is being employed in an area beyond its possible field of application (the adequacy of the regulatory apparatus in the context of human food safety is discussed by Daniel Jones). Thirdly, there is a danger that the rush to establish 'biosafety guidelines' is fuelled not out of a genuine concern with safety but out of a concern that the new technology should gain acceptance, thereby hastening and facilitating its commercial exploitation. What the response also overlooks is the social dimension of terms such as 'risk', as discussed by Andrew Johnson. For example, it matters a great deal to our perceptions of risk by what agency that risk is imposed. People are allergic to risks imposed by others which they would be willing to undergo if they were of their own choosing, and it is such imposed risks which present ethical problems. Who has the right to impose risks on others, and under what circumstances?

A related source of concern further opens up the social dimension of the new technology. Like nuclear technology, biotechnology is extremely sophisticated and has long-term, possibly irreversible effects. Above all, it presupposes an extremely sophisticated social, economic and political infrastructure to support it. For these reasons we see why society as a whole, including societies of the future, have a profound and legitimate stake in its operation. As Paul Thompson urges, society's stake in the new technology is nowhere near exhausted by a calculation of its costs and benefits, for it brings in its wake enormous liabilities, commitments and requirements. We also see why society has every reason to be wary, for even if the science is secure, we cannot assume the same of social, economic and political institutions which are essential to its effective practice. Prudence might suggest that society should invest in technologies which are robust in the face of social change, and it is far from clear that biotechnology answers to this description.

Even if all concerns over the risk attaching to the science and the uncertainties attaching to its social moorings should be satisfactorily resolved, there remains the point that risk as so far construed is primarily a human concern. And while it might be hoped that collective prudence will save humans from such dangers as we can foresee, this will not help the animal subjects of the biotechnology revolution. How we should weigh its effects on them is discussed by Donald Broom, and the increasingly international framework for regulating such effects forms the subject of the chapter by Michael Balls and Donald Straughan.

1.8 WHAT ARE THE LIMITS?

Finally, let us suppose that all the 'continuity' arguments which are advanced to allay public concerns over genetic engineering are successful. It is all very well to be assured that what we are about to do is continuous with what we have just been doing, and that it is all a seamless progression. But such assurances simply invite further questions like: Where will it all end? How much is enough? What story is unfolding? As Michael Banner reminds us, utilitarian considerations can always be used to justify squeezing a little less suffering here, a little more happiness there. So where will the genetic engineering of animals stop? What limits or bounds are envisaged; what sense of proportion is evident? Without at all supposing that there has to be some ultimate destination or final shape to things, it is legitimate to ask what provisional destination or provisional shape of things is being assumed for a society in which the genetic engineering of animals is an accepted practice governed only by the rule to minimize suffering. We have encountered no sensible answer to this question thus far and strongly suspect that none is in the offing. Perhaps, as Jeff Burkhardt suggests, one needs to get into a certain 'way of seeing/way of being'. We also suspect that the endless pursuit of security of resources (mainly human), the alleviation of suffering (animal and human) and the deferring of death (exclusively human), provided only that the utilitarian account book is in surplus, embodies an immature perception of the ends of life, a perception of life without measure, shape or significance. What is amiss may not be the anthropocentric perspective as such, but the pursuit of such an impoverished conception of it. It is difficult to lead our lives without some sense of a narrative shape, both in our individual lives and in the society which provides the context in which they are lived. Moreover, what cannot be denied (except in a desperate bid to save some hypothesis) is the fact of animal subjectivity. And if human subjectivity is so precious, then animal subjectivity cannot be supposed nugatory. As Joyce D'Silva and other pro-animal campaigners urge, surely the role envisaged for the animal community in the human narrative has to be more than that of human resource, or at best – as suggested in a recent pamphlet (Scruton, 1996) – an arena for the exercise of human virtues (and vices). Despite the prevailing utilitarian temper, and borrowing conveniently from contract theory, it is fashionable now to decry the notion of animal rights. Yet the practice of animal biotechnology implies the presumption of a human right – the right to disregard animal subjectivity. We have yet to discern the basis upon which this right is claimed.

REFERENCES

Darwin, C. (1868) *The Variation of Animals and Plants under Domestication*, John Murray, London.

Fisher, J. (1982) *The Origins of Garden Plants*, Constable, London.

Rodman, J. (1977) The liberation of nature? *Inquiry*, **20**, 83–145.

Scruton, R. (1996) *Animal Rights and Wrongs*, Demos, London.

Part One

Scientific procedures and their applications

2

Methods for genetic modification in farm animals and humans: present procedures and future opportunities

Ian Wilmut

2.1 INTRODUCTION

One of the many unregarded miracles of life is the ability of the single-celled, fertilized egg to develop first into a fetus, then a new-born child and finally an adult animal or human being. This involves the formation of a great variety of different tissues at specific stages of development. This miracle occurs because every cell has within its nucleus an instruction manual. Whereas every cell in the body of an animal carries the same blueprint, there are small differences between the blueprints of each animal and these contribute to the fact that each of us is a unique individual. (For a general discussion of the molecular biology of the cell see Alberts *et al.*, 1989.)

The instruction manual, or genome, is written in a language whose alphabet has just four letters. These letters are the four nucleotides Adenylate, Cytidylate, Guanylate and Thymidylate. It is the sequence of these nucleotides in the DNA making up the chromosomes which conveys the information contained in the blueprint. The instruction manual of a mammal contains approximately three billion letters. If each of these were written down on paper so that each page carried 3000 characters, then the manual would occupy 1000 volumes, each of 1000 pages.

The blueprint organizes the development of an adult animal from a fertilized egg by directing the production of combinations of protein in specific tissues at appropriate stages of development. The synthesis of each of these proteins is controlled by a particular 'gene'. A gene, according to one definition, is the unit of information that regulates production of a protein and changes in the protein populations within a cell are the result of differences in gene expression.

There are two aspects to the functioning of each gene. Sequences within the gene determine the structure of the protein to be produced: the coding sequences. In just the same way that the letters within a word convey the meaning of that word, so a sequence of three nucleotides provides the instruction that a particular amino acid is to be incorporated into a protein. Associated with the coding sequences, there are regions of DNA that determine at what stages of development and in which tissues that protein will be produced. Very often, these regulatory elements are situated at the head of the coding sequences of DNA. However, there are cases in which important regulatory elements are to be found within the coding sequences or substantial distances away, either before or after the coding sequence. The way in which the gene is expressed is determined by interactions between the regulatory sequences within the gene and protein factors, known as transcription factors, which bind to the regulatory sequences.

At the present time only a small proportion of the genes within any animal species have been identified and characterized. One particular group of genes, 'housekeeping genes', are active in all cells. By contrast, other genes are only expressed in some tissues and at certain times of development. Genes that direct the production of milk proteins are only active in the mammary gland, and only in late pregnancy and lactation. Similarly, there are genes unique to every other tissue type, such as those of the liver, kidney or brain. In addition to the well-known Human Genome Project, there are others whose aim is to identify the genes of a livestock species. It is to be expected that during the first half of the 21st century most of the genes of livestock species will have been identified. Techniques of genetic modification will make a unique contribution to understanding their role. By being able to prevent the production of a particular protein its role in development and animal physiology will be understood. Similarly, it is only by being able to modify and dissect the regulatory sequences that it will become possible to understand the mechanisms that regulate the functioning of that gene. In turn, this research will reveal new opportunities for the application of techniques for genetic modification in human and veterinary medicine and animal production.

The techniques of genetic modification all depend upon a number of basic molecular biology techniques. It must be possible to identify the gene which is responsible for a particular characteristic. Whereas this is very simple if the protein produced by that gene has been identified, it may be a complex process if it has only been established that the region of a chromosome is associated with a particular phenotype. However, the completion of the mapping projects will facilitate the identification of new genes associated with particular characteristics. Given that a gene of interest has been identified, then it is a simple process to isolate that gene in the laboratory, and to use bacteria in culture to make many copies. Similarly, there are well-rehearsed techniques available for the introduction of precise genetic changes into genes. It is possible to make changes to single letters in the instruction manual, to modify a paragraph or even to change whole pages of the instruction manual. Before the modified gene is used, it is also possible, by simple automated sequencing techniques, to confirm that exactly the change which was desired has been introduced into the copies of the gene.

The objective of this chapter is to review methods of genetic modification of farm animals and humans that are available now or may be established in the near future. Broadly, the genetic modifications are of two types, as they may be made either to the somatic tissues or to all tissues, including the gametes. Changes which are introduced to the gametes, germline transformation, are passed from one generation to another. By contrast, changes specifically introduced into the somatic cells of an animal are not passed on to the offspring of that animal. There are many reasons, ethical, economical and practical, why germline transformations have been restricted to laboratory and farm animals while somatic modifications are more likely to be made to human patients.

The nature of a genetic modification may be either to add an additional gene or to modify a gene which is already present. The addition of a gene, frequently referred to as 'gene transfer', is by far the easier of the two modifications. However, precise modification of an endogenous gene is expected to produce a more consistent effect. The purpose of this chapter is to establish three things: (i) what can be achieved at present; (ii) limitations of the current technique; and (iii) the nature of possible future developments. In assessing the effectiveness of the techniques it is important to remember that the ultimate objective always is to be able to make a precise change without the risk of causing any change to other genes. Later chapters will be concerned with the practical value of these techniques and with their ethical implications. As most of the techniques were developed in the past 10 years and new techniques are being introduced very rapidly, the emphasis of the description will be to give a general impression of what will become possible rather than to be concerned with precise technical details. At appropriate points in the text, the reader is referred to other reviews for information of this nature.

2.2 THE ADDITION OF A GENE IN LIVESTOCK

Addition of a gene into the germline of a mammal has been achieved by several different techniques; however, by far the most frequently employed technique involves the direct injection of several hundred copies of the specific gene into a nucleus in an early embryo. Such 'direct injection' has been used successfully in several species, but by far the greatest number of transfers have been performed in the laboratory mouse (Palmiter and Brinster, 1986).

Direct injection depends upon recovering an embryo a few hours after fertilization. Ideally, it should be recovered at the time when two pronuclei are present. One pronucleus contains the chromosomes derived from the mother, while the other has the chromosomes which were delivered by the sperm. In animals such as the mouse, suitable embryos are recovered from donor females. In the case of farm animals, embryos may either be obtained from such a donor female or by *in vitro* maturation and fertilization of oocytes which are recovered from animals immediately after their slaughter for human consumption.

When embryos are being recovered from donor females, the number of ovulations is increased by administration of hormone preparations with follicle-stimulating properties, and the time of ovulation is regulated. The donor is either mated or inseminated artificially before embryos are recovered a few hours after fertilization. At that time, large pronuclei, late in the first cell cycle, are observed in the centre of the newly fertilized egg.

At the time of recovery, fertilized eggs are located in the oviduct. They are recovered by flushing sterile fluid through the oviduct. General anaesthesia is induced before the abdomen is shaved, washed and sterilized. A cut is made through the abdominal wall to allow the identification of the reproductive tract and its exteriorization. A small tube is inserted into the upper opening of the oviduct to allow passage of fluid. The reproductive tract is returned to the abdominal cavity before closure by suturing the walls of the abdomen with absorbable thread. Recovery from this surgery is very rapid and usually completely uneventful.

The development and continuing improvement of methods for *in vitro* fertilization would provide alternative means of obtaining fertilized eggs at appropriate stages of development. These techniques have so far only been used to produce cattle eggs for direct injection. The ovaries of cattle being slaughtered for commercial reasons each contain a small number of eggs which have almost completed the process of growth and maturation. If these eggs are released from the ovarian tissue and allowed to develop in an appropriate medium for 24 hours, many of them are able to complete this stage in their development

(Gordon, 1994). Under appropriate circumstances, up to 90% of the eggs can be fertilized and begin to develop. If these eggs are cultured overnight, then many of them will have enlarging pronuclei which are suitable for direct injection. At the present time, the disadvantage of using fertilized eggs produced in this way is that a smaller proportion of them are able to develop to become live offspring. While they obviate the need for donor females they unfortunately require a larger number of recipient females than would be required if the injected eggs had been obtained from donors. As research continues it is probable that the efficiencies of the culture techniques will be improved and there will be an increasing use made of embryos produced in culture rather than by recovery from donor animals. However, there will be specific cases where recovery from donor females will still be required. In the case of projects in which the health of the animals is paramount, it would be inappropriate to use material obtained from commercial slaughterhouse animals and it would continue to be necessary to recover eggs from animals of known health status maintained under the strict quarantine conditions. Similarly, there may be occasions when it is essential to know the specific genotype of the resulting fertilized egg and this will only be possible if fertilized eggs are recovered directly from donors, because of the difficulty of tracing ovaries from individual animals.

There are differences between the species in the ease with which it is possible to see the nuclei in an early embryo. Whereas an embryo from a laboratory mouse is translucent and it is possible to see the nuclei very easily, embryos recovered from a pig or cow are very dark, as many granules obscure the view. Dark embryos are centrifuged for a brief period in order to stratify the cytoplasm and reveal the nuclei. Mammalian embryos are all approximately 0.1 mm in diameter. Each embryo in turn is secured by applying suction to a very fine 'holding pipette'. When the embryo has been held in a suitable orientation, a second exceptionally fine pipette is introduced into one of the pronuclei. A small volume of the DNA solution is introduced into the nucleus until it can be seen to swell. After completion of this microinjection the surviving embryos are transferred into recipient foster mothers and are allowed to develop to term.

Only a proportion of the resulting offspring carry the additional genes and it is necessary to identify those 'transgenic' animals. To do this, a sample of DNA is obtained from all of the offspring (Alberts *et al.*, 1989). DNA is extracted from the cells and an analysis (Southern blot) is carried out to discover whether or not the additional gene is present. A number of enzymes have been identified that each cut DNA at specific sequences of nucleotides (letters). If DNA is incubated with one of these enzymes, then it is cut into pieces of a mixture of different sizes, characteristic of that DNA. The pieces of DNA can then be arranged according to their size by a process of electrophoresis. A very thin gel is prepared through which an electric current can be passed. If the DNA pieces are added to that gel at one end and an electric current applied, then the smaller pieces of DNA move furthest through the gel. They can be revealed as a 'ladder' of pieces of DNA of characteristic size.

One of the useful characteristics of DNA sequences is that similar pieces of DNA bind to one another. In the cell, DNA is in a double, twisted strand in which complementary nucleotides bind to one another: adenylate to thymidylate and cytidylate to guanylate. In the laboratory, use is made of this characteristic by adding to the gel a fragment of DNA which is complementary to a part of the injected DNA. The fragment, or probe, is labelled in some way so that its presence can be detected within the gel. Usually a radioactive compound is included within the probe so that when it is added to the gel, and allowed to interact with the samples for a period of time, it has the opportunity to bind to the sample. If injected DNA has been incorporated into the offspring it is detected by exposing a photographic film to the gel. This analysis also provides some indication of the number of copies of the injected gene which have been incorporated into each cell.

The mechanism by which the injected DNA is incorporated into a chromosome is not known. However, it has been suggested that it depends upon the injection of fluid into the nucleus causing breakage in the chromosome (Palmiter and Brinster, 1986). This activates repair mechanisms which inadvertently include copies of the injected gene into a chromosome. Although it has not been established that this is the mechanism of integration, it would certainly account for many of the limitations.

First, the process is inefficient and as a result expensive. The exact proportion of the resulting offspring which carry an additional gene varies from one species to another and indeed from one experiment to another, but it is always very low (1–5%). Second, it seems that the site of integration of the additional gene is very variable. As a result, the integration of the additional gene sometimes causes damage to an existing gene within the chromosome. Detailed studies have only been possible in mice, but in this species it has been shown that a lethal mutation to an existing gene is caused in some 5–10% of lines of transgenic mice which have been made by direct injection. Usually there is no effect in the first generation of animals, because there is a fully functional copy of the gene on the other chromosome of the pair. Lethal mutations are only apparent if systematic breeding is carried out to ensure that some offspring would have copies of the additional gene on both of the pair of chromosomes. In the case of a lethal mutation, animals with two copies of the gene are never found, because of their death during fetal development. It is to be expected that similar frequencies of damage to existing genes occur in all other species. Clearly this test will underestimate the frequency with which genes are damaged because it is only able to detect damage to genes that are essential for normal development.

Third, and also arising from the random site of integration, is the fact that the pattern of expression of the additional gene has been found to be very unpredictable. There are now many studies in which the same gene has been transferred into inbred populations of mice. These are circumstances in which there is the least possible biological variation. Despite this it has been observed on many occasions that the level of expression of the transferred gene is extremely variable. It is believed that the areas of DNA adjacent to the additional gene influence the way in which transcription factors have access to the additional gene and in that way have a great influence on its ability to function.

Finally, this method of genetic modification is only able to add a gene and that gene must function in the presence of all of the existing genes of that animal. These limitations will restrict the potential uses of this approach to genetic modification. The technique is being used for specific pharmaceutical and biotechnological uses; however, it seems unlikely that it will ever be used to modify agricultural aspects of agricultural production. Direct transfer has been used in sheep, goats and cattle to direct the production of proteins needed for treatment of human disease into the milk of these species.

2.3 CHANGES TO EXISTING GENES

The only mammalian species in which it has been possible to introduce precise genetic changes into the germline is the laboratory mouse, by the use of 'embryonic stem cells'. However, exciting new developments suggest that techniques will soon become available for the introduction of precise genetic changes into the germline of many mammalian species by the use of novel procedures for nuclear transfer. The isolation and use of embryonic stem cells will be discussed, before describing the technique of embryo production by nuclear transfer.

In the mouse, methods have been established for the culture of cells derived from the embryo in such a way that they continue to divide – but do not differentiate – embryonic stem cells (Hooper, 1992). Mouse embryos are cultured in the presence of feeder cells in serum-enriched medium. Over a period of a few days the embryos attach to the feeder cells, and the inner cells of the embryo, from which the conceptus would have developed, can be identified as a small clump. If this clump of cells is removed from the culture, separated into groups of a smaller number of cells and then transferred to fresh culture dishes with feeder layers, it is possible in some cases to establish stable cell lines.

If such embryonic stem cells are aggregated with the cells of another embryo they sometimes retain the ability to colonize all of the tissues of the developing conceptus, including the germline. This has usually been achieved by injecting a number of stem cells into the cavity of an embryo at the blastocyst stage. However, there are other strategies in which the cells are aggregated with embryos at earlier stages of development. For convenience, the recipient embryos are usually from a strain of mice with a different coat colour, so that it may be discovered immediately at birth whether or not the resulting offspring have developed from the two different types of cell. While the embryonic stem cells are in tissue culture, it is possible to use 'gene targeting' techniques to introduce precise changes into the genome (see below). In this way it is possible to introduce precise genetic modifications into laboratory mice.

Despite a considerable research effort, embryonic stem cells have been obtained only for a small number of strains of mice and none from any other species. A variety of approaches has been used in attempts to overcome this limitation. Cells which closely resemble embryo-derived stem cells were obtained by culture of primordial germ cells of the mouse. In some cases, these cell lines also have the ability to contribute to all of the tissues of a chimeric offspring. In this way they closely resemble embryonic stem cell lines. Despite the culture of a considerable number of primordial germ cells of livestock species, there are at present no reports that it has been possible to establish stem cell populations in these species.

Recent results suggest that nuclear transfer from established cell lines of a variety of different types may provide an alternative means for the introduction of precise genetic change into mammalian species (Campbell *et al.*, 1996). Two different cells are required for the production of embryos by nuclear transfer. The recipient cell is an unfertilized egg from which the genetic material is removed by micromanipulation, before the genetic material from a donor cell is introduced into the enucleated, unfertilized egg by cell fusion. It is the genetic material of the donor cell that determines almost all of the characteristics of the resulting offspring. It is anticipated that gene targeting procedures will be established for a variety of different cell types, and so allow the introduction of precise genetic change into the genome before the production of new offspring by nuclear transfer.

There are two potential sources of recipient, unfertilized eggs, as was described previously for the technique of direct injection. At the time when the chromosomes are removed from the unfertilized egg they are condensed in a metaphase and so are not visible by ordinary microscopy. In order to be able to confirm that enuclation has been completed, the unfertilized egg is incubated first in the presence of a fluorescent stain which binds specifically to DNA. Local differences in the appearance of the cytoplasm and cell membrane in the region of the chromosomes are used to guide the removal of approximately 20% of the cytoplasm. The potential recipient cell is then removed from the field of view before an ultra-violet (UV) light is switched on and the cytoplasm which has been withdrawn is examined to determine whether or not the chromosomes have been removed.

Nuclear transfer is achieved by fusion of the donor cell to the cytoplasm of the enucleated unfertilized egg. In the mouse, cell fusion has most commonly been induced by exposing the two cells to inactivated Sendai virus. By contrast, in other species fusion is more commonly brought about by exposing the cells to a direct electric current. The donor cell is placed next to the cytoplasm so that the cell membranes are perpendicular to the field of the electric current at the point where the two cells touch. The passage of electric current makes holes in the cell membrane and when these are repaired this is done in such a way that one cell is formed.

In normal circumstances, the arrival of a sperm at fertilization induces the egg to begin to develop (activation) and during nuclear transfer it is necessary to apply an artificial stimulus to the unfertilized egg to achieve the same end. If an electric current is being used to bring about cell fusion, then activation is usually achieved at the same time. However, if other means are being used for cell fusion an activation stimulus must be applied. This stimulus has been provided by electric current or exposure to a variety of agents, including ethanol, strontium or ionophore. Very few of the treatments used to induce activation are able to trigger a completely normal response and it may be that one reason for the failure of development of some of the embryos produced by nuclear transfer is the non-physiological pattern of activation.

The embryos produced by nuclear transfer must be transferred to foster mothers and allowed to develop to term. In livestock species the new embryo develops to the blastocyst stage before transfer to the recipient female. In some cases this initial development has been achieved in a tissue culture system, but in others it occurred in the oviduct of temporary recipient sheep. New research shows that by causing the donor cell to become quiescent before transfer it is possible to obtain development to term following nuclear transfer from cells that have been in culture for prolonged periods. It is to be expected that during the next few years routines will be established for the introduction of precise genetic changes into the genome of such cells before they are used as nuclear donors. This will offer the advantage, for the first time, of being able to introduce specific changes into species other than mice. It may also reduce the number of animals required for the production of each modified line of animals because it will be possible to select only cells that have the desired change.

2.4 GENE TRANSFER IN HUMAN SOMATIC CELLS

Molecular and cellular biology is providing new tools for the treatment not only of genetic diseases, but also for other diseases, as many involve changes in gene expression and it may be possible to use genetic modifications to prevent or cure such diseases. Protocols are being developed for treatment of conditions as varied as cancer and rheumatoid arthritis (Anderson, 1995).

Having identified the target gene the task is either to repair the gene or to introduce an additional copy of the gene into cells that lack the effect of the gene. In either case the gene must be able to function in an appropriate manner and respond to regulatory stimuli. Ideally, the objective is to make a precise change without causing damage to any other gene. Specificity may be achieved by removing cells from the patient and making the change before returning the cells (*ex vivo*). In principle, the opportunity to modify and study cells in the laboratory would appear to offer great advantages, particularly if methods for gene targeting can be established for cell types other than embryonic stem cells. While this approach has great potential advantages it is limited to those tissues that may be isolated and maintained in culture, such as haematopoietic progenitor cells (Dunbar *et al.*, 1994). Alternatively, the change may be introduced in the patient (*in vivo*) as is the case with cystic fibrosis (Boucher, 1996). In these cases a degree of specificity may be provided by localizing treatment, in the case of some treatments for cystic fibrosis to the upper regions of the respiratory tract. However, in other cases the treatment is delivered through the blood supply.

In both cases, the challenge is to be able to make the genetic modification in a sufficiently large proportion of target cells (Hodgson, 1995). There may be three stages: entry into the cell, and then the nucleus, before being included in a chromosome. In achieving these steps, use has been made of the ability of viral agents to enter cells and in some cases to insert their genome into that of the infected cell. Two particular groups of viruses have been the subject of study: adenoviruses and retroviruses (see Alberts *et al.*, 1989). Adenoviruses are the infectious agents responsible for respiratory infections such as the common cold and flu. There are several different types of adenovirus, but most children have been exposed to the more common forms. Adenoviruses enter the cells through receptors, but do not normally become integrated in the patient's genome. By contrast, the life cycle of retroviruses ensures their integration in the genome of the patient. Retroviruses are made of RNA. After they enter the cell, an unusual enzyme directs the production of DNA encoded by the viral RNA – a reversal of the normal pattern of transcription. Enzymes from the virus then direct the insertion of the viral (DNA) genome into any point in a host chromosome. There are advantages and disadvantages of both viral carriers. As the adenovirus is not integrated into the genome, repeated treatment is required, whereas this would not be necessary with integrated retrovirus. However, the act of integration may damage an existing gene or activate one of the cellular genes causing cancer.

Before their use in gene therapy, viral carriers are first modified to prevent their multiplying themselves and causing infection. It is by removing the DNA or RNA sequences encoding the reproductive machinery that space is provided within the restricted genome of a virus to place the gene sequences that are to be carried into the patient. As the modified virus is not able to complete its cell cycle and form infective particles, they must be grown in the presence of helper cells able to provide the missing components. There is an important requirement in all viral protocols, that intact virus must never be present with engineered virus.

Although viral vectors have the great advantage of being able to enter cells efficiently they suffer from some disadvantages. They are only able to carry limited lengths of DNA and in turn this means that there is more difficulty in obtaining efficient expression of the transferred gene. As noted already there is a possibility that the random mechanism of integration into the genome of the patient causes damaging mutations. There is at least a notional risk that a recombination between the viral agent and viral sequences within the cells will lead to the formation of infective viral particles and there is the possibility of an inflammatory response to the treatment. Other, non-biological, means of introducing DNA have been sought with the aim of avoiding these limitations.

Liposomes are made up of fluid mixtures of lipid into which proteins, RNA or DNA can be introduced (Farhood *et al.*, 1996). For the purpose of gene transfer, DNA, which has a negative charge, is complexed with positive lipids to create a slight positive charge. The liposome then has a tendency to bind to and enter (negatively charged) cells before the DNA is released within the cell. This approach avoids the biological risks associated with viral carriers and has not been seen to have adverse immunological responses. However, it provides no measure of specificity.

2.5 *IN VIVO* TRANSFER

Recurrent bacterial infection and subsequent loss of lung function are perhaps the most obvious symptoms of cystic fibrosis (Boucher, 1996). The gene regulates movement of salt and water across the cell membrane of the airways; however, there is still an inadequate understanding of the precise tissues in which the gene acts. Whereas treatment of cells in the upper regions of the respiratory tract might best be with aerosols, a systemic method would be more appropriate for lower regions.

Studies in tissue culture have shown that adenoviral carriers are able to deliver the gene into cells from the airways and that the gene restores salt movement. Initial trials in patients, in which the virus was applied directly to cells within the nose, also showed that there was an improvement in the salt balance. Unfortunately, these encouraging results were not confirmed in later studies. While there are many attractions to the use of adenovirus in aerosols dispersed within the respiratory tract, there are significant limitations. The cells lining the tract are replaced and it would be necessary to repeat the treatment several times each year. Adenovirus stimulates an immune response and in time patients would eliminate the virus very rapidly. Liposomes may offer a means of introducing DNA into cells, without the disadvantage of stimulating an immune response.

2.6 *EX VIVO* TRANSFER

The molecular challenges in modifying cells in culture are similar to those in making similar changes in embryos or embryonic stem cells. The greater limitation is in being able to isolate and maintain in culture a population of cells that retain the ability to colonize those tissue(s) which lack the effects of the deficient gene. The progenitor cells of the marrow are ideal candidates for gene therapy as they are easily obtained, they can be maintained in culture and a small number are able to repopulate the patient (Dunbar *et al.*, 1994). The population has been studied as a potential means to treat cancers of this system, haematological disorders, and immune and metabolic diseases.

A routine procedure has been established in mice such that the genetic modification is present in several different types of cell in almost all recipients. A few days before harvesting of marrow cells, resting progenitor cells are stimulated to resume cell growth and division by using a drug to deplete the population of mature cells. Under the influence of protein factors which stimulate division, the harvested progenitors are cultured for several days in the presence of retrovirus, sometimes in co-culture with cells producing the viral particles. Although a mixture of cells is treated, the genetic modification is present in cells derived from progenitors for long periods after recolonization of mice.

Although this test system is now a routine procedure in mice, development of analogous systems in other species, including human, has proved difficult. Rather than depend upon recovery of bone marrow cells, protocols are being developed for the enrichment of progenitor cells from peripheral blood circulation.

2.7 FUTURE OPPORTUNITIES

Attempts to modify animals genetically began less than 20 years ago and the first treatment of human patients was very recent. The present techniques should be seen as equivalent to the penny-farthing in the development of the bicycle. At some time in the future, means will be available for far more precise genetic modification than is possible at present, although there may be some objectives that prove to be unattainable. While it is certain that great advances will be made, they may not be along any of the present approaches. Similarly, it would be unwise to estimate the time required. This is a very exciting area of science offering great opportunities in medicine, research and agricultural production.

ACKNOWLEDGEMENTS

Many of the ideas discussed in this chapter were developed during research sponsored by Ministry of Agriculture, Fisheries and Food, BBSRC Core Strategic Grant and EU Biotechnology project BIO2CT-920358.

REFERENCES

Alberts, B., Bray, D., Lewis, J., Raff, M., Roberts, K. and Watson, J.D. (1989) *Molecular Biology of The Cell*, Garland Publishing Inc., New York.

Anderson, W.F. (1995) Gene Therapy. *Scientific American*, September 1995, 96–8B.

Boucher, R.C. (1996) Current status of CF gene therapy. *Trends in Genetics*, **12**, 81–4.

Campbell, K.H.S., McWhir, J., Ritchie, W.A. and Wilmut, I. (1996) Sheep cloned by nuclear transfer from a cultured cell line. *Nature*, **380**, 64–6.

Dunbar, K., Bodine, D.M., Sorrentino, B., Donahue, R., McDonagh, K., Cottler-Fox, M., O'Shaughnessy, J., Cowan, K., Carter, C., Doren, S., Cassell, A. and Nienhuis, A.W. (1994) Gene transfer into hematopoietic cells. *Annals of the New York Academy of Sciences*, **716**, 216–24.

Farhood, H., Gao, X., Son, K., Yang, Y-Y., Lazo, S. and Huang, L. (1996) Cationic liposomes for direct gene transfer in therapy of cancer and other diseases. *Annals of the New York Academy of Sciences*, **716**, 23–35.

Gordon, I. (1994) Laboratory production of cattle embryos. *Biotechnology in Agriculture, No. 11*. CAB International, Wallingford, UK.
Hodgson, C.P. (1995) The vector void in gene therapy. *Bio/Technology*, **13**, 222–5.
Hooper, M.L. (1992) *Embryonal Stem Cells: Introducing Planned Changes into the Germline* (ed. H.J. Evans), Harwood Academic Publishers, Switzerland.
Palmiter, R.D. and Brinster, R.L. (1986) Germline transformation of mice. *Annual Review of Genetics*, **20**, 465–99.

3

Animal biotechnology in medicine

Andrew J. T. George

3.1 INTRODUCTION

One of the cardinal features of *Homo sapiens* is our ability to alter other life forms in a deliberate manner. While other animals can have a major impact on other plants and animals whose environment they share, and over the centuries act as a shaping, selective, force in their evolution, only humans have had the ability to decide how they would like a certain animal or plant to be, and then deliberately to plan and execute appropriate strategies to attain their goal. The historic reasons for wanting to alter other life forms are diverse, ranging from the practical one of improving crops and livestock through to the recreational one of developing faster horses or certain features in pet dogs. However, when one looks at a Pekinese dog or a modern farm cow it is clear that the deliberate influence of humans on many species has been massive (Wheeler and Campion, 1993).

In order permanently to alter the characteristics, or phenotype, of an animal or plant it is necessary to alter its genetic makeup, or genotype. If the genetic information encoded by the DNA is not altered then the new features will not be inherited by the next generation; the elaborate hair-cuts of poodles are not passed on to their puppies! Thus, over the millenia, humans have acted as genetic engineers in altering the genotypes of the different species. Until recently, however, we have been unaware of the scientific principles underlying what we have been doing; nonetheless this has not prevented us from using them to great effect!

The genetic engineering practised by the human race in the past has been indirect. What the farmers and racehorse owners have done is to pick certain characteristics in the species that they consider desirable, and then deliberately to breed from animals or plants with those characteristics. Offspring that are not as wanted are killed, and offspring whose phenotype is 'improved' are bred from. Thus, over the years the genotype of the species is altered so as to achieve the desired phenotype. In many ways this process is similar to that of natural evolution; however, the selective force has changed from the 'survival of the fittest' to the 'survival of those that please humans'.

There are disadvantages to this approach to genetic engineering. First, it is slow, being limited by the reproductive rate of the species. Second, it is limited to the genetic material already present in the species (or that which might arise by natural mutation during the course of the breeding process). This sets practical limits on the possible changes. Third, while the genes encoding for a particular aspect of the phenotype may be selected, other genes encoding for other, unwanted, characteristics may be inherited along with the desired genes.

Recently the discovery of the genetic component of inheritance, and the role of DNA in encoding for the genes that make up our genotype, as well as the development of methods for manipulating that DNA, have given us a new method for altering the genotypes of animals. We are now able to practise direct engineering, in which we target alterations in the genetic material of the living organism, in order to effect a specific change in the phenotype. The ability to target the expression of genes has enormous potential, as it overcomes the three problems noted above with conventional breeding; it is relatively fast, genes can be introduced that are not normally expressed in the animals and the alterations in the genotype can be targeted.

3.2 GENETIC MANIPULATION

The vast majority of genetic engineering techniques use bacteria. It is now a totally routine matter to insert novel genes into bacteria, with a view to studying the structure and function of the gene, to monitor the expression or presence of the gene in various cells, to generate large amounts of recombinant protein or to manipulate or alter the gene (for general reviews of many of the procedures involved see Drlica, 1992; George and Hornick, 1996a–c). While the details of such work are outside the remit of this book, this technology has revolutionized the practice of biological and medical science, as well as providing the clinician with a new class of therapeutic agent. Few ethical difficulties are perceived in this sort of work; the main concern of the public (and of the scientific community) is to ensure the safety of the process.

The main concerns of the public rest with the genetic manipulation of animals, including humans. This is increasingly used in medical science, and is likely to become important in clinical practice over the next decade. The three major technologies that merit discussion when considering the medical use of animal biotechnology are: transgenic animals, knock-out animals and somatic gene transfer. These are discussed in more detail in Chapter 2; only a brief overview of the general principles is given here.

3.2.1 Transgenic animals

In this approach a particular gene is injected into the fertilized egg of the animal, which is then implanted into a female animal (Gordon and Ruddle, 1981; Kappel *et al.*, 1994). In a proportion of eggs the gene integrates into the DNA, and the resulting animals have an altered genome that contains the gene. Careful selection, and appropriate breeding, of these animals leads to the generation of a transgenic line of animals carrying the gene. The novel gene can be derived from another animal (or plant or bacterium) or may be artificial. It may also be from the same species but altered so that the expression of the gene becomes abnormal.

3.2.2 Knock-out animals

In knock-out technology a genetic construct that causes the removal of a particular gene is injected into an embryonic cell line (Capecchi, 1989; Koller and Smithies, 1992; Galli-Taliadoros *et al.*, 1995). Cells are selected that have had the gene removed (or knocked out), and are then implanted into blastocysts (early embryos). The embryonic cells mature in the blastocysts, where they form part of the new animal. This animal is a chimera, as it derived from a mixture of the embryonic cell line (containing the knock-out) and the cells comprising the blastocyst. Appropriate breeding can lead to the generation of a 'knock-out' line of animals in which the desired gene is removed.

3.2.3 Somatic gene transfer

The transgenic and knock-out technology described above involved the transfer of DNA into the germline cells (egg or sperm) of the animals. The alterations in the DNA are then inherited from one generation to the next. An alternative approach is to insert genes into somatic cells of animals, that is non-germline cells that comprise the majority of the animal (Verma, 1990; Cournoyer and Caskey, 1993; Kerr and Mulé, 1994). In this technique, a vector – normally a virus – is used to transfer the nucleic acid into the appropriate cells. The genes are expressed only in those cells. Thus, if the vector is targeted to liver cells, only the liver will produce the protein. The expression can be transient; that is, after several weeks the new DNA is thrown out by the cells and they revert to their natural state, or, with some viruses, the new DNA can integrate into the cellular DNA and the expression is permanent. The important distinction between somatic gene transfer and the transgenic or knock-out technologies is that the alterations in the DNA are not present in the germline of the animal. Thus, the animal is unable to pass the new genotype on to its offspring, and when the animal dies so do the alterations in the DNA.

3.3 APPLICATIONS TO MEDICINE

There are four major applications of animal biotechnology in medicine. The first, which to date has been by far and away the most important, has been to use transgenic or knock-out animals to investigate physiological and pathological processes, and to develop model systems of disease that allow development of therapeutic or diagnostic reagents. The second application is to use transgenic animals to produce a recombinant protein. The third application is to produce material that will be useful for therapy; the most notable example of this is to prepare animals for use as donors of organs for transplantation. Finally biotechnology can be used directly as a form of therapy, so-called 'gene therapy'. In these cases somatic gene transfer is used to introduce genes into appropriate somatic cells in the animal, and the expression of the recombinant protein is designed to have a therapeutic effect, either by immunizing the animal, replacing a missing protein, or augmenting a natural physiological process.

3.3.1 Animal biotechnology as a scientific tool

The most common use of animal biotechnology has been for the study of natural physiological processes, and for the pathology that occurs when these processes break down (Jaenisch, 1988; Kappel *et al.*, 1994). This work has immeasurably added to our understanding of many systems in the body, and provided us with the necessary tools to develop and refine new therapies. Indeed, it is increasingly standard practice when studying the function of a gene either to make transgenic animals that express the gene of interest, or to 'knock out' the gene, and determine the effect of these manipulations on the animal.

One case in point is the increased understanding of the role of oncogenes in the development of cancer. Oncogenes are normal cellular or virally derived genes that, either through mutation or inappropriate expression, lead to neoplastic transformation of cells. Expression of oncogenes in an inappropriate manner in transgenic mice, and the subsequent analysis of resulting tumours, has proved invaluable in our understanding of the process of tumorigenesis (Adams and Cory, 1991).

One example is the oncogene *bcl2*, which is involved in the most common chromosomal translocation (t(14:18)) in B-cell follicular lymphomas (Bakhshi *et al.*, 1985; Cleary and Sklar, 1985; Tsujimoto *et al.*, 1985). Expression of *bcl2* in transgenic animals led to follicular proliferation of the B-cells, similar to that seen in human patients (McDonnell *et al.*, 1989). Further analysis of the behaviour of B-cells from these mice showed that the role of *bcl2* was not to increase the proliferative rate of the B-cells, but to prevent their suicide by apoptosis (Wyllie *et al.*, 1980; McDonnell *et al.*, 1989). Further evidence to this effect was derived by knocking out the *bcl2* gene (Veis *et al.*, 1993). In these animals there were several defects, including extensive apoptosis of B-cells. These data helped to demonstrate that the role of *bcl2* is to prevent cell death, and that the development of some malignant states can, in part, be ascribed to the failure of a cellular suicide pathway (Cory, 1995). This has very important implications for the development of novel therapeutic agents for cancer, and, in addition, such animals can provide useful models for the development and testing of such molecules.

3.3.2 Biotechnology to produce recombinant pharmaceuticals

One promising application of biotechnology to medical science is in the production of therapeutic and diagnostic reagents. At the present time many of the proteins used in clinical practice are isolated from material derived from animals (e.g. insulin and most antisera) or humans (e.g. blood products and some antisera). These methods of producing pharmaceutical agents are not without drawbacks; material isolated from humans runs the risk of containing pathogenic viruses and other organisms. This risk is serious: there are approximately 10 000 cases of HIV infection in the USA caused by blood transfusion (Galel *et al.*, 1995), and many haemophiliacs were similarly infected through factor VIII (Darby *et al.*, 1989). In addition, the use of human growth hormone or gonadotrophin prepared from cadavers has been associated with the transmission of Creutzfeldt–Jakob disease (Buchanan *et al.*, 1991; Brown *et al.*, 1994).

The use of material derived from animals reduces this risk. However, it is often different from the equivalent human protein, may not function well in humans, or may be recognized as foreign by the patient's immune system. Thus, patients treated with porcine insulin can develop an immune response against the pig protein, necessitating their conversion to recombinant human insulin.

One approach is to use transgenic animals, in which the gene encoding the desired protein has been inserted, to make large amounts of the protein. A number of animals are being developed as sources of recombinant protein, including pigs, goats, cows and sheep (Gershon, 1991; Velander *et al.*, 1992; Yom and Bremel, 1993; Kumar, 1995). In many cases the gene is modified such that expression is restricted to the lactating breast, and so the protein is secreted into the milk. This has the advantages of reducing potential for harm to the animal, and of greatly simplifying the harvesting of the molecule, using well-established agricultural processes.

Once the transgenic animal has been produced then the goal is to produce relatively cheap protein in large amounts. Currently a number of drugs are being developed for production in this way, including factor IX for blood clotting, α_1 anti-trypsin for emphysema, haemoglobin for the development of blood substitutes, and tissue plasminogen activators for treatment of blood clots (Velander *et al.*, 1992; Yom and Bremel, 1993; Kumar, 1995). If such pharmaceutical drugs can be produced relatively inexpensively, this will increase their availability and use, as well as reducing health care costs.

The major potential problem with this approach is that excessive production of an unnatural protein may disturb the animals' natural physiology, leading to ill health. While no such problems have yet been identified, it is important to continue to monitor such transgenic animals to determine if expression of the protein causes any developmental or chronic health problems. The continued development of controlled expression of the protein (for example in the lactating breast) will further reduce the risk of causing suffering to animals.

3.3.3 Genetically engineered animals as therapeutic tools

The third area in which animals are just beginning to have an impact in medicine is the direct supply of material for treatment of patients. The main example of this is the use of animals as donors for transplantation (Concar, 1994; Dorling and Lechler, 1994; Kaufman *et al.*, 1995). The background to this work has been the massive success of transplantation over the past two decades (George *et al.*, 1995). The transplantation of one organ from a suitable donor to the other has had a major impact on health care in the Western world (Table 3.1). Thus, kidney and heart transplants are the treatment of choice for end-stage renal and cardiac failure, and liver transplants are being used in an increasing number of centres. It has been estimated that, world-wide, about 50 000–60 000 solid organ transplants are carried out each year (Botting, 1995). These transplants have a great effect in reducing both mortality and morbidity; the quality of life following a transplant is normally immeasurably better than before. As well as benefiting the individual, this has a substantial economic benefit to society, particularly as many recipients are young people with families who can return to an active life and so

Table 3.1 Transplants (excluding bone marrow) performed in the UK and Republic of Ireland; 1 January–31 December, 1994*

Organ	*No. of transplants*
Cadaveric kidney	1750
Live related kidney	126
Cadaveric heart	307
Domino heart	23
Heart/lung	52
Lung	115
Pancreas	3
Liver	644
Cornea	2577

*As reported to the United Kingdom Transplant Support Service Authority (UKTSSA). All organs are from cadaveric donors, with the exception of living related renal transplants and domino heart transplants, in which the donor heart comes from the recipient of a combined heart/lung transplant. Not all corneal transplants are reported to UKTSSA. Statistics prepared by UKTSSA from the National Transplant database maintained on behalf of the UK Transplant Community.

not need support. There can also be more direct financial returns; renal transplants are a cheaper form of treatment than long-term dialysis.

The major limitation of transplantation is the supply of organs. In this aspect transplantation has been a victim of its own success. The vast majority of organs donated are cadaveric, that is from dead donors, frequently those who have died as a result of trauma. In a minority of cases the organ is supplied by a living donor, normally a relative. However, in spite of campaigns to increase the donor pool, the supply of organs is insufficient to meet the demand. This is reflected in the increased waiting lists for many transplant operations (as illustrated for kidney transplants in Figure 3.1). Indeed, Figure 3.1 does not show the full picture as there are many low-priority patients who are never placed on waiting lists who, if there were limitless organs, would receive a transplant. In addition, for some organs, such as liver, the waiting time is artificially low as most patients have a very short life expectancy if they do not get a transplant. The shortage of donors is compounded in some countries (notably Asian) by the difficulty in obtaining cadaver organs because of sociocultural beliefs and customs (Woo, 1992).

There are several potential solutions to these problems. The first is to

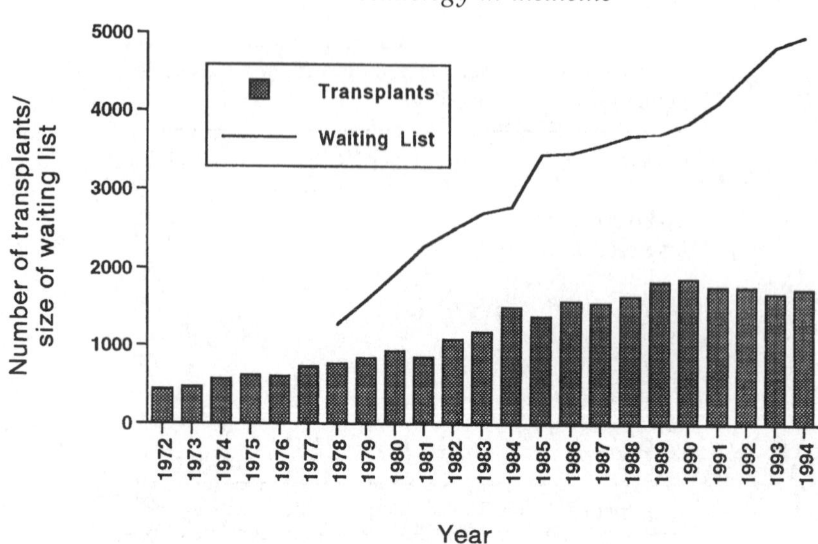

Figure 3.1 Kidney transplants, 1972–1994. The graph illustrates the number of cadaveric renal transplants in the UK and Republic of Ireland notified to UKTSSA 1972–1994, and the number of patients awaiting a kidney at year end, 1978–1994. The number of transplants performed plateaued in the late 1980s, due to shortage of donors, while the waiting list has continued to rise. The waiting list gives a somewhat false impression as potential recipients with low priority for a transplant may not be put on the waiting list. (Data courtesy of UKTSSA.)

increase the number of organ donations. However, while this could be very useful in the near future, it is only a short-term solution; as the transplant criteria widen so the number of people requiring an organ will greatly outstrip the potential supply (Gore, 1993). The second is to develop artificial organs. Dialysis can be used to replace some of the functions of kidneys, and artificial hearts have been used to keep patients alive for short periods (Rowles *et al.*, 1993). Even artificial liver devices (albeit normally incorporating living cells) are being developed (Takahashi *et al.*, 1991; Gibbs, 1993; Sussman *et al.*, 1994). While such machines may have a future role in patient management, particularly in maintaining or preparing a patient for a transplant, they are, with the exception of dialysis, unlikely to be a long-term option for a patient.

The third possibility is to use organs from other animals. This is termed xenotransplantation, as the organ transplanted is xenogeneic with respect to the host. Attempts at xenotransplantation were reported at the turn of the century, though at the time the immunological basis of transplant rejection was not appreciated (Reemtsma *et al.*, 1964b). Following the development of conventional allotransplantation, kidneys from chimpanzees or baboons were transplanted into patients with renal failure (Hitchcock *et al.*, 1964; Reemtsma *et al.*, 1964a,b; Starzl *et al.*, 1964). In most cases the transplanted organ functioned well, in one case for more than 6 months. This gave some patients enough time for a conventional kidney donor to be found. However, if a human donor was not found, the transplants ultimately failed due either to immunological rejection of the organ, which the immunosuppressive drugs failed to control, or to death of the patient from sepsis due to heavy immunosuppression.

These, and similar, attempts at heart and liver xenotransplantation all used primates as donors (Barnard *et al.*, 1977; Bailey *et al.*, 1985; Starzl *et al.*, 1993). This is logical, considering the close evolutionary relationship between humans and other primates, which reduces the immunological difference between the primate donor organ and the human recipient. However, there are major disadvantages to the use of primates; there is considerable public concern over their use, they reproduce slowly in captivity, some are endangered species, and they may contain infectious agents that can cross into the human recipient (Michaels and Simmons, 1994).

The hunt for potential xenogeneic donors has recently concentrated on the pig (Dorling and Lechler, 1994; Kaufman *et al.*, 1995). This animal is consumed as a food in large amounts, reducing the public concern over such an approach [more than 100 million pigs are raised and slaughtered each year in the USA for food (Gundry, 1994)]. In addition, pigs are easy to raise, have big litters, and grow rapidly.

The major trouble with xenotransplantation is the rapid rejection of the organ. Immune rejection of the organ is the major problem in conventional transplantation, and the recipient has to be continually immunosuppressed to prevent the immune system from recognizing the transplanted organ and destroying it. In the case of xenotransplantation the problem is compounded by fundamental incompatibilities between the pig donor and the human recipient. These are capable not only of eliciting an immune response against the donated organ, but also of triggering more basic immunological pathways which lead to violent and rapid rejection of pig into primate grafts (hyperacute rejection) (Calne, 1970). The most important of these pathways is the reaction with pig tissue of natural, pre-existing, human antibodies which bind to the surface of the donor graft, causing its destruction. Most of these antibodies recognize a galactosyl α-1,3 galactosyl carbohydrate structure (known as the α-galactosyl epitope) (Galili, 1993; Sandrin *et al.*, 1993). This epitope is made by an α-1,3 galactosyl transferase enzyme present in all mammals, except humans, Old World monkeys and apes (Galili, 1993; Sandrin *et al.*, 1993). A number of groups are now trying to knock out the transferase enzyme in pigs. Organs from animals lacking the enzyme will not express the α-galactosyl epitope, and so will not undergo rejection by the natural antibodies.

Biotechnology also allows for the development of other strategies to prevent rejection. The major way in which the natural antibodies lead to hyperacute rejection of the organ is through activation of the complement system, a series of proteins present in an inactive form in the serum which is activated, through antibody molecules, to destroy cells. A great deal of effort is now being invested in the production of transgenic pigs expressing human proteins that prevent or reduce the activation of complement (Fodor *et al.*, 1994; McCurry *et al.*, 1995). This strategy should protect the transplanted organs from much of the damage induced by the immune response.

It is likely that we are several years away from the routine use of transgenic pigs as organ donors. There are many barriers to be overcome before the incompatibilities between the pig organs and humans is removed. However, the increased demand for organs will continue to stimulate work in this area.

As well as overcoming the shortage of organs, xenotransplantation offers a number of additional advantages. The use of non-human donors reduces the risk of transferring human viruses to the patient. Second, the genetic modification of the donor may allow us to overcome the normal recognition of the organ by the immune response, or to develop strategies that induce immunological tolerance to the pig, thus preventing or reducing the immune response and lowering the requirement for immunosuppression. Last, but by no means least, the use of animals bred for the purpose of transplantation will reduce the need for distressed relatives to consent to organ donation (Demetrius and Parents of John, 1975; Tymstra *et al.*, 1992).

There are some specific problems that might have to be considered by the public before xenotransplantation is widely accepted (Daar, 1994). Probably one of the strongest is the 'Yuk Factor'. People are uncomfortable with the thought of putting animal organs into humans; it seems an unnatural mixing of species. However, such a response is not new; the introduction of the vaccine for smallpox, derived from cows, initially led to similar concern, as illustrated in Figure 3.2. Similar worries were also expressed following the initial conventional transplants. We should also remember that we already practise xenotransplantation: pig heart valves, which are not recognized by the immune system, are routinely used to replace defective human heart valves.

A second objection that can be raised to the development of transgenic animals for xenotransplantation is that it is wrong to make and raise animals solely for the use of humans. Such an argument has a

Figure 3.2 Early opposition to vaccination. *The Cowpock – or – the Wonderful Effects of the New Inoculation*, by James Gillray for the Anti-Vaccination Society, 1808. There was strong opposition to the introduction of cow-derived vaccines for smallpox. The drastic effects on patients of taking the vaccine pock 'hot from ye cow' illustrate the fears that can be engendered by mixing material from animals with humans. (Figure courtesy of the Wellcome Institute Library.)

certain emotional appeal, one of the possible future scenarios is that people will be able to buy the services of a scientist to create a transgenic pig that is suited to their genetic makeup, and keep that animal until they need it for spare parts (Concar, 1994). There is something inherently unattractive about such a concept. However, logically, creating animals for health care reasons is no different from creating animals for food. If it is acceptable for new animals to have been developed by farmers down the ages for food, then it should be acceptable for scientists to create and use animals for medical use.

3.3.4 Somatic gene transfer

The final application of animal biotechnology is somatic gene transfer (Verma, 1990; Cournoyer and Caskey, 1993; Kerr and Mulé, 1994). As discussed above, this involves the transfer of genetic material into somatic cells of animals, which are not passed on to the offspring. The main use of this approach will be for gene therapy of disease, though it also has a role as a research tool. Historically, somatic gene transfer has been widely used in humans and other animals to immunize; a virus containing genes from a pathogenic organism is injected into the patient where it infects cells, produces protein derived from the pathogen and elicits a protective immune response against the pathogen. Until recently this approach has been restricted to using viruses that naturally contain the gene of interest (e.g. vaccinia for smallpox). Increasingly, genes derived from pathogenic organisms are being inserted into non-pathogenic viruses, with a view to producing a vaccine (Cadoz *et al.*, 1992; Cooney *et al.*, 1993; Graham *et al.*, 1993; Johnson *et al.*, 1994). This approach will allow for the effective vaccination of humans and other animals against a wider range of diseases than is now possible.

The area of gene therapy that is exciting more public concern is the transfer of genes to interfere with physiological or pathological processes. The first use of gene therapy to treat a disease was for adenosine deaminase (ADA) deficiency. This is a disease in which the patient lacks the ability to make the ADA enzyme. As a result a toxic metabolite builds up, which poisons the cells of the immune system. The children are profoundly immunosuppressed and, if not kept in a specialized environment, rapidly die of infection (Cournoyer and Caskey, 1993). The gene therapy trials involve the infection of bone marrow cells from the children with a virus engineered to contain the ADA gene, thereby replacing the missing enzyme. This treatment has caused significant improvement in the immune system of patients (Anderson, 1992). Increasingly such treatments are being devised and tested for a variety of diseases, including genetic disorders such as cystic fibrosis and muscular dystrophy (Kay and Woo, 1994; Kerr and Mulé, 1994), cancer (Culver and Blaese, 1994; Sikora, 1994), neurological diseases (Friedmann, 1994), infectious disease (Gilboa and Smith, 1994) and transplantation (Larkin *et al.*, 1996).

At present, gene therapy of this type is only being considered for human use, though it is being developed in animals. It is likely to be some time before gene therapy, other than for vaccination, has a role in veterinary medicine. It is likely, therefore, that the vast majority of the ethical debate in this area will continue to be dominated by human-centred concerns and needs (Murray, 1991; Muller and Rehmann-Sutter, 1995). Such a debate, while important, is outside the remit of this book.

3.4 DISCUSSION

The use of biotechnology in animals has revolutionized medical science and will continue to do so. However, it is clear that genetic manipulation is the subject of a wide concern, and there are several points that need to be addressed in any consideration of the rights and wrongs of this technology. Should we be allowed genetically to manipulate living creatures? What should the boundaries to this work be? What are the safety implications? Who owns a genetically modified animal? Some of these issues – in particular the last-mentioned – are covered elsewhere in this book, and will not be discussed here.

Although there is an active debate about the use of animals for research, for the purposes of this discussion we shall assume that it is acceptable to use animals in medical experimentation. Obviously if it were not acceptable to use conventional animals for scientific and medical purposes then it would not be acceptable to either use or create genetically modified animals for such purposes. There is not enough space to discuss this topic adequately; however, nearly all medical scientists would believe that – while it is important to restrict the use of animals to the minimum needed, to prevent all unnecessary suffering, and to use alternative methods wherever possible – the use of animals has been, and will continue to be, vital for the progress of medical science (Botting, 1992). There are probably no advances in Western conventional medical science that have not relied on knowledge derived from animal experimentation, direct animal experimentation, reagents derived from animals, or safety testing on animals. For example, in the field of transplantation discussed above, animal experimentation has proved vital for the development of surgical techniques, for the understanding of the immunological processes causing rejection and for the development of immunosuppressive drugs and agents to prevent this rejection (Botting, 1995). Without the use of animals, transplantation operations for humans would not be possible. This and other advances in medicine have caused a tremendous improvement in life expectancy and the quality of life, improvements that we tend to take for granted (Figure 3.3). Both the scientific and the medical community have a moral obligation to continue to develop methods for preventing or treating human disease, and animal experimentation is a vital component in this work. Society, in general, accepts that it is reasonable to breed and use animals for food and clothing and, while questions can arise out of a particular use of an animal, there seems little reason why the preservation of life by medicine should be considered any different from the preservation of life with food or clothes.

3.4.1 Should we be allowed to alter animals genetically?

There is a fear that by tinkering with the genetic makeup of animals we are 'playing God' and interfering in ways that we should not. To some

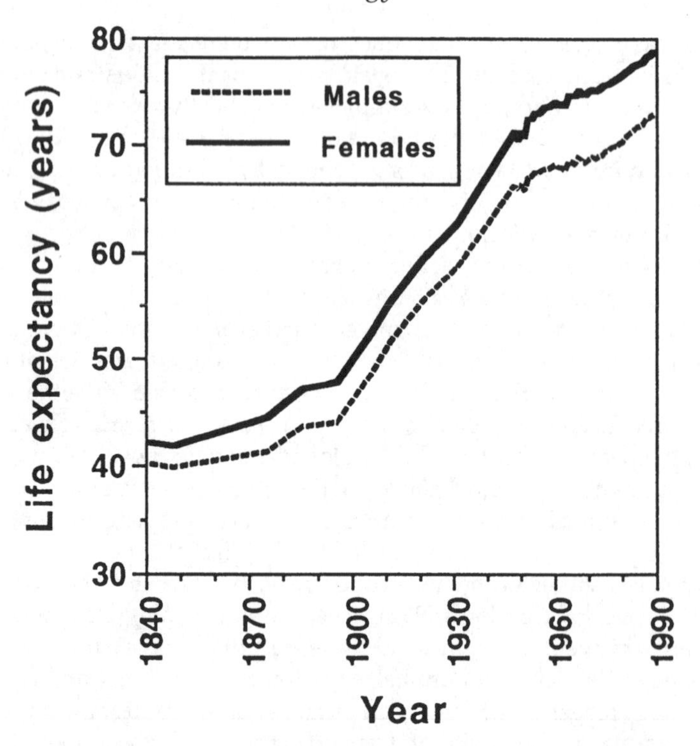

Figure 3.3 Life expectancy, 1841–1989. The graph shows the life expectancy at birth for males and females in England and Wales. Since 1840 there has been a consistent increase in life expectancy for both sexes. (Data taken from Office of Health Economics, 1992.)

extent this fear arises out of a lack of understanding of the reality that humans have continually manipulated other species to obtain 'unnatural' animals by deliberate breeding policies. Some of these animals have been produced in a very uncaring way, such as dog breeds with distressing or painful congenital deformities (Medawar, 1990). Recombinant DNA technology is similar in that it has the potential to be used in an unthinking way, and it is necessary to have proper controls to prevent this. However, the modification of animals by biotechnology is no more 'playing God' than conventional animal husbandry.

It should also be noted that the amount of tinkering with the genetic material of animals that we are able to do using modern biotechnology is very limited. The genome of an average mammal (human) contains about three billion base pairs of DNA, encoding 50 000–100 000 genes (Strachan, 1992). Even the most complex transgenic animal will only have half a dozen or so genes inserted into this genome, probably coded for by less than 30 000 base pairs of DNA, representing up to 0.01% of an animal's genetic material. Compared with the wholesale changes in genetic material caused by breeding, the genetic alterations that we can achieve are therefore minor.

It could be argued that transgenic or knock-out technology represents a qualitative difference in the development of new animals, as it involves the unnatural manipulation of genes. However, theoretically, the equivalent of every transgenic or knock-out animal could be generated by careful selection of the rare natural mutants. For example, one could isolate a pig lacking the transferase enzyme responsible for the α-galactosyl epitope by screening and selecting animals with spontaneous mutations in the enzyme. Human X-linked hereditary diseases, such as Duchenne muscular dystrophy or haemophilia, arise with mutation rates estimated to be 1×10^{-4} and 2×10^{-6} per generation (Vogel, 1990). If similar rates held for the porcine enzyme then the mutation would arise between 2 and 100 times per million piglets. Thus, screening for natural mutants would be a massive task, which might be practicably impossible. However, theoretically, given enough time and resources, it should be possible to obtain animals with any desired phenotype (including those containing genes derived from other species) by selection of spontaneous mutants. Natural mutation is a potent force when linked to an appropriate selection mechanism, and we should not underestimate the ability of the evolutionary process to generate complex and wonderful creatures. The radical difference of biotechnology is that it speeds up and simplifies the process, bringing it from a theoretical possibility to reality.

3.4.2 What boundaries should be set up?

There is considerable debate in the field of gene therapy as to what genes it is reasonable to interfere with, and what genes should not be touched. Currently, the feeling is that work should concentrate on life-threatening conditions, but with more experience, especially with regard to the safety aspects, it is likely that we will be willing to use gene therapy for other serious illnesses. There is agreement that it would be wrong to use gene therapy for cosmetic or trivial reasons (curing baldness or changing hair colour), and it is doubtful that either the public or scientific community would support such use in the foreseeable future. There is also a consensus that, in humans, it would be wrong to alter the germline DNA (Müller and Rehmann-Sutter, 1995), as future generations will not have been able to consent to the alterations in their genome. There is also a worry that this could lead to a new form of eugenics, although it should be remembered that biotechnology of this sort, though possibly able to alter single, but important, features of humans, such as intelligence and height, is too fine a tool to make radical, wholesale, alterations in a person's phenotype. Hitler's terrible goal of producing an Aryan Master Race could not have been achieved by biotechnology, as the mixture of different phenotypic characteristics (and so the relevant genes) is too complex to contemplate producing by current technology. Selective breeding is the only way to achieve that end, if indeed it could be achieved at all. The major worry in eugenic terms with recombinant technology is its potential to screen and select individuals for such breeding programmes. Indeed the ability to screen for genetic diseases is already posing considerable ethical dilemmas, for example relating to health insurance (Murray, 1991).

With non-human animals the debate is on different territory. We already use selective breeding to 'improve' animal strains, and informed consent is not an issue. So, we have to draw up other guidelines. As with animal experimentation and other such moral dilemmas, we need to attempt a cost–benefit analysis, considering several features, including the pain of the animals involved and the benefits to humans and animals of the work being done. This should be done with a proper sense of the importance of the worth of the animals involved. As with conventional animal experiments, painful procedures should not be carried out for trivial reasons, though they may be justifiable if there is the potential for great benefit from them. Biotechnology is no different from conventional animal experiments in this respect.

As with all cost–benefit arguments, there will be arguments as to where to draw the line. The costs and the benefits will be perceived differently by different groups. Some will argue that there are harms that should never be inflicted on animals, and some would include a concept of the integrity of the animal in the equation. The culture in which the argument is taking place will have an impact on the answer and, as society changes, so may the balance of the argument. It is more comfortable to have absolute beliefs and values about these matters, but that may be avoiding the real issues. We should strive to develop an informed debate on these topics to allow society to reach a consensus.

3.4.3 Safety

Recombinant DNA technology has, since its inception, aroused considerable fears over safety, including the anxiety that dangerous organisms will be released into the environment. These public concerns led scientists to adopt in 1974 a voluntary moratorium on certain classes of gene cloning experiments until the risks could be assessed and guidelines drawn up. While our experience with recombinant DNA technology suggests that the risks associated with this technology are very low, and that these guidelines are probably satisfactory, there is understandable concern by the public that the production of novel animals carries a risk. This cannot be denied; while it is difficult to imagine how transgenic and knock-out animals will be dangerous, it is impossible to predict with total accuracy the consequences of every action. There is risk in all human activity; science and medicine are no exception. Clinicians and scientists have made mistakes in the past, and no doubt will do so in the future. It is therefore necessary to carry out risk assessment analysis to determine both the possibility of something going wrong and the possible consequences of any such error. Such assessments have to be carried out in the UK before any recombinant DNA experiment to identify the appropriate safety precautions for that work.

In coming to a more general risk assessment of this technology, it is important to have a realistic idea of the nature of the dangers and the risk that they might occur. A Frankenstein-created monster is a powerful image; however, it is not a realistic reflection of the real dangers! The most likely risk is environmental, that the release or escape of a bioengineered animal will disrupt a local ecosystem, by competing with or destroying other species. Similar problems have frequently been noted, for example by the introduction of mammals into Australasia or in the escape of mink into British waterways. It is clear that the introduction of any genetically engineered animals into the wild, or into a confined habitat from which they might escape, needs careful environmental analysis. It is worth reiterating the point that the alterations made to animals using recombinant techniques are relatively minor, and that the introduction of a 'natural' animal into a new habitat that does not contain the animal will probably cause a greater environmental upheaval than the introduction of a gene-modified animal into an environment already containing its wild-type counterpart.

The safety of biotechnology used in medicine will need continual evaluation, and the guidelines will need to be revised as we gain more experience. Work to date has suggested that the actual risks are low and acceptable, especially when compared with the greater dangers inherent in other human activities, such as power generation. However, this is an area where there is no room for complacency on the part of the scientific community.

3.4.4 Wider ethical issues arising from genetic manipulation

There are wider social issues arising out of the use of biotechnology in medicine. Most of these are not unique to medicine and science, but they are probably among the most important issues that need to be raised.

Biotechnology, similar to much high-technology medicine, is very expensive. Pharmaceutical companies concentrate their development efforts on products that have the potential for profit, and so will develop drugs (and genetically engineered animals) for the treatment of diseases where there is a customer who can pay. As a result, most research is into cancer, heart disease, allergies and other diseases of the developed world. Relatively little attention is paid to parasitic and infectious diseases that affect developing nations. This problem is compounded by the fact that much university research is funded by charities (in the UK the majority) which, quite reasonably, support research with potential application to their fund-raisers. The result is to bias medical provision ever further to the treatment of diseases of the developed world, while relatively little attention is paid to the diseases of developing countries. Naturally, a great number of scientists, charities and drug companies do have an interest in diseases of the developing world. However, biotechnology has the potential to increase the polarization in medical provision by providing care to the minority of people with the money to pay for it.

There are also monopolistic concerns in the development of biotechnology that mirror the situation seen in other industries, such as the media or telecommunications. For example, the company, or consortium of companies, that will own the rights to pigs used for xenotransplantation will have a very powerful market position. They will be able to influence medical provision and research in a large number of countries, yet they will be answerable only to their shareholders. It is dangerous to believe in conspiracy theories, and to assume that the companies plan to do anything more than make some money and produce useful pharmaceutical agents; however, the concentration of power may be a subject for concern. Once again this problem is not one of biotechnology *per se*; nonetheless, many of these general questions pose more ethical dilemmas than particular issues raised by the use of genetically engineered animals.

3.5 CONCLUSION

The use of genetic engineering of animals has already greatly influenced medical science, in providing model systems to study physiological and pathological processes. Such technology is likely to have an increasing role in providing useful therapeutic reagents. In addition it will be used directly in the treatment of disease, both by the provision of engineered organs or tissues, and by the direct gene therapy of disease. There are still many ethical questions that need to be considered, but few of these questions are unique to genetic engineering of animals. Indeed, humans have, albeit unconsciously, been carrying out highly effective genetic engineering of animals for several millenia by selective breeding programmes.

ACKNOWLEDGEMENTS

I would like to thank Mr Phil Hornick, Revd Julian Reindorp, Dr Mark Matfield, Dr Graham Dwyer and Revd Malcolm Strange for critical reading of this manuscript, Dr Gisela Field for advice and Zhao An-zhu for secretarial assistance.

REFERENCES

Adams, J.M. and Cory, S. (1991) Transgenic models of tumor development. *Science*, **254**, 1161–7.

Anderson, W.F. (1992) Human gene therapy. *Science*, **256**, 808–13.

Bailey, L.L., Nehlsen-Cannarella, S.L., Concepcion, W. and Jolley, W.B. (1985) Baboon-to-human cardiac xenotransplantation in a neonate. *Journal of the American Medical Association*, **254**, 3321–9.

Bakhshi, A., Jensen, J.P., Goldman, P., Wright, J.J., McBride, O.W., Epstein, A.L. and Korsmeyer, S.J. (1985) Cloning the chromosomal breakpoint of t(14;18) human lymphomas: clustering around J_H on chromosome 14 and near a transcriptional unit on 18. *Cell*, **41**, 899–906.

Barnard, C.N., Wolpowitz, A. and Losman, J.G. (1977) Heterotopic cardiac transplantation with a xenograft for assistance of the left heart in cardiogenic shock after cardiopulmonary bypass. *South African Medical Journal*, **52**, 1035–8.

Botting, J.D. (ed.) (1992) *Animal Experimentation and the Future of Medical Research*, Portland Press, London.

Botting, J. (1995) The contribution of animal experiments to kidney transplantation. *RDS News*, April, 8–14.

Brown, P., Cervenáková, L., Goldfarb, L.G., McCombie, W.R., Rubenstein, R., Will, R.G., Pocchiari, M., Martinez-Lage, J.F., Scalici, C., Masullo, C., Graupera, G., Ligan, J. and Gajdusek, D.C. (1994) Iatrogenic Creutzfeldt–Jakob disease: an example of the interplay between ancient genes and modern medicine. *Neurology*, **44**, 291–3.

Buchanan, C.R., Preece, M.A. and Milner, R.D. (1991) Mortality, neoplasia, and Creutzfeldt–Jakob disease in patients treated with human pituitary growth hormone in the United Kingdom. *British Medical Journal*, **302**, 824–8.

Cadoz, M., Strady, A., Meignier, B., Taylor, J., Tartaglia, J., Paoletti, E. and Plotkin, S. (1992) Immunization with canarypox virus expressing rabies glycoprotein. *Lancet*, **339**, 1429–32.

Calne, R.Y. (1970) Organ transplantation between widely disparate species. *Transplantation Proceedings*, **2**, 550–6.

Capecchi, M.R. (1989) Altering the genome by homologous recombination. *Science*, **244**, 1288–92.

Cleary, M.L. and Sklar, J. (1985) Nucleotide sequence of a t(14;18) chromosomal breakpoint in follicular lymphoma and demonstration of a breakpoint-cluster region near a transcriptionally active locus on chromosome 18. *Proceedings of the National Academy of Sciences of the USA*, **82**, 7439–43.

Concar, D. (1994) The organ factory of the future. *New Scientist*, **142**, 24–9.

Cooney, E.L., McElrath, M.J., Corey, L., Hu, S.L., Collier, A.C., Arditti, D.,

Hoffman, M., Coombs, R.W., Smith, G.E. and Greenberg, P.D. (1993) Enhanced immunity to human immunodeficiency virus (HIV) envelope elicited by a combined vaccine regimen consisting of priming with a vaccinia recombinant expressing HIV envelope and boosting with gp160 protein. *Proceedings of the National Academy of Sciences of the USA*, **90**, 1882–6.

Cory, S. (1995) Regulation of lymphocyte survival by the bcl-2 gene family. *Annual Review of Immunology*, **13**, 513–43.

Cournoyer, D. and Caskey, C.T. (1993) Gene therapy of the immune system. *Annual Review of Immunology*, **11**, 297–329.

Culver, K.W. and Blaese, R.M. (1994) Gene therapy for cancer. *Trends in Genetics*, **10**, 174–8.

Daar, A.S. (1994) Xenotransplantation and religion: the major monotheistic religions. *Xeno*, **2**, 61–4.

Darby, S.C., Rizza, C.R., Doll, R., Spooner, R.J.D., Stratton, I.M. and Thakrar, B. (1989) Incidence of AIDS and excess of mortality associated with HIV in haemophiliacs in the United Kingdom: report on behalf of the directors of haemophilia centres in the United Kingdom. *British Medical Journal*, **298**, 1064–8.

Demetrius, R.M. and Parents of John (1975) Transplantation: the relatives' view. *Journal of Medical Ethics*, **1**, 71–2.

Dorling, A. and Lechler, R.I. (1994) Prospects for xenografting. *Current Opinion in Immunology*, **6**, 765–9.

Drlica, K. (1992) *Understanding DNA and Gene Cloning. A Guide for the Curious*, 2nd edn, John Wiley and Sons, New York.

Fodor, W.L., Williams, B.L., Matis, L.A., Madri, J.A., Rollins, S.A., Knight, J.W., Velander, W. and Squinto, S.P. (1994) Expression of a functional human complement inhibitor in a transgenic pig as a model for the prevention of xenogeneic hyperacute organ rejection. *Proceedings of the National Academy of Sciences of the USA*, **91**, 11153–7.

Friedmann, T. (1994) Gene therapy for neurological disorders. *Trends in Genetics*, **10**, 210–14.

Galel, S.A., Lifson, J.D. and Engleman, E.G. (1995) Prevention of AIDS transmission through screening of the blood supply. *Annual Review of Immunology*, **13**, 201–27.

Galili, U. (1993) Interaction of the natural anti-Gal antibody with α-galactosyl epitopes: a major obstacle for xenotransplantation in humans. *Immunology Today*, **14**, 480–2.

Galli-Taliadoros, L.A., Sedgwick, J.D., Wood, S.A. and Körner, H. (1995) Gene knock-out technology: a methodological overview for the interested novice. *Journal of Immunological Methods*, **181**, 1–15.

George, A.J.T. and Hornick, P. (1996a) Molecular biology: part I. *Surgery*, **14**(5), 108a–8d.

George, A.J.T. and Hornick, P. (1996b) Molecular biology: part II. *Surgery*, **14**(6), i–iii.

George, A.J.T. and Hornick, P. (1996c) Molecular biology: part III. *Surgery*, **14**(7), 156a–6c.

George, A.J.T., Ritter, M.A. and Lechler, R.I. (1995) Disease susceptibility, transplantation and the MHC. *Immunology Today*, **16**, 209–11.

Gershon, D. (1991) Biotechnology. Will milk shake up industry? *Nature*, **353**, 7.

Gibbs, W.W. (1993) Deliverance. Medicine closes in on an artificial liver device. *Scientific American*, **269**(6), 14–15.

Gilboa, E. and Smith, C. (1994) Gene therapy for infectious diseases: the AIDS model. *Trends in Genetics*, **10**, 139–44.

Gordon, J.W. and Ruddle, F.H. (1981) Integration and stable germline transmission of genes injected into mouse pronuclei. *Science*, **214**, 1244–6.

Gore, S.M. (1993) The shortage of donor organs: wither, not whether, *Xeno*, **1**, 23–4.

Graham, B.S., Matthews, T.J., Belshe, R.B., Clements, M.L., Dolin, R., Wright, P.F., Gorse, G.J., Schwartz, D.H., Keefer, M.C., Bolognesi, D.P., Corey, L., Stablein, D.M., Esterlitz, J.R., Hu, S.-L., Smith, G.E., Fast, P.E., Koff, W.C. and The NIAID AIDS Vaccine Clinical Trials Network (1993) Augmentation of human immunodeficiency virus type 1 neutralizing antibody by priming with gp160 recombinant vaccinia and boosting with rgp160 in vaccinia-naive adults. *Journal of Infectious Diseases*, **167**, 533–7.

Gundry, S.R. (1994) Is it time for clinical xenotransplantation (again)? *Xeno*, **2**, 60–1.

Hitchcock, C.R., Kiser, J.C., Telander, R.L. and Seljeskg, E.L. (1964) Baboon renal grafts. *Journal of the American Medical Association*, **189**, 158–61.

Jaenisch, R. (1988) Transgenic animals. *Science*, **240**, 1468–74.

Johnson, R.P., Hammond, S.A., Trocha, A., Siliciano, R.F. and Walker, B.D. (1994) Induction of a major histocompatibility complex class I-restricted cytotoxic T-lymphocyte response to a highly conserved region of human immunodeficiency virus type 1 (HIV-1) gp120 in seronegative humans immunized with a candidate HIV-1 vaccine. *Journal of Virology*, **68**, 3145–53.

Kappel, C.A., Bieberich, C.J. and Jay, G. (1994) Evolving concepts in molecular pathology. *FASEB Journal*, **8**, 583–92.

Kaufman, C.L., Gaines, B.A. and Ildstad, S.T. (1995) Xenotransplantation. *Annual Review of Immunology*, **13**, 339–67.

Kay, M. and Woo, S.L.C. (1994) Gene therapy for metabolic disorders. *Trends in Genetics*, **10**, 253–7.

Kerr, W.G. and Mulé, J.J. (1994) Gene therapy: current status and future prospects. *Journal of Leukocyte Biology*, **56**, 210–14.

Koller, B.H. and Smithies, O. (1992) Altering genes in animals by gene targeting. *Annual Review of Immunology*, **10**, 705–30.

Kumar, R. (1995) Recombinant hemoglobins as blood substitutes: a biotechnology perspective. *Proceedings of the Society for Experimental Biology and Medicine*, **208**, 150–8.

Larkin, D.F.P., Oral, H.B., Ring, C.J.A., Lemoine, N.R. and George, A.J.T. (1996) Adenovirus-mediated gene delivery to the corneal endothelium. *Transplantation*, **61**, 363–70.

McCurry, K.R., Kooyman, D.L., Alvarado, C.G., Cotterell, A.H., Martin, M.J., Logan, J.S. and Platt, J.L. (1995) Human complement regulatory proteins protect swine-to-primate cardiac xenografts from humoral injury. *Nature Medicine*, **1**, 423–7.

McDonnell, T.J., Deane, N., Platt, F.M., Nunez, G., Jaeger, U., McKearn, J.P. and Korsmeyer, S.J. (1989) *bcl*-2-immunoglobulin transgenic mice demonstrate extended B-cell survival and follicular lymphoproliferation. *Cell*, **57**, 79–88.

Medawar, P. (1990) Animal experimentation in a medical research institute, in *The Threat and the Glory* (ed. P. Medawar), Oxford University Press, Oxford, pp. 243–52.

Michaels, M.G. and Simmons, R.L. (1994) Xenotransplant-associated zoonoses. Strategies for prevention. *Transplantation*, **57**, 1–7.

Müller, H. and Rehmann-Sutter, C. (1995) Gentherapie und ethik. *Schweizerische Medizinische Wochenschrift*, **125**, 34–41.

Murray, T.H. (1991) Ethical issues in human genome research. *FASEB Journal*, **5**, 55–60.

Office of Health Economics (1992) *Compendium of Health Statistics*, p. 17.

Reemtsma, K., McCracken, B.H., Schlegel, J.U., Pearl, M.A., DeWitt, C.W. and Creech, O., Jr (1964a) Reversal of early graft rejection after renal heterotransplantation in man. *Journal of the American Medical Association*, **187**, 691–6.

Reemtsma, K., McCracken, B.H., Schlegel, J.U., Pearl, M.A., Pearce, C.W., DeWitt, C.W., Smith, P.E., Hewitt, R.L. and Creech, O., Jr (1964b) Renal heterotransplantation in man. *Annals of Surgery*, **160**, 384–410.

Rowles, J.R., Mortimer, B.J. and Olsen, D.B. (1993) Ventricular assist and total artificial heart devices for clinical use in 1993. *ASAIO Journal*, **39**, 840–55.

Sandrin, M.S., Vaughan, H.A., Dabkowski, P.L. and McKenzie, I.F. (1993) Anti-pig IgM antibodies in human serum react predominantly with Gal(α 1-3) Gal epitopes. *Proceedings of the National Academy of Sciences of the USA*, **90**, 11391–5.

Sikora, K. (1994) Genes, dreams, and cancer. *British Medical Journal*, **308**, 1217–21.

Starzl, T.E., Marchioro, T.L., Peters, G., Kirkpatrick, C.H., Wilson, W.E.C., Porter, K.A., Rifkind, D., Ogden, D.A., Hitchcock, C.R. and Waddell, W.R. (1964) Renal heterotransplantation from baboon to man: experience with 6 cases. *Transplantation*, **2**, 752–76.

Starzl, T.E., Fung, J., Tzakis, A., Todo, S., Demetris, A.J., Marino, I.R., Doyle, H., Zeevi, A., Warty, V., Michaels, M., Kusne, S., Rudert, W.A. and Trucco, M. (1993) Baboon-to-human liver transplantation. *Lancet*, **341**, 65–71.

Strachan, T. (1992) *The Human Genome*. βios Scientific Publishers: Oxford.

Sussman, N.L., Gislason, G.T. and Kelly, J.H. (1994) Extracorporeal liver support. Application to fulminant hepatic failure. *Journal of Clinical Gastroenterology*, **18**, 320–4.

Takahashi, T., Malchesky, P.S. and Nose, Y. (1991) Artificial liver. State of the art. *Digestive Diseases and Science*, **36**, 1327–40.

Tsujimoto, Y., Gorham, J., Cossman, J., Jaffe, E. and Croce, C.M. (1985) The t(14;18) chromosome translocations involved in B-cell neoplasms result from mistakes in VDJ joining. *Science*, **229**, 1390–3.

Tymstra, T., Heyink, J.W., Pruim, J. and Slooff, M.J.H. (1992) Experience of bereaved relatives who granted or refused permission for organ donation. *Family Practice*, **9**, 141–4.

Veis, D.J., Sorenson, C.M., Shutter, J.R. and Korsmeyer, S.J. (1993) Bcl-2-deficient mice demonstrate fulminant lymphoid apoptosis, polycystic kidneys, and hypopigmented hair. *Cell*, **75**, 229–40.

Velander, W.H., Johnson, J.L., Page, R.L., Russell, C.G., Subramanian, A., Wilkins, T.D., Gwazdauskas, F.C., Pittius, C. and Drohan, W.N. (1992) High-level expression of a heterologous protein in the milk of transgenic swine using the cDNA encoding human protein C. *Proceedings of the National Academy of Sciences of the USA*, **89**, 12003–7.

Verma, I.M. (1990) Gene therapy. *Scientific American*, **263**(5), 34–41.

Vogel, F. (1990) Mutation in man, in *Principles and Practice of Medical Genetics*, 2nd edn, vol. 1 (eds A.E.H. Emery and D.L. Rimoin), Churchill Livingstone, Edinburgh, pp. 53–76.

Wheeler, M.B. and Campion, D.R. (1993) Animal production – a longstanding biotechnological success. *American Journal of Clinical Nutrition*, **58**, 276S–81S.

Woo, K.T. (1992) Social and cultural aspects of organ donation in Asia. *Annals of the Academy of Medicine of Singapore*, **21**, 421–7.

Wyllie, A.H., Kerr, J.F.R. and Currie, A.R. (1980) Cell death: the significance of apoptosis. *International Review of Cytology*, **68**, 251–306.

Yom, H.C. and Bremel, R.D. (1993) Genetic engineering of milk composition: modification of milk components in lactating transgenic animals. *American Journal of Clinical Nutrition*, **58**, 299S–306S.

4

Biotechnology in animal agriculture

George E. Seidel, Jr

4.1 AGRICULTURE AND HUMAN PROGRESS

I will start with working definitions of biotechnology and animal agriculture. Such definitions are always stipulative; for purposes of this chapter, I define biotechnology as 'application of scientific and/or technological procedures and principles to biology'. Some may protest that this definition is too broad, but in my opinion even simple selective breeding is biotechnology. For example, producing ponies and draft horses from the same equid stock, already accomplished centuries ago, had more dramatic end results than the minor tinkering we have done by applying artificial insemination and embryo transfer technology in the past few decades. Another striking example of selective breeding is crossing species barriers resulting in sterile animals, a practice used since biblical times to produce mules, and used currently to cross cattle and bison.

Defining animal agriculture is problematic. Appropriate definitions might vary between developed and developing countries (which have the majority of agricultural animals) and among cultures within countries (particularly developing ones), and could include or exclude aquaculture, invertebrates, less prevalent species such as emus, research animals, and recreational animals like racing camels and some horses. For this chapter, I will define animal agriculture as 'use of birds and mammals to support material needs'. This, of course, includes animal use by the owner/manager directly, e.g. for food, fibre, power, and fuel, or selling the animals or their products (including recreation) to someone else and using the proceeds for living expenses, recreation, accumulating resources for retirement, etc. Note that most cultures use agricultural animals themselves as sources of prestige (e.g. owners of prize fowl, sheep, horses, etc.), reservoirs of wealth, and/or as a medium of exchange, including dowries.

From these definitions, applications of biotechnology to animal agriculture could range from estimating weights of two male water buffaloes in Pakistan to decide which male to castrate and use for power and which to leave intact as a breeding bull, to developing a strain of disease-resistant turkeys by adding a computer-generated DNA sequence to turkey embryos using a viral vector.

4.1.1 Return to less efficient agriculture is unlikely

In most cultures, the pace of change is accelerating. This reinforces the natural and sensible tendency to embrace the known and familiar, especially when it is perceived to be founded on good and solid principles that have proven successful. These tendencies are particularly and appropriately strong with agricultural animals. However, many trite examples exist about how we cannot go backwards with animal agriculture. For instance, sufficient farmland is not available to feed enough horses to replace automobiles, and a corollary, the problems with the horse manure that would be produced in cities. More meaningful statistics also abound: horses were (and are) much more dangerous modes of transportation than automobiles, mass transit, or planes, particularly on a per distance basis, but also on a per unit time basis.

A point not readily evident is that, with the exception of poultry for meat, there are far fewer farm animals in Europe and North America than there were a few decades ago (Seidel, 1986), and these numbers continue to decline, particularly on a per capita basis. For most species, the amount of food produced per animal is 50–150% greater than 50 years ago. Interestingly, most of this change is not due to larger animals, but rather due to an increase in efficiency. In fact, most farm animals are getting smaller (Seidel, 1986).

To illustrate efficiency, if higher pregnancy rates and fewer losses due to predation and disease enable production of 85 calves per 100 cows per year, rather than 65, it will require many fewer cows and bulls to produce the millions of calves that are marketed annually. If one adds artificial insemination, the number of mature bulls required decreases considerably. There also are statistical aspects to such data that are more complicated to appreciate. If it takes 18 rather than 36 months to grow a calf to slaughter weight, at any time there will be only half as many animals present in this phase of agriculture to produce the same amount of meat. Of course, the animals live only half as long, a fact that might be perceived as good or bad by about equal numbers of people.

In my opinion, return to the less efficient (and in many cases less humane) animal husbandry practices of the past will not occur. It would require many more animals and more arable land than exists, would be ecologically unmanageable, and would greatly increase the price of food and lower the standard of living of the poorest people. Of course, the major reason that more efficient food production is essential on a global basis is growth in population.

In a few instances, there are sensible changes of practices to those of earlier times, including less use of pesticides and increased use of crop rotation. However, it is more correct to think of these changes as appropriate use of the accumulated armamentarium of biotechnology than going backwards in time. In some cases, agricultural practices are aimed at non-agricultural needs. Some years ago I suggested that animal agriculture in parts of Europe might be more accurately described as a zoo or museum rather than as agriculture. In retrospect, I find this description insensitive. For some of these farms, the monetary governmental subsidy is of considerably greater magnitude than the value of the farm produce on the world market. Clearly, it may be very appropriate to provide such large subsidies for social, ecological, or economic reasons such as promoting tourism. Although maintaining quaint agricultural practices which are non-profitable without huge subsidies appears to be reversing time, or standing still technologically, this perhaps should no longer be defined *primarily* as production agriculture; starvation would occur on a huge scale if similar agricultural approaches were used universally.

The prison farms that evolved in the United States are an interesting variation on subsidised agriculture. They provided meaningful work for prisoners, wholesome food, and excellent test sites for some kinds of experiments. In recent years, most such farms have been phased out because they were perceived as too expensive to operate.

4.1.2 Anthropomorphic arrogance in animal husbandry

Almost every husbandry practice, whether of a biotechnological nature or not, and including possessing any domesticated animals, is offensive to one person or another. Note that there is some evidence that domestication of dogs and likely other species might most accurately be described as evolutionary adaptation of animals to people rather than active domestication (Curtis, 1993; Morey, 1994). Practices such as confinement (some degree of confinement is essentially universal), castration, dehorning, tail docking, freeze or hot-iron branding, and implanting agricultural animals with growth promotants are examples of controversial practices. Curiously, vaccination, hoof trimming, and hair clipping are considered more natural. Studies about the animal's perspectives on these practices are difficult, infrequent, and often ignored if they contradict pre-conceived notions, both for and against the practices. In many cases it does not seem inappropriate to decide for the animal, in the same way that we do for children with circumcision, vaccinations or surgery. Decisions to dock a lamb's tail to decrease parasitic disease or dehorn calves to prevent injury to others are made to benefit the individual at a later age or the herd as a whole. Similar public health decisions are made for the human population. For example, each year a number of young boys suffer due to complications from rubella vaccination (Preblud *et al.*, 1980), even though the vaccination is of no value to the boys themselves: the main justification of vaccination is to prevent birth defects when pregnant women become infected, as they would with higher frequency if males remained unvaccinated, and thus served as a reservoir for the disease.

One issue not widely researched is the extent to which various husbandry practices, like tail docking in sheep, are traumatic intrusions on animals' lives, relative to everyday vagaries such as being butted around by other animals. A different issue is the extent to which loss of a tail or horns or testes deprives an animal of happiness. Anthropomorphic arrogance, presuming to know what practices are most desirable for domesticated animals without studying them, should be recognized and dealt with. These issues are relevant to biotechnological interventions including genetic engineering, since one can breed for lack of horns (polled), develop methods of immunocastration, and select for short tails or even shorten tails by transgenic procedures (tailless animals sometimes occur spontaneously as do human babies with tails).

4.1.3 Anticipating consequences accurately

We cannot accurately anticipate all consequences that will occur years in the future for any action (or inaction), especially concerning genetic alterations of animals (Rollin, 1996). Many authors point out examples of problems without acknowledging examples of success. Two generalizations summarize the situation:

1. Every action or change has costs and benefits.
2. Some fraction of innovative changes will have decidedly negative, even disastrous net consequences (and, of course, the converse).

Every action is an example of the former statement; for instance, at one time it could be said that penicillin saved more lives than any single drug, and killed more reasonably healthy people than any drug due to allergic and anaphylactic reactions. It is unclear if this is true in veterinary medicine. A more benign example is that when animal productivity increases, e.g. due to improved health resulting in less morbidity and mortality, short-term prices and profitability frequently decrease due to surplus production.

Examples of well-known, unanticipated disastrous consequences include introducing rabbits to Australia (sheep and cattle also were introduced) or feeding animal by-products to cattle in the United Kingdom, clearly leading to more nutritious rations, but likely resulting in bovine spongiform encephalopathy (BSE) when practices of heat-treatment of offal changed. Clearly, no one can anticipate all the things that can go wrong or right with agricultural biotechnologies (in part because there are infinite permutations and combinations). This, however, does not justify abandoning research or refusing to implement improvements in animal husbandry. Nor does it excuse responsibility for serious thought and research into possible untoward consequences of new approaches. One reasonably robust approach is carrying out limited-scale field trials before promoting wide-scale adoption of a new practice.

4.1.4 Changing needs/wants for agricultural animals

A myriad of factors affect demand for agricultural animals and products, including market components such as cultural inertia; price, availability and convenience; and advertising. In addition, government policies such as support of research as well as taxes, laws and regulations can have considerable impact. A broad generalization seems to be that per capita use of animal products is slowly decreasing in so-called developed countries, and slowly increasing in developing countries, with the situation less clear in the least-developed countries. Although use of animal products in China and India is growing rapidly, in part due to the introduction of new biotechnologies, I would not expect convergence with per capita use in northern Europe. On the other hand, the dairy industry in India has grown to be one of the largest in volume for a single country, and likely has contributed immeasurably to the health and well-being of its citizens, particularly women and children. Even sporadically consumed small portions of nutrient-dense yoghurts, cheeses, ice cream, etc., mostly from high-fat water buffalo milk, provide considerable nutritional benefits (Gupta, 1983). A second major benefit of the dairy industry in most countries, but particularly in India, was to create a cornucopia of reasonable paying jobs for transporting, processing and marketing milk products, manufacturing related equipment and providing veterinary services, etc. These jobs are diffused throughout the country while providing a steady source of income to (mostly) small farmers. Biotechnology has had major impacts in the form of artificial insemination, vaccine development, and new bacterial cultures for fermented milk products.

A very different analysis of needs/wants for animal products emerges from surveys in the United States. One reads much about decreased demand for animal fat. This has affected the dairy industry in that premiums paid to farmers for milk fat are lower than a few years ago, but there is still a premium paid since milk fat is needed for cheese and ice cream production as well as for most other dairy products. Selective breeding is used very effectively to change milk composition to meet market demands.

There is tremendous verbiage about the need for lean beef, yet at this time in North America, lean carcasses are discounted 20% relative to fat carcasses (select versus choice grades). Although there clearly are niche markets for low-fat beef, current market signals to cattle breeders and feeders are decidedly in the other direction. Note that the market does discount the layers of waste fat outside of the meat by a second measure of carcass quality, the yield–grade system.

One curious example of a lack of concordance between what consumers say they want and what they in fact purchase is a recent study of people who described themselves as vegetarians (Schweitzer and Young, 1995). It emerged that what self-reported vegetarians in fact ate was very similar to what non-vegetarians ate, even meat; this study included 4700 people.

My conclusion from these admittedly anecdotal observations is that the true test of demand for animal products will usually be the market. Although the market can be influenced by factors such as promotion, education, advertising and price, the driving force is in fact what consumers purchase. The market also has its unpredictable elements, frequently with no proven scientific basis, such as panic about contaminants and 'politically correct' sequelae. Agricultural biotechnology will affect the market via price as well as factors such as safety, uniformity within products, and variety of products.

4.2 ANIMAL WELL-BEING

4.2.1 Identification and records

Identification of individual animals is of importance for establishing ownership, health purposes (both for animals themselves and for public health), and husbandry, including animal breeding records for biotechnology applications. Procedures to establish identity are usually invasive to the animal. Some countries now require 'permanent' identification of certain agricultural animals by law. Methods of identification include hot- or cold-iron branding, photographs, nose prints, tattoos, ear notching, ear tags, leg bands and implantable electronic devices. For most purposes, easy readability from a distance is essential.

Biotechnology will increasingly be applied to animal identification. Blood typing has been used for years for certain niche applications, but it may be replaced by DNA typing, which is much more capable of discrimination among animals. DNA typing can be used on milk, blood or any other body tissues. Special applications include determining parentage of offspring in difficult situations such as multi-sire mating pastures. This approach circumvents misidentification due to new-born switching at birth, incorrectly tagging animals that lost identification, and matings by males that breach fences. These kinds of misidentification occur on the order of 5% of the time in pure-bred herds of cattle, which presumably have better management than grade (non-pure-bred) herds. Interestingly, the highest error rates in identification as verified by blood typing occur with natural breeding, less so with artificial insemination, and least with embryo transfer and related technologies.

An interesting – some would say insidious – aspect of records used for routine purposes such as health decisions and ownership is that they can also be used for biotechnology purposes. One of the first major applications of high-speed (for those days) computers was analysis of milk production records of dairy cows, primarily for purposes of calculating the fat content and quantity of milk. This enabled farmers to make intelligent husbandry decisions, for example when to replace unprofitable cows. When information on parentage was added, it became possible to use this information for very sophisticated selective breeding approaches. The needs of this application even resulted in more rapid development of certain fields of mathematics (Henderson, 1953), which in turn have been applied to other areas of animal breeding.

The application of such information can greatly speed up genetic changes in populations. As with any information, it can be put to good or careless use. For example, one might select for rapid growth but in the process decrease fertility. One greatly valued current application is to use computer-generated predictions of birthweights of a particular bull's progeny to select service sires for primiparous pregnancies, and thus decrease the incidence of traumatic births. Information generated in this way is much more accurate than with more primitive methods (e.g. just using the service sires' own birthweights), and the information is widely used.

The use of records will increase greatly as DNA typing is used more widely. The term 'marker-assisted selection' describes one area of such applications. Direct molecular selection such as for B alleles for casein is now being used to select bulls whose daughters' milk will result in higher yields of cheese.

4.2.2 Minimizing surgical and related interventions

Practices including dehorning, castration, ovariectomy, tail docking, debeaking, etc., are controversial. Of course, some of these practices can be omitted completely if one is willing to accept more disease and injury to animals; elimination of others is more problematic, for example castration of male pigs. Without castration, pig meat is unpalatable due to boar taint. Virtually all of these procedures can be replaced with less intrusive manipulations using biotechnology. Already mentioned were taking advantage of naturally occurring mutations such as lack of horns or tails and breeding those characteristics into the population. Angus cattle represent a breed in which the polled (hornless) condition has been fixed for more than a century. A related example is selection of pigs for lack of tusks. The genes (and alleles) for many of these traits are now being identified, which can greatly increase rates of introducing such traits into populations, for example distinguishing homo- and heterozygotes by DNA typing, or even by using transgenic approaches. Failing to use heterozygotes that carry deleterious recessive alleles for breeding purposes is an indirect, but powerful, use of biotechnology to decrease animal suffering.

Another example is vaccines for immunocastration. There is no doubt that castration can be mimicked by vaccination if sufficient investment is made. While I do not wish to delve into the ethics of animal castration, it is clear that males of many domestic species are dangerous to other animals and humans. In most years of this century, more people in the United States were killed by bulls than all other species of animals combined, except *Homo sapiens* (mostly males too!). Death rates of dairy farmers and their family members due to bulls have plummeted since widespread application of artificial insemination, because of decreased bull numbers and housing of bulls at well-constructed sites.

In the past, it was common to spay (ovariectomize) heifers to improve feedlot performance and minimize bruises to carcasses due to injuries from homosexual mounting during oestrus. For most heifers currently fattened in North America, this problem has been solved by feeding the orally active progestin, melengesterol acetate, which is chemically similar to one component of oral contraceptives used by women; this compound inhibits sexual behaviour in cattle and has slight growth-promoting attributes. Interestingly, this drug is approved for use by the US Food and Drug Administration only for its growth-promoting characteristics.

A functional variation on this theme is castrating bulls at a young age when it is less painful (and psychologically less invasive?) and replacing the growth-promoting properties of testicular hormones with an anabolic ear implant. These biotechnological substitutes for painful or invasive practices should be encouraged in my opinion.

Many people have no quibble about surgery and related interventions as long as appropriate anaesthesia and/or analgesia are used. However, in some instances these ameliorations may be worse than the procedure itself. For example, catching an animal to give anaesthesia, keeping it caught while the anaesthetic takes effect, then docking its tail, then catching it half a day later to give an analgesic may be more difficult for the animal than catching it, docking the tail instantly, and letting it get on with life. Some years ago, I had a finger tip instantly amputated in an agricultural accident; the pain of amputation was nil, but the embarrassment substantial! It seems to me that the appropriate studies need to be carried out rather than simply minimizing what is aesthetically displeasing.

Bloodless approaches are frequently used because they *seem* more humane, e.g. caustic dehorning pastes or elastrators (strong rubber bands) for castration or tail docking. In such cases, blood supply is compromised, and the testes or tails simply whither and drop off in a month or so. Although bloodless, these may result in some pain and considerable itching, and also might be replaced with genetic or other suitable biotechnological approaches.

For future farm animals, I expect routine biotechnological circumvention of numerous current practices such as those just described. In many cases, the practices will be profitable and will improve animal well-being. Cost–benefit ratios of any mandated practices should be considered carefully to avoid undesirable sequelae such as accelerating the demise of small farms. New practices that decrease animal well-being substantially should be introduced cautiously and only for justifiable reasons.

4.2.3 Health considerations

One of the greatest biotechnological interventions of all time is vaccination. Yet there is still considerable variability in effectiveness of vaccination programmes, especially for farm animals. However, due to molecular biology studies, the physiology of the immune system is increasingly understood. Also, recombinant antigens are being used to produce much improved vaccines. Even so, it will never be practical to vaccinate against every imaginable disease in livestock for a number of logistical reasons.

Many people feel that the biotechnological approach of choice for dealing with diseases is to breed for, or add, genes for generalized or specific disease resistance. This is clearly possible and desirable, but only to a point. First, selection for increased immunity to one disease will likely result in decreased immunity to other diseases, and not in a predictable way. Second, if one selects for too much immunological activity, there will likely be an increase in autoimmune diseases in addition to interference with normal functions such as reproduction.

Because of the complexities described, management of vaccination and other aspects of health programmes will always be a balancing act of selective breeding and selective vaccination combined with minimizing exposure to disease-causing microorganisms, including some degree of confinement, and even quarantine. Improved diagnostic and therapeutic methods will continue to minimize animal suffering when illness does occur if they are affordable.

4.2.4 Behavioural aspects

Animal mental health is exceedingly difficult to study, and involves considerable anthropomorphic baggage. Even so, this is no excuse for inaction. Some sensible measures of animal well-being include normal growth, reproduction and resistance to disease. Another set includes measuring preferences, although, as with children, just because an animal prefers something does not necessarily mean it is good for it. Obviously, absence of aberrant behaviours such as pacing, self-mutilation, cannibalism, and lack of parental behaviour constitutes other appropriate measures.

A group of measures that I have some difficulty with are species-specific behaviours (see Chapter 12). For example, if a pig does not act like a pig but instead like a dog, its *telos* may be violated. Negative examples abound; for example, carnivores kill things to eat in the wild, so carnivores not permitted this behaviour are treated unfairly, or ruminants that are fed diets which lead to little rumination are being deprived, despite the fact that they prefer grain, or use of artificial insemination deprives animals of sexual pleasure. It seems incoherent to make the latter argument at the same time as promoting neutering of pet animals.

Obviously, these are not 'either/or' issues, but rather a question of striking a balance. I support considering *telos* as one component of evaluating biotechnological practices. I also support additional thinking on the subject. For example, one striking way to alter behaviour is to wean young from their mothers at a few days of age and have humans bring them milk, which is how dairy calves have been raised for years. Such calves fix on humans, similar to goslings in the classic studies of Lorenz (1979). Unless maltreated at a later time, such calves come running toward people whenever they appear. In contrast, calves that nurse their mothers and are raised extensively tend to regard humans suspiciously, and run away rather than toward them, at least in the absence of their becoming conditioned to come for feed. These behaviours were demonstrated most dramatically when applying embryo transfer technology by having beef animals as the surrogate mothers for embryos of dairy breeds. Resulting calves did not behave like dairy heifers. A related example is stallions fixing on people when treated over-affectionately as foals, and showing little sexual interest in mares. With wild animals such as bears, deer, mountain lions and wolves, this type of behaviour usually leads to disaster for the animals. However, should behaviours such as domestic animals enjoying being around humans be considered inappropriate or aberrant?

4.3 ANIMAL MANAGEMENT

4.3.1 Housing and environment

There is hardly a more controversial subject than what is appropriate for the immediate physical environment for domestic animals, particularly the degree of confinement. It is clearly possible to breed animals to fit particular environments; for example, *Bos indicus* cattle are more heat-tolerant than *Bos taurus*. Animals selected for optimum performance in a relatively constant environment are unlikely to adapt readily to environmental extremes, and animals that have the genetic makeup to adapt to extremes likely will not perform as well as the specialized animals selected for constant conditions.

Most genetic changes can probably be accelerated as we learn more about the genes involved. For example, heat-shock protein genes are transcribed when body temperature is higher than normal, and much is being learned about regulation of those genes. Similar advances are being made concerning genes controlling stress, appetite, aggression, sexual behaviour, etc. (Gross and Siegel, 1985). For me, changing animals genetically so that they are appropriate for a particular environment seems sensible. This has obviously already occurred for most domestic animals to a considerable extent.

4.3.2 Reproductive technologies

A truly remarkable menu of reproductive biotechnologies has become available, as discussed by Ian Wilmut in Chapter 2. These technologies increase rates of change due to selective breeding. To the extent that the genetic goals are desirable, this can be very useful. However, these biotechnologies transcend selective breeding applications. This is perhaps best illustrated by cryopreservation of gametes and embryos. Although frequently a by-product of animal breeding programmes, many millions of doses of semen and hundreds of thousands of embryos are stored in liquid nitrogen, likely never to be thawed for production purposes because they have become genetically obsolete. A corollary is that many strains of domestic animals, for example numerous breeds of cattle, are dying out because their maintenance can no longer be justified economically. However, while maintained in liquid nitrogen, the frozen sperm and embryos will probably remain viable for a few thousand years. Much of this genetic material is appropriately discarded each year, but some remains in inventory indefinitely. Programmes to conserve such material deliberately are under way in several countries. Although such conservation may have limited justification for animal agriculture, it is a form of insurance, and has cultural value. Note that such programmes are completely dependent on biotechnologies such as semen collection, cryopreservation, artificial insemination and embryo transfer.

Another example of an indirect benefit of reproductive biotechnology is use of the embryo (frozen or not) as a means of moving genetics from country to country, or from herds with infectious diseases to those free of the disease without introducing additional diseases. The reason for this fortuitous situation is that it is possible to sanitize individual embryos for most diseases (Wrathall, 1995). In addition, extreme dilution of fluids surrounding embryos in the course of collection and transfer adds another layer of safety. There is not one documented case of disease being transmitted from one country to another in the course of international exchange of hundreds of thousands of embryos (Wrathall, 1995). The process is inherently safer in this respect than moving genetic material via semen or live animals. The stress of shipping animals and holding them in quarantine for months also is circumvented. Although semen and live animals can be moved internationally without moving diseases, the process is more complex and costly than with embryos.

An example of an unanticipated problem with a reproductive biotechnology is the occurrence of abnormalities in lambs and calves when early-stage embryos are cultured with certain *in vitro* procedures for 5 or more days or when certain methods of cloning by nuclear transplantation are used (Garry *et al.*, 1996). At birth, some of these calves and lambs are metabolically abnormal, and some are up to 50% larger than normal. Neonatal mortality of such offspring is higher than normal, primarily due to the metabolic abnormalities. These kinds of abnormalities occur in about 1% of calves produced by natural mating, and the incidence seems to increase to about 10–15% with long-term *in vitro* culture of embryos, and close to 50% with animals produced via cloning by nuclear transplantation. Because of these problems, cloning by nuclear transplantation is currently not used as an agricultural production biotechnology, but is still strictly experimental.

A striking feature of these problems is their transient nature. If animals are given intensive care for the first day or two after birth, they become essentially normal; furthermore, the abnormalities do not appear to be passed on to the next generation, and thus are not genetic. This is an example of an unexpected problem from a biotechnology that was detected before widespread application. It will likely be used as a model to study the cause of the abnormalities, which may eventually be applied to circumventing the tens of millions of neonatal deaths annually due to this problem in the 'normal' population.

For most livestock enterprises, approximately one- to two-thirds of the costs are for feed, and the remaining costs for everything else include housing, labour, veterinary care, purchase of animals, interest, etc. Thus, managing feed costs and getting the maximum nutrition from feed will continue to receive attention from farmers. Another important concept is that nutritional status is closely related to every other component of animal husbandry, including health, reproduction and growth rate.

Various biotechnologies are closely tied to nutrition. For example, swine rations are formulated with vitamins and amino acids, as opposed to the older, absolute requirements for animal products such as blood meal and meat scraps to meet vitamin needs. Although ruminants do not require vitamins as feed supplements, health and performance improve with the addition of such supplements in many situations.

There are many components of nutritional biotechnology involving use of growth hormones. While there is much controversy over their use medically as well as in animal husbandry, it is clear that supplemental growth hormone builds upon normal physiology. For example, the body uses growth hormone to direct nutrients to milk production rather than to body fat during lactation, and animals that naturally grow faster or produce more milk than average have higher concentrations of growth hormone in their bodies (Figure 4.1). A good demonstration of the adage that 'you don't get something for nothing' is that these higher-producing animals, whether from increased natural growth hormone from their own genetics or from supplemental growth hormone, require more careful management, including more feed, than average-producing animals. Use of supplemental growth hormone clearly is inappropriate without adjusting nutrition to match needs.

Appropriate experiments also can demonstrate when a biotechnological product or approach is inappropriate. For example, about 40 years ago it was shown that milk production can be increased dramatically with thyrotropin, a product that increases metabolic rate by stimulating the thyroid gland. However, this was not a normal physiological mechanism for increasing milk production, and there were long-term

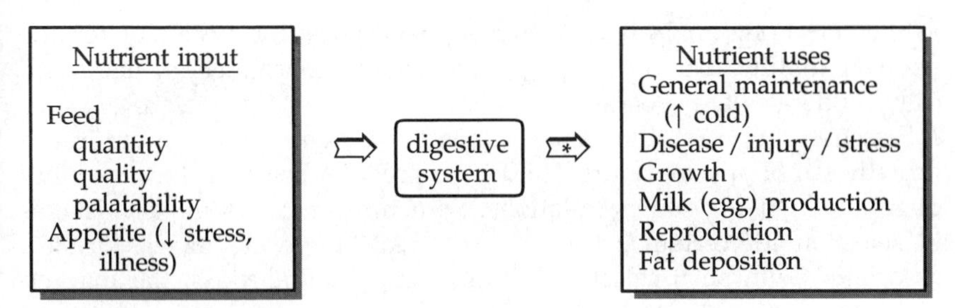

= main point of regulation; for example, growth hormone causes nutrients to be used for growth in young animals and milk production in older animals; low growth hormone results in fat deposition.

Figure 4.1 Nutrient partitioning.

detrimental effects on animals, including excessive weight loss and decreased reproductive function.

Anabolic agents, including steroids, represent another controversial application of nutritional biotechnology. Most of these build on natural physiological principles such as increased muscle mass in most male mammals due to testosterone. Another mechanism exploited is to change the microbial physiology in the rumen with ionophores so that less energy is belched into the atmosphere by ruminants, primarily in the form of methane. It also is clear that such agents can be abused by inappropriate application or diversion to illegal human use.

4.4 SELECTIVE BREEDING

4.4.1 Principles

Conventional selective breeding is a variation on evolution, and many principles of evolution apply, such as 'survival of the fittest'. Fittest of course is defined on the basis of the animal's environment, both in a physical sense, e.g. climate, altitude, etc., and a biological sense, e.g. forest or grassland, availability of food, and effects of predators. The characteristics of those animals that adapt – especially in the sense that they reproduce successfully – end up as the characteristics of the species.

In the same sense, those animals that adapt to environments of domestication and reproduce give rise to the characteristics of domestic animals. This undercurrent of 'natural' selection is automatically present when one superimposes 'artificial' selection such as for greater wool production, faster racing horses or more protein yield in milk.

While selection can lead to extreme phenotypes, the necessity for the animals to function normally, particularly in reproduction, automatically moderates the process.

Selection imposed by *Homo sapiens*, for example for more rapid growth, is not any more profound than predator selection for capability of rapid flight in wild populations. Sometimes farmers make mistakes in selection for certain extremes, or at least there are increased costs associated with such selection. While it may sound crass, the marketplace is a potent moderating force in selection, since agricultural animals need to be profitable to continue to evolve. Animals that die because they are too large at birth, or are unhealthy because of selection for too large a mammary system (or for any other reason) will be unprofitable, and their lines will die out. Nature works similarly via evolution. For example, it has been postulated that the great Irish deer died out some millenia ago because antlers on the males became so large as to make these animals poorly functional (but see Gould, 1980, p. 90) (Short, 1976). These evolutionary dead ends could clearly be thought of as mistakes of nature. Even before the appearance of *Homo sapiens*, many more species became extinct than survived. Note that this does not justify extinction of species due to human activities.

Another example of mistakes of nature is that embryonic death occurs in a high percentage of pregnancies. This is best documented in domestic animals, with 20–30% embryonic death and *Homo sapiens*, with in excess of 50% embryonic death. Much of this is due to genetic mistakes, and embryonic death is, in part, a correction mechanism for these mistakes. A broader perspective on how nature works can be gleaned from the classic paper of the Nobel laureate, Jacob (1977), 'Evolution and tinkering'.

4.4.2 Transgenic technology – a special case of selective breeding

Transgenic technology can be defined in many ways, for example as deliberate alteration of the sequence of nucleotides in DNA of gametes or early embryos. Such alterations include insertions, deletions, substitutions, and rearrangement of nucleotides. These alterations also occur naturally in all organisms with considerable frequency, and are termed mutations. Many of these are repaired, and many result in cell death. Others result in genetic diseases, and still others help maintain the natural genetic variation that makes each animal different. Such modifications are the very basis of all genetic variation and the sequelae of evolution, including speciation and natural selection.

The natural causes of DNA alteration include sunlight, retroviruses, and mistakes during DNA synthesis and cell division. Most of these occur in somatic cells, but the more protected germ cells also are affected. The overwhelming majority of these alterations in DNA have no measurable effect on multicellular organisms such as farm animals because the cells die out or are not used. As an example, only some dozens of the trillions of sperm produced by a male bird or mammal ever fertilize an ovum.

Natural selection and selective breeding take advantage of the genetic variation described. Although transgenic technology is a very potent experimental tool, and will eventually be applied for production in farm animals, I can think of no aspect that does not occur naturally. DNA is even exchanged between species in nature, for example via viruses. What is different is that there is less randomness to transgenic DNA changes than in nature, because there usually is some designed DNA alteration for a desired purpose.

Most natural mutations that are present in early embryos are detrimental, resulting in fertilization failure, embryonic death or congenital abnormalities. However, some are beneficial in some respects and detrimental in others; rarely they may be strictly beneficial. With the current state of the art, the same holds for transgenic alterations; that is, most are detrimental, and others have advantages and disadvantages, e.g. increased disease resistance but lower growth rates, or increased growth but lower fertility. There are several exceptions to the above generalizations; for example, transgenic technology has been used to change detrimental mutations back to normal.

Although I expect a few spectacular successes within the next decade with transgenic technology for agricultural purposes – for example, resistance to specific diseases – the logistical complexities are so enormous that transgenics are unlikely to become a routine component in breeding agricultural animals for some time. Low success rates, long generation intervals, huge expense, lack of basic genetic information, lack of inbred lines, regulatory costs, and social concerns are examples of impediments to applying this technology. Also, as discussed above, most transgenic modifications, as with naturally occurring modifications (mutations), will result in abnormalities (many of which result in embryonic death).

Another major problem is to make agricultural animals homozygous for a transgene. This is rather simple to do in mice via mating between relatives or using embryonic stem cells. Such cells are not currently available for farm animals, and matings between relatives results in inbreeding (not a problem with inbred mice), which depresses phenotypic performance, making it difficult even to evaluate such animals.

A special case is creating transgenic animals to produce pharmaceuticals in milk, blood or eggs. This is likely to be reasonably successful, but will involve very few animals, and applications will be more industrial than agricultural enterprises. Note that there have been numerous examples of using non-transgenic animals for pharmaceuticals, for example collecting urine from hundreds of pregnant mares to produce conjugated oestrogens for treatment of post-menopausal women. Another special case of transgenics is to add the human lactoferrin gene to cattle to produce improved milk for human infants (Krimpenfort *et al.*, 1991).

While I see only limited application of transgenic methods to production agriculture in the near future, I envision excellent opportunities to use transgenic technology in research. With regulations already in place, I see no difference in potential for abuse than with a myriad of other research tools and models. Furthermore, some kinds of experimental questions can be answered with many fewer transgenic than non-transgenic animals. In the sense that transgenics will provide basic information about disease, growth, milk production, reproduction, and nutrition, this technology will be very important to production agriculture as that information is applied.

4.4.3 Selection for specific traits

Many aspects of animal agriculture can be manipulated by selective breeding. Table 4.1 presents examples of needs that are amenable to selective breeding as well as to other approaches. It is possible to select for almost any trait that can be measured, from fertility to tail length. Most traits have both genetic and environmental components. For example, size depends both on level of nutrition and the genetics of the individual. Selective breeding is obviously more effective when traits have a large genetic component, or high heritability, for example

Table 4.1 Agricultural animal biotechnology needs

Amenable to selective breeding	*Requires strategies other than selective breeding*
Lack of horns (polled)	Early pregnancy tests
Circumvent seasonal breeding	Precise timing of ovulation
Lactation without pregnancy	Pre-conception sex selection
Chickens that lay one egg each day	Immunocastration (♂ and ♀)
Decreased feet and leg problems	Precise timing of birth
Decrease deleterious alleles	*In vitro* gametogenesis
Improved passive immunity transfer to neonates	Easy identification
Disease resistance	On-the-farm diagnostic kits for disease

percentage of fat in milk, and less effective when the environmental component dominates, e.g. for fertility. Traits selected in dairy cattle include: amount of milk, percentage of protein in milk, size of cow, shape of mammary system, size of teats, milking speed, docility, shape of the foot, etc. Most of these traits are moderately heritable, which means that rates of change are modest, but substantial over many generations. Note that as selection is applied to more traits in a population, the rate of change for each trait decreases.

One other important concept is heterosis, which results in improved traits by crossing different lines, breeds, or even subspecies. Such cross-breeding increases heterozygosity and is especially useful for improving 'survival' traits of low heritability such as fertility, disease resistance, neonatal survival and mothering ability. A huge problem is that relatively pure lines of parents must be maintained to obtain the advantages of hybrid vigour resulting from cross-breeding. Biotechnology has a significant impact because few males need to be retained if artificial insemination is used. Since cross-bred animals will not breed true, cloning ideal individuals by nuclear transplantation may eventually be used extensively to circumvent this problem.

4.5 CONCLUSIONS

Almost every aspect of animal agriculture involves biotechnology of some form. Biotechnologies are tools that can be applied in beneficial or detrimental ways. In my opinion, failure to use available tools when there are pressing problems to be solved is as morally indefensible as is using biotechnology inappropriately, for example by increasing animal suffering. Clearly, some amount of regulation of biotechnologies is appropriate, even though regulations can be used for unintended purposes such as delaying work for political purposes. Dialogue is one of the more important tools for maintaining a healthy animal agriculture.

REFERENCES

Curtis, S.E. (1993) Variations in U.S. animal production systems: current trends and their impacts on animal well-being and the economics of production, in *Food Animal Well-being*, Purdue University, West Lafayette, IN, pp. 55–69.

Garry, F.B., Adams, R., McCann, J.P. and Odde, K.G. (1996) Postnatal characteristics of calves produced by nuclear transfer cloning. *Theriogenology*, **45**, 141–52.

Gould, S.J. (1980) *Ever Since Darwin*, Pelican, Harmondsworth.

Gross, W.B. and Siegel, P.G. (1985) Selective breeding of chickens for corticosterone response to social stress. *Poultry Science*, **64**, 2230–3.

Gupta, P.R. (1983) *Dairy India 1983*, Dairy Yearbook India, New Delhi.

Henderson, C.R. (1953) Estimation of variance and covariance components. *Biometrics*, **9**, 226–52.

Jacob, F. (1977) Evolution and tinkering. *Science*, **196**, 1161–6.

Krimpenfort, P., Rademakers, A., Eyestone, W., van der Schans, A., van den Droek, S., Kooiman, P., Kootwijk, E., Platenburg, G., Pieper, F., Strijker, R. and de Boer, H. (1991) Generation of transgenic dairy cattle using 'in vitro' embryo production. *Biotechnology*, **9**, 844–7.

Lorenz, K. (1979) *The Year of the Greylag Goose*, Harcourt Brace Jovanovich, New York.

Morey, D.F. (1994) The early evolution of the domestic dog. *American Scientist*, **82**, 336–47.

Preblud, S.R., Serdulla, M.K., Frank, J.A., Jr, Brandling-Bennett, A.D. and Hinman, A.R. (1980) Rubella vaccination in the United States: a ten-year review. *Epidemiologic Reviews*, **2**, 171–94.

Rollin, B.E. (1996) Bad ethics, good ethics and the genetic engineering of animals in agriculture. *Journal of Animal Science*, **74**, 535–41.

Schweitzer, C. and Young, M.K. (1995) Eating in America today: nutrition implications for the meat industry. *Proceedings of Conference: Demand Strategies*, **1**, 17–21.

Seidel, G.E., Jr (1986) Characteristics of future agricultural animals, in *Genetic Engineering of Animals* (eds J.W. Evans and A. Hollaender), Plenum Publishing Corp., New York, pp. 299–310.

Short, R.V. (1976) The origin of species, in *Reproduction in Mammals 6. The Evolution of Reproduction* (eds C.R. Austin and R.V. Short), Cambridge University Press, Cambridge, pp. 110–48.

Wrathall, A.E. (1995) Embryo transfer and disease transmission in livestock: A review of recent research. *Theriogenology*, **43**, 81–8.

5

The effects of biotechnology on animal welfare

Donald M. Broom

5.1 SUMMARY

Some effects of biotechnology on animals are obvious but many require careful scientific study to evaluate properly. This chapter is about what should be done – for little has been done.

Techniques for the scientific assessment of the welfare of animals have developed rapidly in recent years and many of these can be applied to animals which are genetically modified, or treated with biotechnology products. Each modified strain or treated animal should be compared with unmodified or untreated animals using measures of physiology, behaviour, anatomy, immune system function, pathological change, growth, reproduction and longevity. Using a wide range of measurements, any increased levels of pain, fear or distress should be revealed. These measurements show how poor welfare is, but other studies can indicate the extent to which the welfare of such individuals can be good. These methods should also be used to assess the effects of embryo transfer.

A potential problem in using some welfare assessment techniques is that an animal may be affected by the genetic modification in a way which alters the aspect of its biology which is being measured. For example, if a preference is tested but the relevant sensory functioning has genetically altered, or if an adrenal response is to be measured and adrenal functioning has been changed in the modified animals, then the measurement procedure would be invalid. Such problems must be considered wherever the welfare of transgenic animals is to be assessed.

Every person who works with transgenic or treated animals should be aware of how to assess their welfare and should act so as to avoid or minimize poor welfare. When a transgenic animal has been developed, details of any effects of the genetic manipulation on the welfare of the animal should be part of the specification available for potential users. While there is some legislation concerning the welfare of animals which are part of experiments, in most countries the only legislation relevant to their welfare after this is of a general nature – for example, that concerning cruelty to animals. Such legislation is not adequate for transgenic animals or animals treated with biotechnology products. The legislation should stipulate that no genetically modified or treated animal should be permitted to be used commercially until comprehensive studies of the welfare of the animal have been carried out during two generations and continuing for maximum commercial life. The decision as to whether the use of the modified animal is permitted should depend upon whether there is a net benefit for the welfare of all animals, including humans. A commercial profit is not sufficient justification for modifying an animal in such a way that its welfare is poor.

5.2 INTRODUCTION

If animals are to be produced as a consequence of transgenic procedures or treated with biotechnology products, two important questions which need to be answered are: (i) whether or not there are positive or negative effects on welfare; and (ii) the magnitude of those effects. Hence, it is essential to use a definition of welfare which allows scientific measurement. The welfare of an animal is its state as regards its attempts to cope with its environment (Broom, 1986). This state refers to the amount of difficulty which the individual has in trying to cope with its environment and the extent to which it is failing to cope. When it fails to cope, or seems likely to do so, it is said to be stressed. The state of the animal includes the feelings of the individual, which may be good feelings or suffering (Broom, 1996). In order for it to be a useful scientific concept we must be able to think of welfare as varying over a range from very good to very poor.

There is a rather small range of measures which give us information about welfare at the good end of the scale (Table 5.1) but a much longer list of measures which can tell us about how poor the welfare of the animal is (Table 5.2). All of these measures and the concept of welfare are explained in detail by Broom (1991) and Broom and Johnson (1993).

Measurements of animal welfare should be made in an objective, scientific way. Once they are made, moral judgements concerning what is acceptable can be made more easily. However, the process of scientific evaluation should be kept separate from the moral judgement.

Table 5.1 Measures of good welfare

- Variety of normal behaviours shown
- Extent to which strongly preferred behaviours can be shown
- Physiological indicators of pleasure
- Behavioural indicators of pleasure.

From Broom and Johnson (1993).

Table 5.2 Measures of poor welfare

- Reduced life expectancy
- Reduced ability to grow or breed
- Body damage
- Disease
- Immunosuppression
- Physiological attempts to cope
- Behavioural attempts to cope
- Behaviour pathology
- Self-narcotization
- Extent of behavioural aversion shown
- Extent of suppression of normal behaviour
- Extent to which normal physiological processes and anatomical development are prevented.

From Broom and Johnson (1993).

5.3 BREEDING AND WELFARE

Conventional breeding methods can change animals in such a way that they have more difficulty in coping or are more likely to fail to cope (Broom, 1994, 1995). One example of such an effect is the sensory, neurological or orthopaedic defect found commonly in certain breeds of dog. Others are the effects of the genes promoting obesity in mice, double muscling linked to parturition problems in cattle, and many examples of selection promoting fast growth and large muscles in farm animals. Modern strains of pigs have relatively larger muscle blocks, more anaerobic fibres and smaller hearts than have the ancestral strains (Dämmrich, 1987). They are more likely to die or to become distressed during any activity. Modern broiler strains grow to a weight of 2–2.5 kg in 37 days as compared with 12 weeks 30 years ago. Indeed, the maturation age is decreasing by one day per year at present. Their muscles and guts grow very fast but the skeleton and cardiovascular system do not. Hence, many of the birds have leg problems, such as tibial dyschondroplasia or femoral head necrosis, or cardiovascular malfunction such as that which gives rise to ascites.

It is clear that the welfare of meat-producing animals which are growing too fast for their legs and heart is becoming poorer and poorer because of this genetic selection and that the continuation of this trend is morally wrong. The competitive nature of the industry makes it difficult for individual producers to take action to reverse the trend, but many of them are now breaking the cruelty laws. This point is made here because there is pressure on those concerned with genetic engineering to make such animals grow even faster.

5.4 THE WELFARE OF TRANSGENIC ANIMALS

Transgenesis can result in better welfare, in no change from the average for unmodified animals, or in poorer welfare. Some of the points concerning welfare assessment are explained in more detail later in this chapter.

Some genetic manipulations can be beneficial to the modified animals. For example, in the work on avian leucosis virus resistance, if genes conferring disease resistance are inserted into the genome of an individual, then the welfare of the modified individual is better than that of the unmodified individual. If the animal can cope with disease challenge better, then its welfare is slightly improved for most of the time and very much improved in the circumstance where disease challenge occurs.

When the transgenic animal is modified so that it can produce a novel protein in its blood or milk, there may be no effect at all on its welfare. However, there could be some adverse effect, and the predictability of that effect will vary according to the precision of the transgenesis procedure. Gene transfer by introducing embryonic stem cells into a blastocyst is more predictable in its effects than the introduction of genetic material by microinjection.

The production of disease-susceptible animals by transgenesis, so that the animals can be used in medical research, will result in poorer welfare whenever the gene is expressed. The extent of the poor welfare will differ considerably according to the level of expression and the disease state.

If, as discussed in the preceding section, the animals produced as a result of transgenesis were modified in a way which increased their growth rate, or the growth of a particular organ, or differential growth in such a way that an already productive genetic strain was made even more productive, there is a serious risk that the welfare of the animals would be worse as a direct consequence of the manipulation. Those carrying out such work should consider whether the animals are already close to some biological limit to adaptability before proceeding.

Genetic manipulation could affect sensory functioning, the structure of bones or muscles, hormone production, detoxification ability, neural functioning, etc. The question which must be considered is not whether or not there is a change but whether there is a change which affects the animal's welfare. In some cases, any effects of the genetic modification on the welfare of other individuals must be considered, for example if the modified individual were made more aggressive.

In a study of the effects on welfare of transgenesis or treatment with biotechnology products, control animals which have not been modified or treated should also be used. A wide range of measures of welfare are necessary because the actual effects on the individual will seldom be known and also because species and individuals vary, both in the methods which they use to try to cope with adversity and in the measurable signs of failure to cope. A simple welfare indicator could show that welfare is poor but absence of an effect on one indicator of poor welfare does not mean that the welfare is good. For example, if the major effect of a manipulation was a behavioural abnormality or an increase in disease susceptibility but only growth rate was measured, a spurious result could be obtained. The choice of measurements should include the main methods of assessing poor welfare which are mentioned here but often it will be obvious from a preliminary study of morphology, or a clinical examination, which measurements of function or of pathology will be most relevant.

The effects of genetic manipulation or treatment with biotechnology products may not be apparent at all stages of life, so the animal must be studied at different stages, including the oldest age likely to be reached during usage. Some effects may be evident in the second generation but not in the first, so modified animals should be studied for two generations.

5.5 THE WELFARE OF ANIMALS TREATED WITH BIOTECHNOLOGY PRODUCTS

Biotechnology products could be identical to naturally occurring chemicals such as hormones. However, since they are often produced by bacteria they may not be identical. For example, recombinant bovine somatotrophin (BST) differs slightly from the natural BST. Some of such products may be completely different from any chemical normally found in the species. In addition to this possible difference, the quantities of the products which can be given to animals are often much greater than normal physiological levels. As a consequence of these important possibilities for difference, the effects of biotechnology products on welfare should be assessed in the same way as the effects of transgenesis and should be subject to the same legislative controls.

5.6 THE EFFECTS OF EMBRYO TRANSFER ON WELFARE

There are two areas for investigation in relation to embryo transfer. The first comprises the immediate effects of the procedures themselves and the second comprises the effects during pregnancy, at parturition and soon afterwards.

The collection of eggs and the insertion of eggs into another female animal can be carried out without the necessity for surgery in a large animal such as a cow. However, in animals of the size of sheep or pigs or smaller, an incision must be made in the abdominal cavity to carry out the procedures. The effects of these procedures can be monitored in the same ways as those described for transgenic animals.

When the insertion of an egg into a female mammal results in the growth of a fetus which is larger or of a different shape from the fetus that the mother would produce after mating with a male of similar type, problems may occur during pregnancy and at parturition. Some problems during pregnancy – and most problems at parturition – result in poor welfare in the mother, the young animal or both.

5.7 MEASURES OF WELFARE

5.7.1 Preference studies

As listed in Table 5.1, an important technique in welfare research is the measurement of the strength of animal preferences. Studies of positive preferences involve choice tests, often with some operant technique being used to indicate how hard the individual will work to obtain a particular resource or have the opportunity to carry out a certain behaviour (Dawkins, 1983; Arey, 1992; Manser *et al.*, 1996). A possible problem which must be considered when using such methods is that the sensory or motor ability of the animal might be altered by the transgenesis. Positive preferences could on occasion give ambiguous results, but in general it would be expected that what is important to normal animals would also be important to transgenic animals or animals treated with biotechnology products. Studies of aversion and its strength would be of value in studies of transgenic animals. If, for example, the modified animal were changed so that bright light was aversive, the extent of the aversion could be measured in studies of actual movement away from light, of reluctance to be moved towards a well-lit place or of some specific task which had to be performed in order to avoid the onset of bright light.

5.7.2 Reproductive success

Some zoo animals cannot breed, when potential breeding partners are present, because of an inadequacy in their environment. The welfare of these animals is less good than that of animals which can breed. Inability to reproduce would be an indicator of poor welfare in transgenic or treated animals.

5.7.3 Growth, weight loss, mortality and life expectancy

If control animals can grow or maintain weight in a given situation but modified animals fail to grow or lose weight, this would indicate poorer welfare in the latter. Abnormal weight gain could also indicate a problem. It is important to use a biologically relevant control in such studies. An animal could be losing weight because it is lactating or is a reproductively active male, like a red deer in rut. On the other hand, an animal which is in the pre-hibernation condition could put on a great deal of weight.

Measures of mortality rates have long been used in studies of the effects of housing conditions or management methods on animal welfare. As Hurnick and Lehman (1988) have pointed out, a housing condition, management method or treatment which resulted in the animal having a lower life expectancy indicates poorer welfare in that condition or with that treatment. Indeed, a human who died early because of some form of self-abuse or an energetic lifestyle would be considered to have been under greater stress than a similar but longer-lived person. Other examples include cetaceans which die early in poor zoo conditions and dairy cows which do not live as long under the very high production conditions of recent years as they did when their metabolic pace of life was lower (Agger, 1983; Broom, 1993b).

5.7.4 Physiological measures

Aspects of normal physiological functioning, e.g. of the kidneys, could be affected in some genetically modified or treated animals. Some of the abnormalities would be detected by clinical examination, but others require specific tests to be carried out for their detection.

Several physiological measurements are of value in assessing the extent to which emergency responses have been used by an individual. When there is a short-term problem, the individual may increase its heart rate and adrenal activity. Modified or treated animals could be tested in situations in which control animals would show a known mean level of physiological response in order to ascertain whether or not those situations caused them more problems. It might also be useful to investigate longer-term usage of adrenal cortex responses by means of dexamethasone and adrenocorticotrophic hormone challenge tests (Dantzer and Mormède, 1983; Mendl *et al.*, 1992).

A further method of coping with adversity is to use endogenous opioids in the brain to self-narcotize. The welfare of individuals which have to do this is poorer than that of those who do not. The measurement of levels of plasma opioids appears to give little information about this coping method and the experimental use of opioid antagonists is difficult to interpret. However, studies of opioid receptor density may prove useful in welfare assessment (Zanella *et al.*, 1991, 1996).

5.7.5 Measures of immune system, disease and injury

When animals show substantial adrenal cortex responses, this is often associated with some degree of immunosuppression (Kelley, 1985; Siegel, 1987). There are also other mechanisms by which difficult conditions lead to impairment of immune system function. Measurements of immunosuppression include antigen challenge tests, *in vivo* lymphocyte stimulation tests, *in vitro* lymphocyte proliferation tests and specific tests of natural killer cell or macrophage efficacy (Broom and Johnson, 1993). If a genetically modified animal had less efficient immunological defences than an unmodified control, then it would be coping less well with its environment, so its welfare would be poorer. Disease always indicates some effect on welfare, so if that animal was also diseased and suffering then its welfare would be considerably worse. One of the first steps in assessing the welfare of a modified animal is to carry out a thorough clinical examination.

Injury also means poor welfare, the extent depending on the magnitude of the injury and the amount of associated suffering. A predisposition to injury because of weakness of some kind also indicates reduced ability to cope with the environment and hence poor welfare. Hens in battery cages (Knowles and Broom, 1990; Norgaard-Nielsen, 1990) and sows in stalls (Marchant and Broom, 1994, 1996) have weak bones because they get insufficient exercise. If a modified or treated animal had thin skin, weak bones or some other effect which predisposed the individuals to injury, then its welfare would be poorer than that of controls.

5.7.6 Behavioural measures

Abnormalities of behaviour are often the easiest indications of poor welfare to recognize and are an integral part of a proper clinical examination. However, careful behaviour recording is also important in welfare assessment and no attempt to assess welfare would be complete unless such recording were carried out. In order to recognize problems in carrying out normal movements, the observer must first establish which movements occur and with what frequency in normal individuals. When Andreae and Smidt (1982) wanted to assess the extent of abnormality of standing and lying movements occurring in young bulls kept on slippery slats they compared these movements with those of bulls on non-slippery floors. In a study of the extent of walking abnormalities in broiler chickens Kestin *et al.* (1994) classified locomotor ability by its difference from normality and reported that the majority of birds had some locomotor problem before they reached slaughter age.

In studies of the effects of inadequate housing conditions where the animal has insufficient control over important events in its life, stereotypies are sometimes shown. These repeated, relatively invariant movements with no obvious function are readily recognized. Examples are route-tracing in zoo animals, water spout circling in laboratory rodents and crib-biting in horses. Other abnormalities of behaviour include self-mutilation, excessive aggression, unresponsiveness, and attention to localized sources of irritation or pain. [For further details, see Broom and Johnson (1993).]

5.8 PROGRESS IN WELFARE ASSESSMENT OF TRANSGENIC ANIMALS

Some clinical examination will have been made of most transgenic animals and in the more extreme cases where welfare is obviously poor, for example the Beltsville pigs (Pursel *et al.*, 1989), the experimental study will have been terminated (van der Wal *et al.*, 1989). A report on the behaviour of sheep genetically modified to produce human blood clotting factor in their milk reveals no problems (B.O. Hughes, personal communication) but no comprehensive study of the welfare of a transgenic animal has been published. This represents a serious failing on the part of researchers, administrators and governments who have allowed developments to proceed to the point where some of these animals are being used commercially or in medical research. The results of studies of the welfare of the animal should be put in the specification of the animal prepared for subsequent users.

5.9 PROGRESS IN THE ASSESSMENT OF THE WELFARE OF ANIMALS TREATED WITH RECOMBINANT DNA PRODUCTS

Work on the effects of recombinant bovine and porcine somatotrophin injection has also been directed almost entirely towards finding out how to improve productivity in dairy cows and pigs. Any results which indicate what the effects on the welfare of the animals might be have been derived largely as an incidental by-product of the main study. This rather short-sighted approach to the testing of BST and PST and lack of concern for the animals has been one of the causes of public disquiet about the use of these products.

Since BST occurs naturally, low levels of it are unlikely to have any adverse effects on welfare, but even at low levels the effects need to be checked because each of the different forms of recombinant BST available has some differences in amino acid sequence from the natural form. BST injection results in increases in the amount of insulin-like growth factor-1 (IGF-1) in the blood and in milk (Prosser and Mepham, 1989; Prosser *et al.*, 1989, 1991). These increases can be substantial and it has been shown that high levels of IGF-1 can affect rat bone growth (Juskevich and Guyer, 1990). Low levels of IGF-1 are likely to have no adverse effect, but it is a potent mitogen and the effects of high-level intake on the cow, on the calf which consumes the milk, or indeed on the people who consume the milk, are unknown (Mepham, 1991).

The most clearly documented effects of BST and PST are on disease incidence and on reproduction (Broom, 1993a; Simonsen, 1993; Willeberg, 1993). The effects of BST injection are similar to changes which occur during the rising phase of lactation and high-yielding cows which are not treated with BST are particularly susceptible to disease at this time. Kronfeld (1988) states that high levels of BST result in subclinical hypermetabolic ketosis which can lead to reduced reproductive efficiency and a higher incidence of mastitis and other production-related diseases. However, studies reviewed by Phipps (1989) provide no evidence for increased incidence of ketosis following BST treatment. Several of the studies of cows treated with BST so that milk yields are particularly high report that the incidence of mastitis can increase. There are also some reports of increased incidence of lameness (Phipps, 1989; Craven, 1991). A general survey of mastitis incidence following BST use (Phipps, 1989) makes it clear that there have been several studies in which BST use did not result in a greater likelihood of mastitis. However, high production levels are associated with greater incidence of both mastitis and lameness (Broom, 1994), and BST use can result in high production levels, so the discrepancies in research results in the effects of BST on mastitis may depend upon how great were the maximal production levels using BST. Increases in disease following BST use may be directly related to the metabolism associated with high production levels, but welfare is obviously poorer if mastitis and lameness occur, whatever the exact reason for it.

Surveys of the results of several studies of BST-treated animals by Epstein (1990) and Epstein and Hardin (1990) showed that the conception rates of control cows were reduced after treatment from 89% to 59% and 95% to 50%, respectively. Assuming that the attempts to get the cows to conceive were equivalent, these results also indicate poorer welfare in BST-treated cows. Phipps (1989), in reviewing the evidence for effects of BST on reproduction, distinguishes: (i) between the use of BST early in lactation and late in lactation; and (ii) between higher and lower doses of BST. If the BST is administered early in lactation and at higher dose levels, the reductions in pregnancy rate reported by Epstein can be produced. However, it seems that administration of lower dose levels of BST later in lactation are less likely to have any adverse effects on welfare.

A further point, which may be very important to the cows, is that each injection has some effect on a cow, and repeated injections may cause swollen and tender injection sites (Comstock, 1988). More general effects of BST use are, firstly, that higher mastitis incidence may result in more antibiotic treatment and greater risk of the development of pathogen resistance and, secondly, that the possible change from smaller to larger dairy farms which could result from widespread BST usage could lead to poorer average stockmanship and less individual care of cows.

5.10 CAN WE PRODUCE NEW ANIMALS WHOSE WELFARE IS NEVER POOR?

Domestication involved selection of genetic lines which adapted well to human proximity. The widespread existence of poor welfare in domestic animals, however, shows that there are limits to how much animals can adapt to conditions imposed on them by humans. Genetic engineering could change animals further than has been possible so far with conventional breeding in this same direction. However, there will always be limits to change in animals which we require to feed themselves actively and otherwise regulate their interactions with their environment.

If tissue culture were to be used, animal cells might be cultured without the need for a nervous system and supracellular regulatory systems, so that there would be no poor welfare.

5.11 LEGISLATION REQUIRED

In the European Union there is legislation about animal experimentation which requires that some account should be taken of the animal's welfare during experimentation on transgenesis, or on treatment with biotechnology products. Research workers need to consider the welfare of the animal carefully and should be able to justify all of their actions to a member of the general public. However, after the animal ceases to be experimental, or if a genetically modified animal or product of biotechnology for treatment of animals is brought in from another country, the animals are not covered by the animal experimentation legislation.

It will not be adequate to depend upon the moral consciences of those who use transgenic animals, and specific legislation is needed concerning testing before usage. There is EU legislation relating to human health and preservation of the environment. There should also be legislation requiring that no genetically modified animals or animals treated with biotechnology products should be used commercially unless their welfare has been assessed using an adequate range of measures at suitable intervals throughout life and on through the next generation. If there is a net benefit for the welfare of animals, including humans, then the genetic manipulation should be permitted. In this assessment, benefits for humans would have to be direct and would not include increased monetary profit. This is a stricter criterion than just to say that any harm to the animal must be weighed against any benefit, because this latter criterion could allow severe effects solely for financial gain. Modifications of animals which are carried out for commercial purposes only, but which result in poor welfare, should not be permitted. There is legislation in the Netherlands stating that genetically modified animals cannot be used unless specific permission is given. The EU and other countries should be following that lead. If such action does not occur quickly it will become more difficult as economic pressures build up.

REFERENCES

Agger, J.F. (1983) Production disease and mortality in dairy cows. Analysis of records from disposal plants from 1969–1982, in *Proceedings of the 5th International Conference on Production Diseases in Farm Animals*, Uppsala, pp. 308–11.

Andreae, U. and Smidt, D. (1982) Behavioural alterations in young cattle on slatted floors, in *Disturbed Behaviour in Farm Animals* (ed. W. Bessei), Eugen Ulmer, Stuttgart.

Arey, D.S. (1992) Straw and food as reinforcers for prepartal sows. *Applied Animal Behaviour Science*, **33**, 217–26.

Broom, D.M. (1986) Indicators of poor welfare. *British Veterinary Journal*, **142**, 524–6.

Broom, D.M. (1991) Animal welfare: concepts and measurement. *Journal of Animal Science*, **69**, 4167–75.

Broom, D.M. (1993a) Assessing the welfare of modified or treated animals. *Livestock Production Science*, **36**, 39–54.

Broom, D.M. (1993b) A usable definition of animal welfare. *Journal of Agricultural and Environmental Ethics*, **6**(suppl. 2), 15–25.

Broom, D.M. (1994) The effects of production efficiency on animal welfare, in *Biological Basis of Sustainable Animal Production Proceedings of 4th Zodiac*

Symposium, EAAP Publ. 67 (eds E.A. Huisman, J.W.M. Osse, D. van der Heide, S. Tamminga, B.L. Tolkamp, W.G.P. Schouten, C.E. Hollingsworth and G.L. van Winkel), Wageningen Press, Wageningen, pp. 201–10.

Broom, D.M. (1995) Measuring the effects of management methods, systems, high production efficiency and biotechnology on farm animal welfare, in *Issues in Agricultural Bioethics* (eds T.B. Mepham, G.A. Tucker and J. Wiseman), Nottingham University Press, Nottingham, pp. 319–34.

Broom, D.M. (1996) Animal welfare defined in terms of attempts to cope with the environment. *Acta Agricultura Scandinavica Section A. Animal Science*, **27**, 22–8.

Broom, D.M. and Johnson, K.G. (1993) *Stress and Animal Welfare*, Chapman & Hall, London.

Comstock, G. (1988) The case against bGH. *Agriculture and Human Values*, **5**, 36–52.

Craven, N. (1991) Milk production and mastitis susceptibility: genetic relationships and influence of bovine somatotropin treatment, in *Mammites des Vaches Laitières* (ed. J. Espinasse), Polygone, Toulouse, pp. 55–9.

Dämmrich, K. (1987) Organ change and damage during stress – morphological diagnosis, in *Biology of Stress in Farm Animals: an Integrated Approach* (eds P.R. Wiepkema and P.W.M. van Adrichem), Martinus Nijhoff, Dordrecht, pp. 71–81.

Dantzer, R. and Mormède, P. (1983) Stress in farm animals: a need for re-evaluation. *Journal of Animal Science*, **57**, 6–18.

Dawkins, M. (1983) Battery hens name their price: consumer demand theory and the measurement of animal needs. *Animal Behaviour*, **31**, 1195–205.

Epstein, S.S. (1990) Potential public health hazards of biosynthetic milk hormones. *International Journal of Health Services*, **20**, 73–84.

Epstein, S.S. and Hardin, P. (1990) Confidential Monsanto research files dispute many bGH safety claims. *The Milkweed*, **128**, 3–6.

Hurnik, J.F. and Lehman, H. (1988) Ethics and farm animal welfare. *Journal of Agricultural Ethics*, **1**, 305–18.

Juskevich, J.C. and Guyer, C.G. (1990) Bovine growth hormone: human food safety evaluation. *Science, New York*, **249**, 875–84.

Kelley, K.W. (1985) Immunological consequences of changing environmental stimuli, in *Animal Stress* (ed. G.P. Moberg), American Physiological Association, Bethesda, Maryland, pp. 193–223.

Kestin, S.C., Adams, S.J.M. and Gregory, N.G. (1994) Leg weakness in broiler chickens, a review of studies using gait scoring, in *Proceedings of the 9th European Poultry Conference, Glasgow, 7–12th August 1994. Volume II*, WPSA, pp. 203–6.

Knowles, T.G. and Broom, D.M. (1990) Limb bone strength and movement in laying hens from different housing systems. *Veterinary Record*, **126**, 354–6.

Kronfeld, D.S. (1988) Biologic and economic risks associated with use of bovine somatotropin *Journal of American Veterinary Medical Association*, **1921**, 1693–6.

Manser, C.E., Elliott, H., Morris, T.H. and Broom, D.M. (1996) The use of a novel operant test to determine the strength of preference for flooring in laboratory rats. *Laboratory Animals*, **30**, 1–6.

Marchant, J.N. and Broom, D.M. (1994) Effects of housing system on movement and leg strength in sows. *Applied Animal Behaviour Science*, **41**, 275–6 (abstract).

Marchant, J.N. and Broom, D.M. (1996) Effects of dry sow housing conditions on muscle weight and bone strength. *Animal Science*, **62**, 105–13.

Mendl, M., Zanella, A.J. and Broom, D.M. (1992) Physiological and reproductive correlates of behavioural strategies in female domestic pigs. *Animal Behaviour*, **44**, 1107–21.

Mepham, T.B. (1991) Bovine somatotrophin and public health. *British Medical Journal*, **302**, 483–4.

Norgaard-Nielsen, G. (1990) Bone strength of laying hens kept in an alternative system, compared with hens in cages and on deep litter. *British Poultry Science*, **31**, 81–9.

Phipps, R.H. (1989) A review of the influence of somatotropin on health, reproduction and welfare in dairy cows, in *Use of Somatotropin in Livestock Production* (eds K. Sejrsen, M. Vestergaard and A. Neimann-Sorensen), Elsevier, London, pp. 88–119.

Prosser, C.G. and Mepham, T.B. (1989) Mechanism of action of bovine somatotropin in increasing milk yield in dairy ruminants, in *Use of Somatotropin in Livestock Production* (eds K. Sejrsen, M. Vestergaard and A. Neimann-Sorensen), Elsevier, London, pp. 1–17.

Prosser, C.G., Fleet, I.R. and Corps, A.N. (1989) Increased secretion of insulin-like growth factor I into milk of cows treated with recombinantly derived bovine growth hormone. *Journal of Dairy Resources*, **56**, 17–26.

Prosser, C.G., Royale, C., Fleet, I.R. and Mepham, T.B. (1991) The galactopoietic effect of bovine growth hormone in goats is associated with increased concentrations of insulin-like growth factor I in milk and mammary tissue. *Journal of Endocrinology*, **128**, 457–63.

Pursel, V.G., Pinkert, C.A., Miller, K.F., Bott, D.J., Campbell, R.G., Palmiter, R.D., Brinster, R.L. and Hammer, R.E. (1989) Genetic engineering of livestock. *Science, New York*, **244**, 1281–8.

Siegel, H.S. (1987) Effects of behavioural and physical stressors on immune responses, in *Biology of Stress in Farm Animals* (eds P.R. Wiepkema and P.W.M. van Adrichem), *Current Topics in Veterinary Medical Animal Science*, Martinus Nijhoff, Dordrecht, pp. 39–54.

Simonsen, H.B. (1993) PST treatment and leg disorders in growing swine: a welfare hazard? *Livestock Production Science*, **36**, 67–70.

van der Wal, P., Niewhot, G.J. and Politiek, R.D. (eds) (1989) *Biotechnology for Control of Growth and Produce Quality in Swine: Implications and Acceptability*, Pudoc, Wageningen.

Willeberg, P. (1993) Bovine somatotropin and clinical mastitis: epidemiological assessment of welfare risk. *Livestock Production Science*, **36**, 55–66.

Zanella, A.J., Broom, D.M. and Hunter J.C. (1991) Changes in opioid receptors of sows in relation to housing, inactivity and stereotypies, in *Applied Animal Behaviour: Past, Present and Future* (eds M.C. Appleby, R.I. Horrell, J.C. Petherick and S.M. Rutter), Universities Federation for Animal Welfare, Potters Bar, pp. 140–1.

Zanella, A.J., Broom, D.M., Hunter, J.C. and Mendl, M.T. (1996) Brain opioid receptors in relation to stereotypies, inactivity and housing in sows. *Physiology of Behaviour*, **59**, 769–75.

Part Two

The social context

6

Why biotechnology?

Caird E. Rexroad, Jr

6.1 INTRODUCTION

Technology can be defined variously as 'applied science', 'a scientific method of achieving a practical purpose' or 'the totality of the means employed to provide objects necessary for human sustenance and comfort'. The first two definitions point clearly to roots of technology in science. Animal biotechnology then refers to technologies that derive from the application of the sciences of animal biology. Science can be defined as 'knowledge covering general truths or the operation of general laws especially as obtained and tested through the scientific method' (Webster). A justification for animal biotechnology then must speak to both science and technology. In this brief chapter, I will explore the rationales both for conducting scientific research and for the application of its findings in the form of biotechnology.

6.2 THE NECESSITY OF SCIENCE

The desire to understand ourselves and the world about us has been a characteristic of humanity throughout recorded history. Humans have explained their origin and nature in many ways, giving rise to the creation myths of each civilization. Western Judeo-Christian tradition sees the biology of humans as resulting from creative acts of God (*The Holy Bible*, Genesis, Chapters 1–3). Modern cosmologists see the existence of the universe and humans in it as deriving inexorably from the forces that came into existence shortly after the 'Big Bang' (Hawking, 1988). These explanations differ in detail, but point with certainty to one conclusion: humans have an innate desire to understand their nature and origin. Recognition that events in nature are repeatable led to the conclusion that the material world, including humans and animals, can in some sense be understood.

Since the 'enlightenment', the primary method for understanding the material world has been experimentation to test hypotheses, thus ever increasing the complexity of our understanding. In the recent past, human understanding has progressed rapidly from the 'cell theory' to 'introns and exons', and from lenses to lasers. Scientists have at the same time developed ever more sophisticated tools, from balances to restriction enzymes. These great successes of the scientific method have not been achieved without the payment of a price for challenges to existing dogma: consider Galileo.

One important area of science that supports animal biotechnology is genetics. Practitioners of the science of genetics, in its many forms, seek to understand: 'How does life persist and adapt? Why do organisms grow, age, and die? Why do or don't offspring look like parents?' The goal of the first part of this chapter is to look at the science of genetics to address the question: 'Is such scientific inquiry necessary, affordable, and ethical?'

The question of the necessity of the science of genetics is the question: 'What need is met by this science?' The introduction to this chapter directs the answer: The need is human understanding of the biological world. Without the studies and the theories of Gregor Mendel and Charles Darwin, we would not understand much of the biological world about us. For instance, we would not know that many diseases in humans and animals arise from variations in the genetic constitution. Features of our world view change also. Scientific explanation of diseases suggests that we no longer need to invoke supernatural explanations for diseases. It becomes clear that understanding, based on science, does more than fulfil simple curiosity; it also provides a basis on which to structure philosophy.

Are there necessities for science beyond understanding? While it may have utility for philosophy and technology, the necessity of science is simply understanding.

6.3 THE AFFORDABILITY OF SCIENCE

The next question, 'Is the science of genetics affordable?', has no single answer but depends on the economic and philosophical models proposed to test such questions. What costs should we count to determine the affordability of understanding genetics? If we count dollars, pounds, francs, marks, yen or any other currency, the answer is not clear. Governments rarely spend money purely for the sake of understanding. They spend money because they expect utilitarian results, i.e. technology. An alternative question exists: 'What is the price of ignorance of genetics?' Some of the price is superstition and treatments of disease based on unsupported postulates about how diseases arise. The answer to the question of the affordability of a science such as genetics is tied to the value of knowledge, as a basis for understanding and behaviour.

6.4 ETHICAL CONCERNS CAUSED BY SCIENCE

Is the conduct of the science of genetics ethical? The impact of knowledge is not neutral. The ethical impact of new knowledge clearly depends on the nature of the knowledge and the ethics of the recipient of such knowledge. Research in genetics allows us to know to a limited degree the genetic composition of humans and animals. Should a society be able to make decisions on the fitness of humans based on such knowledge? Should male fetuses be preferentially aborted in a dairy herd? Is the trait for red hair linked in cattle to undesirable traits? It appears that the utilization of knowledge (i.e. technology, philosophy) causes ethical concerns, rather than the knowledge itself. An ethical question remains: 'Is it ethical to prohibit the discovery of new knowledge that might arise from the science of genetics?' Few societies prevent geneticists from discovering mechanisms that affect physiology, or determine the basis for disease. The support for this kind of research suggests that it is ethically acceptable to understand the world and to act in some manner based on the knowledge provided by research.

6.4.1 Animal research

The scientific method demands manipulation of the physical object of study. In genetics, that object is biological and often animal. Ethical concern about genetic experimentation on animals is mostly a special case of ethical concern about animal experimentation in general. Some philosophers view animal experimentation as acceptable if its utilitarian value is high. A less prevalent but strong view is that such experimentation is never acceptable. Special concerns arise in genetic experimentation because attempts to understand genetics may involve altering an animal's genetic composition, as is the case with transgenic animals.

6.4.2 Ethics of transgenesis

The argument that transgenic animals present a special ethical concern suggests that the reviewing ethical system views genetic constancy as a unique ethical value or that genetic changes imposed by humans differ qualitatively from other changes. There appears to be no widespread ethical system that places special value on unchanging genetics for any species.

Humans have practised genetic selection that has resulted, for instance, in both poodles and pomeranians deriving presumably from a wolf-like ancestral stock. Humans have selected, for utilitarian reasons, animals genetically changed to have 'domestic behaviour', copious milk production, unique colours, etc. Genetic constancy has been seen as good only from the utilitarian perspective of having a 'pure-bred'. Even the desire for genetic constancy would contradict any ethical view based on 'nature'. In nature, gene pools change based on environmental pressure and random events. The genome of a species is not stable; it is an ever-shifting composite of changing genes.

6.4.3 Ethics of DNA

The argument that there are 'special' ethical implications to genetic insertion experiments suggests that the information-providing chemicals (DNA, RNA) that constitute the genome differ in quality from other chemicals of life. Physiological systems are hierarchical and interactive with many effectors, including proteins, sugars, fats, and minerals at the top. Beneath this level are the proteins that synthesize, modify, transport or contain the top level of effectors. At the bottom of this hierarchy is the genetic code implemented in DNA and effected usually as RNA that generates the higher orders of information specified as enzymes, membranes, receptors, etc. Expression of the genetic code is not a constant for a given animal and is variably expressed depending on the presence or absence of higher level effectors in the hierarchy. The history of biological sciences has been to study each level of the physiological hierarchy as tools become available to make that level accessible to manipulation. Growth was studied first by modifying nutrient intake with increasing sophistication from modifying forages to purified additives to vitamins, etc. The recognition that food utilization and carcass composition were altered by signal peptides led to experimental manipulation using peptides and other hormones, viz. growth hormone, oestrogens and thyroxine. Transgenes reflect intervention at the next lowest level of that hierarchy. The possibility of a clear ethical distinction for the practice of science among these various hierarchical levels seems to be an unacceptable restriction of accumulation of knowledge.

6.4.4 Risks of genetics research

A question which is separate from the necessity, affordability and ethical nature of genetics research is the risk of such genetics research. The release of a new virus into the wild as the unpredicted consequence of genetic research would be an example of a potential risk.

One cannot postulate that all genetics research brings equal risks. Research with viral transgenes that might recombine with an endogenous retrovirus to make a new retrovirus may carry risks or even benefit. Insertion of most mammalian structural genes into an animal poses little or no risk to society. Risks have to be assessed based on knowledge, i.e. prior scientific research, or otherwise research should be conducted under risk-reducing constraints. Society has the right and duty to be advised of genetics research-related risks and to be given the opportunity to make decisions about the conduct of particular research programmes. The self-restriction on recombinant DNA research in the United States reflected a risk-reducing approach to genetics research while gathering knowledge that would enhance future risk assessment. Destruction of world stocks of smallpox virus is a risk-based decision indicating that the risk of infection to a world no longer vaccinated was greater than the utility of additional knowledge of the virus.

6.5 THE NECESSITY OF BIOTECHNOLOGY

Biotechnology as based on the science of genetics can also be inspected for its necessity, affordability, ethical implications and associated risks.

Technology, as indicated, is the application of science. Unlike science, which seeks to identify 'laws' and thus should converge into a unique set of knowledge, technology based on science can take on many alternative forms. Technology based on the wheel has provided ox carts, automobiles, trains, turbines and merry-go-rounds. Necessity for technology is defined by utility. Utility however can be broad in scope – some technologies may merely amuse (the hula hoop is a wheel) while others save lives (peristaltic pumps for dialysis use wheels). Other chapters in this book detail the utility of genetic biotechnology. Lives will be saved or made liveable by medicines produced by animal biotechnology. Humans will have more affordable foods or foods that have compositions better suited to human consumption. Who should dictate the development of genetic science into particular biotechnologies in those cases where alternative possibilities might exist?

Recognizing that alternatives to some genetic technologies may not exist, the suitability of biotechnologies can be investigated from the point of view of affordability. Affordability is perhaps the most easily investigated question. Affordability is determined by the economic decisions of those willing to invest in biotechnologies. In the short run, affordability is dictated by willingness of investors to capitalize research and development. In the long run, affordability decisions will be based on the history of prior success or failure, remembering that failure includes societal rejection of offered technologies. While this assessment of affordability is simplistic, it is functional in a market economy that permits open discussion of issues but may not suffice in economies where governments or interest groups are seen as having the wisdom to make economic decisions.

6.6 ETHICAL CONCERNS ABOUT BIOTECHNOLOGY

6.6.1 The ethical environment

Ethical concerns about the implementation of the findings of genetic science into biotechnology are broader than ethical concerns about genetic science. Ethical questions, for instance, arise about changing the relationship between man and animals when genetic biotechnologies are put into practice. Is the use of genetically modified animals for milk production different from the use of a highly selected Holstein? If so, how is it different? The genetically modified animal may not be different so much as people's attitudes towards the animal. Unfortunately, the development of ethical criteria to evaluate this question is difficult because the question is posed at a time in the development of the western world when many people do not appreciate animals as sources of food, fibre and pharmaceuticals. In such a world, ethical decisions relating to the use of animals by humans may be made in a vacuum without scientific and philosophical underpinnings.

6.6.2 Transgenesis and '*telos*'

One suggested ethical problem with transgenic animals is that gene insertion violates the '*telos*' of the species. Such arguments are not difficult to challenge because the definition of a quality such as *telos* is subjective and depends on acceptance of a particular philosophical system. Because science deals with the material world, we might argue that humans already have changed the *telos* of many species by selection with resulting acceptable changes at multiple genetic loci. The genome of higher vertebrates consists of 50 000–100 000 genes and numerous modifiers of their action. Current genetic modification technologies may add a single or a few genetic elements. On the other hand, an animal can be viewed as a set of genetic potentials (genes). The actions of those potentials is manipulated via feeding and the environment provided by man. Do these practices impinge on a species' *telos*?

Ethical questions arise because philosophical systems differ in their resolution of questions about biotechnologies. Some ethicists claim to speak for the animals (to protect their *telos*?): presumably concerned with 'Who grants informed consent for genetic modification?' While such a question has value in debate, it has little practical effect, as humans clearly have assumed responsibility for domestic and wild animals to the point of spending great sums to assure their survival. The more important question seems to be the underlying question of 'Which group of humans (what set of values) will determine the practices acceptable?' Ethical values arise from consensus in societies. Scientists should play a role in reaching social consensus regarding use of genetically modified animals as an informed decision.

6.7 THE RISKS OF BIOTECHNOLOGY

The risks associated with the application of animal biotechnology probably cannot be generalized. A longstanding genetic biotechnology is artificial insemination with cryopreserved sperm. From the outset, the benefits were clear. Rapid genetic progress in traits such as milk production is evident. The genetic risks that arise are specific: reduced allelic variation from dependence on fewer bulls. There were also potential societal risks as the number of farmers needed to produce milk decreased. Individual risks were reduced: farmers are safer because fewer bulls are housed on dairy farms.

The illustration of sperm cryopreservation and artificial insemination makes a strong argument for the pursuit of animal biotechnology. While there are societal costs, such as potential loss of rural populations, its overall benefit to a world that is industrialized and increasing in population is clear. This technology meets the need to feed people healthy, affordable foods. The associated sciences that provided this technology continue to benefit people: infertile couples now can produce offspring through sperm cryopreservation, embryo cryopreservation and *in vitro* fertilization. The population genetics studies that made it clear that artificial insemination would benefit agriculture have also provided explanations of the origin of the human race and the movements of disease around the world. It should be expected that enormous benefits will continue from research on biotechnology-related sciences and their implementation in new technologies.

REFERENCES

Hawking, S. (1988) *A Brief History of Time*, Bantam Press, London.
The Holy Bible, Genesis, Chapters 1–3.
Webster's Ninth New Collegiate Dictionary, Merrian-Webster, Inc., Springfield, MA.

7

Campaigning against transgenic technology

Joyce D'Silva

7.1 THE WELFARE OF FARM ANIMALS

Animal welfarists take a dim view of animal biotechnology, not because they are Luddite or anti-innovation in itself, but because they have yet to see any benefits to animals themselves resulting from the new technologies, and are aware of ample evidence of actual harm being caused to the creatures involved.

Their cause for concern is amplified by reflection on existing technologies such as selective breeding, which have given us physiologically crippled animals which generally survive just long enough to get to their desired slaughter weight. These animals have been deliberately bred for speedy growth and high yields of meat or milk, but as a result, enormous stresses have been placed on their metabolisms.

Let us take a look at some examples of selective breeding as they have impacted on farm animal health and welfare.

Take the chickens we rear to eat, the broiler chickens. Here, we have a telling example of how selective breeding to increase productivity can have disastrous results on animal welfare. Broiler chicks now grow to slaughter weight, i.e. 2 kg, in just 42 days. That is half the time it took them 30 years ago. By genetic selection and the use of growth-promoting antibiotics routinely incorporated in their feed, the chickens now put on muscle at a phenomenal rate. However, their skeletal structure has not developed to cope with this weight increase and as a result most of the birds are crippled to a degree in the last week or two of their very short lives. As it is, Bristol University scientists found that 90% of broilers had abnormality in their gait of varying degrees of severity (Kestin *et al.*, 1992). Statistically their research would mean that, in the UK, 24 million birds a year can only walk with difficulty when strongly motivated, and 12 million can only crawl on their shanks or struggle to walk using their wing tips as crutches to stop them falling over. But not only are these chickens growing too fast for their legs, their cardiovascular system also cannot cope (Webster, 1992). About 6 million a year die from ascites – pooling of body fluids in the abdomen caused by congestive heart failure. So selective breeding has given us baby birds which die due to heart failure.

The modern turkey has also been selectively bred for massive muscular (meaty) development. The result is that breeding males can no longer mount the females and virtually all reproduction takes place by artificial insemination. The overweight turkeys themselves suffer from severe joint problems in their legs as their inadequate bone structure fails to support the enormous weight of meat it carries.

Having been selectively bred for productivity and indoor rearing, today's pigs also suffer similar conditions to broiler chickens' heart failure and leg problems – too heavy a body on legs that cannot stand the strain and which find no respite on the cold concrete and metal slats of the average pig pen.

Take the case of the dairy cow. The ancestor of the modern Western dairy cow had small udders, slung at the highest point on the belly. Genetic selection has changed all that. The modern high-yielding cow may produce 10 times as much milk as her calf would suckle from her – if it were permitted to do so. Her udder is large and pendulous, often hanging so low that new-born calves have difficulty locating the teats. As her hind legs have to change position to accommodate the bulging, swinging udder, she frequently develops gait problems which result in lameness. The udder may even hang so low that painful teat injuries occur. As Professor Webster (1987) has put it succinctly, 'The modern dairy cow is not a satisfactory shape'. Although causes of lameness are varied, it is now recognized that feeding and housing are major causal factors. However, as 70% of foot damage is to the hind feet and over 70% of these incidences affect the outer claw, genetic selection must take the primary blame (Greenhalgh *et al.*, 1981). Cows cannot stand or walk normally because their huge udders will not let them. Too much weight is then put on the outer claw.

Genetic selection has given us crippled chickens, painfully lame cows, turkeys unable to breed, pigs and chickens that die from heart failure in their infancy – no wonder the concerned welfarist takes a decidedly cautious view of a new technology which can create in a year what may have taken the selective breeders decades to achieve.

These welfare disasters have resulted from utilization of selective breeding technologies which have been driven by the desire to maximize profits. This profit motive is often couched in more honour-able-sounding phrases such as 'maximizing efficiency', 'achieving the animal's genetic potential' – even 'the necessity to produce food for the hungry/starving nations of the world'.

In the UK, much of the pig and poultry sectors are vertically integrated and farmers rear the type of animals and in the manner dictated by the supplier/buyer. The manner of rearing is of course the manner of the factory farm. Indeed, today's Ross/Cobb broiler chicken or Large White pig is not even suited physiologically for outdoor, free-range systems. In other words, these animals have been mutilated genetically to the point where a 'natural' existence, as lived by their biological ancestors, could be too much of a shock for their systems.

Proponents of genetically engineered animals are quick to tell us that biotechnology has been around for centuries and that man has been 'genetically manipulating' domestic animals for nearly as long: as if a linear pedigree has some moral worth in itself. The welfarist would answer that the evidence of harm caused by that 'genetic manipulation' is all around us to see – or perhaps, more truthfully, is hidden away behind the closed doors of the factory farm. The only pedigree may be a pedigree of pain.

7.2 WHY TRANSGENIC FARM ANIMALS?

Let us look at the proposed motivation for creating transgenic farm animals:

1. Disease resistance
2. Increased productivity
3. Production of pharmaceutical proteins in milk/eggs
4. Provision of organs for transplant to humans

Not one of these motives is welfare-led.

7.2.1 Disease resistance

Even disease resistance is not being sought for welfare motives. Yet on the face of it this sounds benign enough – who wants animals to suffer diseases? But, what diseases are we considering? In nearly all cases the diseases are those endemic to the factory farm. Animals or birds cooped up in their hundreds or thousands indoors are incredibly vulnerable to the rapid spread of disease. Hence the battery of vaccinations and prophylactic medication commonly utilized on our intensive farms today. It is little wonder that those in the business would be happy to reduce production costs by rearing animals resistant to disease in the first instance.

But to the animals concerned there seems to be little benefit. A different method is used to prevent disease. Yet the same boring, frustrating and stressful rearing conditions continue – simply propped up by the microinjection rather than the hypodermic or spray. This can hardly be seen as a positive welfare move.

7.2.2 Increased productivity

Increased productivity, as we have seen with selective breeding, has already brought untold misery to millions of farm animals and poultry. There is no reason to believe that genetic engineering of farm animals would produce any better consequences. In fact, the scientific literature to date is almost a blow-by-blow account of experimental disasters.

United States Department of Agriculture researchers at Beltsville put human and bovine growth hormone genes into pigs – aiming at increased growth rates and leaner meat. They achieved these ends but at what cost? They report 'several health problems in transgenic pigs due to the excessively high levels of growth hormone. Pigs expressing high levels of growth hormone tend to be lethargic, they exhibit indications of muscle weakness, and some are susceptible to stress. Others tend to lack coordination in their gait, probably because they are rather tender on their feet. So far, all of our transgenic gilts that express the growth hormone transgene have been anoestrus. Their reproductive tracts are infantile. Boars also tend to lack libido, but can be used for breeding with the help of electro-ejaculation and artificial insemination. A number of transgenic expressing pigs have died from gastric ulcers before they reach one year of age. Others have had lesions on the stomach lining when slaughtered for carcass evaluation. Some of the pigs show evidence of arthritis' (Bolt *et al.*, 1988). The authors conclude with frankness, 'to date, the technology has not produced an animal with a beneficial transgene'.

Similar experiments with adding growth hormone genes to lambs resulted in degeneration of the liver and kidneys which may have reflected diabetes-associated degenerative changes (Rexroad *et al.*, 1990).

A recent experiment in microinjecting 2555 calf embryos with a gene which is known to promote muscle development resulted in the birth of just one transgenic bull calf. The calf grew normally for 8 weeks and then began to build up muscle. By 10 weeks he started showing signs of weakness and rapidly became unable to stand up. The researchers wrote, 'At 15 weeks of age, it became obvious that the muscle weakness was not a transient phenomenon, and humane considerations led us to euthanize this animal' (Bowen *et al.*, 1994).

Such experiments continue. Already a patent has been applied for on a transgenic chicken containing an additional growth hormone gene. The patent claims that the chickens get to slaughter weight sooner (as if

42 days was not soon enough) and that males kept for breeding purposes produce sperm at an earlier age. Already, 100 genetically altered pigs are undergoing commercial trials in New South Wales and the Australian authorities have granted a patent on these animals. If the trials are successful, the pigs will be marketed.

Of course normal animals can also have their metabolisms adjusted by having biotech products injected into or implanted in their bodies. Thousands of dairy cows world-wide are now being given fortnightly injections of recombinant bovine growth hormone or bovine somatotrophin (BST). Research has shown that not only does BST increase milk yield, but it also doubles the period of catabolic stress which normally follows calving. In addition, BST use is associated with an increase in clinical mastitis incidence of between 15% and 45% (Willeberg, 1993). Mastitis is an extremely painful infection of the udder/teats. When the first BST product, Posilac, went on sale in the United States in February 1994, the company literature which arrived with it on farmers' doorsteps warned of a host of possible adverse side effects from treatment, including 'an increase in digestive disorders such as indigestion, bloat and diarrhoea ... cows injected with Posilac have increased numbers of enlarged hocks and lesions of the knee ... and second lactation or older cows have more disorders of the foot region ... use has been associated with reductions in haemoglobin and haematocrit values during treatment'. This tallies with the Report of the Committee on Veterinary Medicinal Products to the European Commission (CVMP, 1993) which referred to 'the observed anaemia ... due to a slowing down, caused by the treatment, of the physiological compensation of the constitution of the blood during lactation'.

Analysis of detailed reports (February, 1995) from the US Food and Drug Administration on problems arising in cows on farms where BST is being used shows that out of 9509 individual cows, over 2000 were reported with mastitis, nearly 2000 with reproductive disorders such as premature birth and abortions, well over 1000 with digestive disorders such as anorexia and weight loss, and over 1600 with foot or leg problems. A further report (August, 1995) shows similar incidences of problems – with one major change. The monthly rate of digestive disorders reported in the first 6 months of the second year of BST use shows an increase of 579%. This could suggest that cows enduring a *second* lactation on BST are experiencing more digestive problems (CVM, 1994-5).

The recombinant pig growth hormone gene, porcine somatotrophin, or rPST, is now licensed for use in Australia and may soon be licensed for commercial use in the USA. Injected regularly, PST increases the growth rate in pigs. At higher doses, growth rate is maintained, although appetite and feed intake are depressed. Body fat decreases, protein increases and the result is a dramatically leaner animal. However, PST injections cause body organs such as liver, heart and kidneys to increase in size, and bone strength to be reduced with an increase in the incidence of lameness. PST also appears to inhibit ovary development in gilts (Bryan *et al.*, 1989). When PST was administered to lactating sows near the end of gestation, they developed respiratory distress and haemorrhaging ulcers, resulting in the death of some sows (Smith *et al.*, 1991).

7.2.3 Production of pharmaceutical proteins in milk/eggs

Production of pharmaceutical proteins in the milk of transgenic animals is often hailed as one of the truly beneficial uses of transgenic livestock. The reconstructed human genes are targeted so that they are only expressed in the animal's mammary gland. In addition, the genes chosen to pioneer this work have perhaps been carefully selected for maximum public acceptability. We all feel sympathy for the innocent haemophiliac or diabetic, or the elderly emphysema sufferer. How can we not want to help them? But good PR does not a medicine make. In spite of massive media hype about the human blood-clotting Factor IX project at Edinburgh, nothing has been heard of this for some time. It would be interesting to know if problems have arisen with the transgenic animals.

The same research team and company (PPL Therapeutics plc) are now putting their efforts into alpha-1 antitrypsin. This product was originally 'hyped' as being a cure for emphysema sufferers (who tend to be elderly, and often smokers). More recently, the selling point has been as a potential cure for cystic fibrosis sufferers (children and young people with a life-threatening inherited disease). With clinical trials as yet incomplete, we do not know whether the product will turn out to be either safe or effective. Sheep, the animals being used, are of course as disease-prone as any animal species and in certain parts of the world are prone to scrapie – the spongiform encephalopathy credited with initiating bovine spongiform encephalopathy (BSE), and closely related to human spongiform encephalopathies such as Creutzfeldt–Jakob disease. This as yet unknown disease agent is notoriously resistant to normal sterilization procedures.

As yet, long-term effects on the transgenic animals have not been analysed and morbidity/mortality rates of the transgenic laboratory animals used in the development of this technique have not been published in the scientific literature. Why not?

Even if long-term analysis shows up no welfare or health problems with each type of molecularly pharmed transgenic animal, Compassion in World Farming (CIWF) fears that such animals will be kept, once commercial production gets under way, in ultra-sterile conditions to maximize hygiene. Contact with faeces/other animals/humans will be minimized and the likely scenario is isolation in stainless steel cages with mesh/slatted flooring. As sheep are essentially flock animals with a strong herd instinct, and are natural ruminants, this type of confined, isolated existence can only be seen as cruel: factory farming to the 'nth' degree.

Several years of research have been undertaken by Pharming Health Care Products, in an attempt to produce cattle which are transgenic for human lactoferrin. The gene was microinjected into target embryos in April 1990 and one transgenic bull, Herman, has been bred from. Pharming Health Care Products have been able to produce animals which express the transgene in their milk although it is not yet clear if it is a human-identical lactoferrin.

In the meantime, scientists in Nebraska have incorporated the human lactoferrin gene into a plant tissue culture, and the transgene is expressed by the transformed cells. It is entirely possible that molecular pharming in transgenic animals will prove to be a blind alley, and that production of the required proteins in plant tissue cultures will prove to be quicker, simpler, cheaper, and cruelty-free.

7.2.4 Provision of organs for transplant to humans

Creating transgenic animals to facilitate organ transplants (xenografting) has many similar disadvantages to molecular pharming. The compatibility of the product may be hard to achieve – researchers are still only at the stage of congratulating themselves when a heart from a transgenic animal continues beating for 60 days after transplant into a primate. What was not at first widely publicized was the fact that these primates retained their own hearts and were treated with levels of immunosuppressive drugs that would be toxic in humans.

As with the transgenics in molecular pharming, those transgenic organ 'donors' will be kept in sterile conditions and will in all likelihood be born by specific pathogen-free (SPF) methods to maximize disease-free status. SPF pigs are obtained by anaesthetizing the pregnant sow shortly before she farrows. The entire uterus containing the piglet embryos is removed in a sterile 'bubble'. The piglets are reared in isolation conditions and the sow is slaughtered.

In January 1997, the government published a report by The Advisory Group on the Ethics of Xenotransplantation (known as the Kennedy Report). The Report recommended that primates should not be used as source animals for transplants as to do so would 'constitute too great an infringement of their right to be free from suffering' (4.2). Sadly, the Report went on to accept the use of pigs as source animals whilst admitting that the animals 'may be exposed to harm' (4.3) and declaring 'we regret that animal suffering is caused' (4.8) (Kennedy, 1997).

7.3 REPRODUCTIVE TECHNOLOGIES IN PRACTICE

Transgenic animals are initially created in the laboratory. To get the egg cell or embryo to the laboratory requires its removal from an animal (probably a super-ovulated animal) and to grow the manipulated embryo to term requires its transfer into another 'surrogate' mother.

To induce super-ovulation in farm animals they are subjected to hormone treatment to stimulate the production of more ova than would naturally be produced. In cattle, this is achieved through a series of prostaglandin and follicle-stimulating hormone injections – sometimes up to 10 injections may be given. Embryos are collected about a day after insemination. In both sheep and pigs the embryos are removed via surgery; in cattle they may be 'flushed' by insertion of a flexible rubber tube made rigid with a metal stilette which is passed through the cervix into the uterus. Alternatively, the animal may be killed or castrated.

Once the eggs have been microinjected with the desired gene, then they must be transferred into a recipient animal. Sometimes this is an intermediate animal such as a rabbit – a technique used to screen out non-viable embryos and minimize use of expensive large mammal recipients. If cattle are being used, the embryos are transferred to the uterus of the recipient mother via the cervix. This takes place at a time when the recipient cow's cervix is closed. Although in the UK epidural anaesthesia is required at this point, non-veterinary surgeons can administer it. In many other countries no anaesthesia at all is required for this invasive and painful procedure.

7.3.1 Cloning

One fast growth area for the future is that of cloning. A herd of cloned animals will all reach slaughter weight at the same time with identical carcasses. This is a supermarket buyer's paradise. As Professor Peter Street of Reading University's Department of Agriculture succinctly puts it: 'In effect, with this implanted, designed embryo, if we then are able to manipulate the feeding system, we can design the whole carcass, if you like, from embryo to plate to meet a particular market niche' (Street, 1992).

The current most common cloning technique is embryo splitting, but nuclear transplantation and embryonic stem cell techniques hold greater potential for multiplication of large numbers of animals. It has been estimated that just five generations of cloning could produce 100 000 viable clones (Stice, 1992).

Although cloning is essentially a laboratory technique and hardly appears to raise welfare problems, such problems begin when the clones are implanted and brought to term by surrogate mothers. For example, many cloned calves are abnormal and grow to twice the usual size at birth. Birth is therefore inevitably by caesarean, with all its attendant health and welfare problems for the cow. Some 10% of cloned calves have other abnormalities such as joint problems.

In 1996, scientists at the Roslin Institute, Edinburgh, proudly announced in the journal *Nature* that they had produced a number of cloned sheep (Campbell *et al.*, 1996). Much media hype followed, including an article in *The Independent* which described the experiment as 'a technical tour de force in mammalian embryology' (Wilkie, 1996). Three of the five lambs died shortly after birth. Later, it was discovered that all but one of the lambs were larger than normal at birth, and post-mortems on the dead lambs showed congenital abnormalities in their kidneys and cardiovascular systems. The scientists had failed to mention any of this in their published paper!

One year later, the Roslin scientists produced 'Dolly', the cloned sheep, the first animal cloned from a cell taken from an adult (Wilmut *et al.*, 1997). This new experiment has been lauded and deplored, mainly because of its implications for the cloning of humans. What have been overlooked are the animal welfare implications. To achieve 'Dolly', a ewe was given hormones to stimulate super-ovulation; she had her egg cells removed surgically, and the cloned embryo was placed, surgically, into the ligated oviduct of another ewe. After 6 days, this ewe was killed and the embryo placed, surgically, into the uterus of its final surrogate mother.

Cloning has huge implications for transgenesis. It is of course by means of embryo transfer (ET) that genetically engineered embryos can be grown to term, and the fastest way to multiply a new transgenic line will be by cloning. Already, the scientists from Roslin are talking about 85% of our dairy cows being clones within 20 years (Bulfield, 1997).

The other great danger in cloning is that the production of genetically identical animals means that such animals are not only identically super-fast growing, super-lean, etc., but also identically vulnerable to the same pathogens. Just one strain of disease to which all the cloned animals were highly vulnerable could wipe out an entire herd.

7.5 CONCLUSIONS

What we have before us is a battery of techniques for manipulating animals' breeding patterns, genomes and, ultimately, their lives. These techniques range from the comparative sophistication of bovine embryo transfer to the as yet haphazard methods of gene sequence selection and microinjection.

We have yet to see any benefit for human or beast, though great things are promised.

What we do see though – writ clearly across the whole gamut of these technologies – is the human desire to maximize profits and to extend human life at any cost.

We see no concern for animal welfare. Animal health is viewed as a means to an end (profits) rather than as a positive state in itself. We see no recognition of the biological integrity of species – or respect for the past and future natural evolution of species. We see no acknowledgement of animals' sentiency or the individuality of each animal.

What appears to dominate the biotech-vision is a currently dominant species – human – exercising its technical prowess over other equally sentient species, to its own financial advantage, and possibly as a cheap aid to the health and longevity of its own kind. We can only regret that the technological progress and power of the human species has apparently not been matched by any evolution of compassion or of appreciation of the 'sacredness' and uniqueness of the animals with whom we happen to share our planet.

Campaigning against Transgenic Technology? Well, what else could we do? And Compassion in World Farming will continue to say No to all laboratory procedures, breeding processes and farming systems which place a heavy burden on the genome, physiology or psychology of farm animals.

REFERENCES

Bolt, D., Pursel, V., Rexroad, C., Jr and Wall, R.A. of USDA (1988) Improved animal production through genetic engineering: transgenic animals, in Proceedings of forum, *Veterinary Perspectives on Genetically Engineered Animals*, sponsored/published by the American Veterinary Medical Association.

Bowen, R.A. *et al.* (1994) Transgenic cattle resulting from biopsied embryos: expression of c-ski in a transgenic calf. *Biology of Reproduction*, **50**, 664–8.

Bryan, K.A. *et al.* (1989) Reproductive and growth responses of gilts to exogenous porcine pituitary growth hormone. *Journal of Animal Science*, **67**, 196–205.

Bulfield, G. (1997) *The Daily Telegraph*, 7th March, p. 10.

Campbell, K.H.S. *et al.* (1996) Sheep cloned by nuclear transfer from a cultured cell line. *Nature*, **380**, 64–6.

CVM (1994–5) Updates from the USFDA Centre for Veterinary Medicine, October 3rd 1994, March 14th 1995 and October 12th 1995.

CVMP (1993) Final Scientific Report of the Committee for Veterinary Medicinal Products on the application for marketing authorization submitted by the Monsanto company for somatech, dated 27/1/93, Commission of the European Communities.

Greenhalgh, P.R., McCallum, F.J. and Weaver, A.D. (1981) *Lameness in Cattle*, 2nd edn, Wright Scientechnica, Bristol.

Kennedy, I. (1997) A report by the Advisory Group on the ethics of xenotransplantation of animal tissues into humans. Chairman, Professor Ian Kennedy, Department of Health, HMSO, London.

Kestin, S.C., Knowles, T.G., Tinch, A.E. and Gregory, N.G. (1992) Prevalence of leg weakness in broiler chickens and its relationship with genotype. *Veterinary Record*, **131**, 190–4.

Rexroad, C.E. *et al.* (1990) Insertion, expression and physiology of growth-regulating genes in ruminants, in Proceedings of the Second Symposium on Genetic Engineering of Animals. *Journal of Reproduction and Fertility*, supplement no. 41, 119–24.

Smith, V.G. *et al.* (1991) Pig weaning, weight and changes in hematology and blood chemistry of sows injected with recombinant porcine somatotropin during lactation. *Journal of Animal Science*, **69**, 3501–10.

Stice, S.L. (1992) Multiple generation bovine embryo cloning. Proceedings, Symposium on Cloning Mammals by Nuclear Transplantation, Colorado State University, 1992.

Street, P. (1992) Dept. of Agriculture, Reading University, in *Horizon* programme, *Fast Life in the Food Chain*.

Webster, A.J.F. (1987) *Understanding the Dairy Cow*, BSP Technical Books, Oxford.

Webster, A.J.F. (1992) quoted in *Horizon* programme, *Fast Life in the Food Chain*.

Wilkie, T. (1996) Do not cower from science. *The Independent*, 8th March 1996.

Willeberg, P. (1993) Bovine somatotropin and clinical mastitis: epidemiological assessment of the welfare risk. *Livestock Production Science*, **36**, 55–66.

Wilmut, I. *et al.* (1997) Viable offspring derived from fetal and adult mammalian cells. *Nature*, **385**, 810–13.

8

Consensus conferences and technological animals

Lars Klüver

'It is ethically acceptable to use technological animals, if the primary purpose is to develop new treatments for diseases, which otherwise cannot be cured. To cure cancer is a very important goal. To reach that goal we can accept that animals suffer.'

Teknologinævnet (1993)

This was said unanimously by a panel of lay people in a consensus conference in November, 1992. When lay people are given the opportunity to make up their minds on their own premises, they are open-minded, responsible and pragmatic. They do as fine a job as an ethicist could, and besides the comprehensiveness of their statements, they have something which a single academic can never get – large political credibility.

The objective of the consensus conference in 1992 was to assess developments in the field of 'technological animals' by taking up questions related to ethics, health, economy and environment. In addition, the conference was intended to diffuse interest and knowledge in order to stimulate public debate on technological animals.

The consensus conference on technological animals was held in Copenhagen, September 23–25, 1992. The conference was arranged by The Danish Board of Technology in cooperation with the Research Committee of the Danish Parliament, Folketinget.

The heart of a consensus conference is a panel of 12–16 lay people – the jury – who decide on the questions which the conference is to answer. After listening to the answers of a panel of experts and interrogating the experts, the lay panel writes the final document of the conference in which they – in effect – answer their own questions.

There are two noteworthy aspects of the consensus conference on technological animals. First, the democratic aspect – why is it relevant to involve lay people in ethical and democratic debate, as is done in a consensus conference? Second, the content – what did they come up with at this specific conference?

8.1 VALUES, KNOWLEDGE AND ATTITUDES

It has often been said that if people knew more about biotechnology, their attitude towards it would be more positive. However, compared with people of other European countries, Danes are among those who have most knowledge about biotechnology, but who are also most sceptical towards its benefits (Eurobarometer, 1991, 1993).

Danish consensus conferences have shown that during the process, the laymen in the jury undergo an evolution, moving from initial standard responses to being nuanced in their argument, ready for negotiations and able to explain their opinions. However, they do not show much change in their fundamental attitude. Increased knowledge is thus primarily used to clarify one's own attitudes, and only to a lesser degree to make a basic change in attitudes (Rienecker and Erichsen, 1990).

Values thus seem to be firmly inter-connected as a rather stable complex. There is a tendency for a person who holds a positive attitude towards biotechnology also to emphasize economic growth and military security, whereas the person who is sceptical towards biotechnology will emphasize environmental protection and social services (Borre, 1989, 1990).

Thus, one's attitude towards biotechnology is created from the personal set of values which one usually attempts to live up to. Anyone trying to form an opinion on a new technology will evaluate how this technology fits with his or her personal set of values, and attitudes will be formed in that light.

If you are member of a lay panel in a consensus conference on 'Technological animals', you will probably learn a lot when you listen to the experts – but it is very unlikely that the knowledge gained on different biotechniques and their effects on animal welfare will change your ideas about good/bad and right/wrong in this world. It takes more than superior knowledge about a closed problem to change basic values.

If you have a less than average knowledge of a certain subject, additional knowledge up to a given level may take away misunderstandings and myths that you have formed or accepted because you did not have a reference point. More knowledge may make you understand the motives and attitudes of other people, because you are now able to respond with some certainty concerning your own standpoint. As a result, you may add nuances to your opinions.

Apart from its effect in narrowing the 'don't know' group, added knowledge and openness in debate, such as is found in a consensus conference, make people move from 'yes!' to 'yes, but...' or from 'no!' to 'no, unless...'. And that is actually a good starting point for a consensus-making process in society and for a constructive use of a new technology.

In other words, while knowledge does not change our basic attitudes, knowledge – as a layer on top of our values – may provide us with an ability to understand the attitudes of others, to make our own attitudes more nuanced and to form opinions (Klüver, 1995).

8.2 ETHICS AND POLICY

In democratic terms, the fact that basic attitudes are stable and based on a broad set of values in itself makes it important to respect the opinion of each individual – irrespective of how illogical or emotional this opinion may seem to a professional analyst. This may seem irresponsible to some people, but it is not possible in a democracy – or in any other form of decision making, incidentally – to demand from decision makers that they know everything about anything. Decision making must be based largely on values and on knowledge only to the extent that it is present, communicated and understood; that is, to a limited extent.

Ethics as a political matter is more complex than ethics as an academic discipline. It involves the individual set of values each member of society has. It involves a search for coherent argumentation, such as we see in the discipline of ethics. And the resulting reasoning has to work in a mixed society with built-in conflicts of power and interests. If ethics is to be not merely an academic discipline, isolated from actual decision making in society, but also politically applicable in a democracy, we will as a consequence have to broaden the concept of ethics from analytical criteria such as stringency and consistency in argumentation, towards procedural criteria, such as democracy and responsible pragmatism.

Politicians often meet claims for ethics to be drawn into the decision making process. But democracy is – in an ethical context – an imperfect construction, because it has to harmonize a multitude of different ethical standpoints.

8.3 THE ROLES OF LAY PANEL AND EXPERTS

Democracy can be regarded as a procedural agreement – the rules of a game. There can be no doubt that the imperfection of democracy contains conflicts: Should power be given to the pragmatic element (the citizens) or to the coherent element (the experts)? There is no ultimate answer to that question, but at least, in order to find a balance between coherent reasoning and democratic pragmatism, the rules and the roles have to be clear. In a consensus conference, they are.

The key group in a consensus conference is the lay panel. A lay person is in this context defined as a citizen who is not a professional in the relevant field and who cannot be regarded as stakeholder.

Besides their capacity as ambassadors for public opinion, lay people have a number of relevant qualifications as political assessors:

- Motivation: they and their children are the ones who have to live with the decisions. And they frequently feel themselves isolated from decision making that often has profound effects on their daily life.
- Relevant expertise: they are experts on living daily life in society. As assessors they have the necessary knowledge and social framework.
- Legitimacy: they have their constitutional rights to speak as free individuals and citizens in society, and they are as citizens the object of democracy.
- Independence: they are non-stakeholders, only representing themselves. They do not have to take orders from anyone.
- Neutrality: they have no reason to cause harm to anyone and no reason to cover up anything.
- Credibility: they try to live up to the values they usually believe in.

Although most of these characteristics are ideal in the sense that they are fulfilled only to a certain degree, they clearly sketch the features of lay people that make their assessments so important.

Experts cannot live up to at least the last four of these qualities. Experts have a profound knowledge in a limited field which makes them relevant as resources for the political process. Further, experts are trained in being analytical – a tool which is needed in any serious debate. But they cannot claim to have the necessary qualities as spokesmen in a democratic debate.

Just to mention a few problems, they usually come from a scientific discipline and are usually dependent upon grants from scientific funds, which are steered by other scientists. At least in Denmark, a growing proportion of publicly employed scientists now and then change scientific focus, when the issues of research and development programmes change. Although these experts have a high professional moral standard, these facts undermine their political credibility.

Citizens are well equipped to take on democratic responsibility as assessors in technology assessment. The role of the experts, and among them the ethical experts, must be to supply knowledge and advice to citizens and politicians. A consensus conference is built upon that framework (Klüver, 1995).

8.4 WHAT IS A CONSENSUS CONFERENCE?

In broad terms, a consensus conference is a democratic process, involving dialogue between experts and citizens, but on the premises of the citizens. It is a method of technology assessment in which experts give an input of knowledge and citizens draw up the conclusions. The final document reflects how far the panel can go regarding consensus – it can be interpreted as the highest possible common denominator in the panel.

Topics to be taken up at a consensus conference should be of current political interest, require expert knowledge, be possible to delimit and not too abstract, and finally involve unresolved issues and conflicts. In other words, the topic should be well suited to be on the public agenda.

The organization of a consensus conference is best done by an independent institution because of the need for a democratic process. Any motivation among the organisers to manipulate the panel in a certain direction will erode the credibility of the results.

Advertisements for panellists are published in local newspapers and a panel is selected by the planning group. The panel should be mixed regarding relevant demographic characteristics (sex, age, location, education, occupation – in certain cultures religion or race could be included) in order to ensure that different values meet in the panel. But no efforts are made to make the panel statistically representative.

During two preparatory weekends, the lay panel prepares for the conference, for example by brainstorming concerns they have, and turning them into main and sub-questions that the conference should answer. The planning group then puts together an expert panel able to cover the questions posed.

The actual conference is usually a 4-day event. Expert presentation takes place at the first full day of the conference. On the first half of the second day, supplementary questions are put forward by the lay panel and there is a dialogue between experts, lay panel and the audience. During the rest of that day and the next day the lay people write the final document which is presented to the public on the fourth day of the conference.

The results of a consensus conference, as judged from the reactions of Danish politicians, are very useful to politicians as they give insight into the common opinions behind the public debate. Politics are not consensus-based and as a consequence these conferences should only be seen as a contribution to the political processes. But it is valuable for politicians to know in which directions and how far the citizens can go along the road of consensus (Grundahl, 1995; Klüver, 1995).

8.5 BACKGROUND TO THE CONSENSUS CONFERENCE ON TECHNOLOGICAL ANIMALS

In April 1987, the first Danish consensus conference, which involved lay people in the jury of the conference, was launched. It was about 'Gene Technology in Industry and Agriculture', and the lay people concluded that none of them could accept that gene technology be performed on animals (Teknologinævnet, 1987). This conference directly resulted in a Parliament decision concerning the Danish Biotechnology R&D Programme 1987–90: no research on transgenic animals was going to be funded by the programme.

At that time, practical use of these kinds of animals seemed to be a vision only – or to some people a picture of a frightening future. Some experts found it hard to believe that transgenic animals would be a realistic technology inside a 20-year time span. Others found the decision of the Danish Parliament appalling – Denmark would in a few years lose the chance of taking part in this important biotechnological development.

In 1992 it was clear that transgenic animals would be technologically feasible before the end of the decade. In order to have a Danish debate in due time, The Danish Board of Technology together with the Research Committee of the Danish Parliament arranged a consensus conference on manipulation of germ cells of animals. The popular title of the conference became 'Technological Animals'.

8.6 THE FINAL DOCUMENT

The full conclusions of the lay panel assessment in the 1992 consensus conference on 'Technological Animals' is printed in the final document (Teknologinævnet, 1993). The following is a summary.

8.6.1 Administration and regulation

The jury found that interest groups and lay people should, to a greater extent, be represented in fora where the circumstances for research are discussed.

It has been argued that national restrictions on the use of technological animals are irrelevant because of international regulations. This should not prevent us from taking a national standpoint, the panel argued.

8.6.2 Economics

Economic considerations seem to be a most important regulatory mechanism through priority setting in research policy and through market mechanisms. This led the jury to conclusions about impacts on different economic policies.

In future uses of technological animals, for example in the food industry, it may turn out to be unavoidable to pay a license fee to a patent owner since the cost of research and development of the technological animals must be covered. Paying these fees may have a negative impact on the financial situation of the individual producer of food products.

The established and important principle of 'farmer's privilege', which allows the farmer to keep seed from his own crop for his own purposes, would be broken if license money had to be paid for the product in question.

Our competitiveness will be weakened only if national restrictions are imposed and the consumer accepts transgenic animals. If the consumer does not accept them, our competitiveness will be weakened by investment in these animals. That is, investment in a controversial technique is not necessarily to the good.

An alternative to development of technological animals for agriculture is increased research and development efforts in organic farming. Although the panel did not manage to get exact figures from the experts, the general impression was that very little is invested in organic farming compared with modern biotechnology.

The economic resources which have been reserved for risk assessment and ecological understanding of gene technology seem very small.

It must be expected that companies will be inclined to carry out risk analyses themselves if consumers are conscious about their choices and demand healthy and ecologically sound products. So, consumers have a responsibility of their own.

Through the introduction of the 'polluter pays principle' we would be protected against short-term economic decisions taken by the companies.

8.6.3 Patents

The lay panel found that by patenting animals we run the risk of regarding animals as things.

The lay panel did not understand the logic in the field of patenting: you cannot patent a gene because it is a discovery and not an invention, but by synthesizing an identical gene, patenting becomes feasible. Thus, a discovery becomes an invention.

The Danish Parliament has already decided to prohibit the patenting of animals. Based on the view that patenting of life is not acceptable, this decision should be maintained and promoted in negotiations for international patent regulations.

8.6.4 Risk analysis

The possible danger of releasing technological animals into nature could be that the genetically manipulated animals may dominate or out-compete the naturally occurring species. Furthermore, technological animals could transfer unwanted new characteristics to wildlife species.

Because of this, the panel found that there are limits to what can be accepted regarding deliberate release of genetically manipulated animals.

The risks from releasing larger animals, for example cows on a fenced field, seem negligible. But it must be considered irresponsible to release smaller technological animals such as, for example, genetically engineered fish, insects, etc. which are difficult to control in nature. Fish farms in dams or at sea impose risks of accidental releases, which makes it irresponsible to produce transgenic fish.

The use of technological animals entails risks of reduction of the gene pool, for example when cloning is used in farming. Compared with traditional selective breeding this poses a problem. On the other hand, the panel found that the same technologies make it possible to maintain biodiversity by freezing eggs and semen.

As scientists are able only to a limited degree to foresee the different risks, risk assessments should be carried out as thoroughly as possible, case-by-case and step-by-step. Risk assessments ought to be made with regard to experiments as well as when seeking permission for production.

Despite the assurances of the experts, the panel did not feel it was safe to eat genetically manipulated animals. The long-term consequences of consuming foods produced by technological animals are unknown, they argued.

Labelling of food products from technological animals is a practicable demand and therefore is only a question of will. The panel found that labelling should be mandatory.

8.6.5 Ethics, values and nature

We have unlimited possibilities for using animals as we want to. But there are limits to our rights in this respect.

That the animals do not suffer does not in itself make it responsible to use technological animals. Suffering is an important criterion, but not the only one, the panel said.

The goal does not justify the means in any situation. The means should always be seen in relation to the goals in question. It is for example totally unacceptable to produce 'new' pets.

On reasonable grounds, however, it must be considered ethically acceptable to produce technological animals to be used for purposes such as developing new treatments against serious diseases, such as cancer.

However, it is considered unethical to produce technological animals like cows or pigs on the grounds that they may become better adjusted to withstand the existing methods of production in agriculture. If the existing production methods give problems, they should be corrected instead of the animals.

We are making animals into things through the way we produce food. We have done that for a long time and it is unethical. We should not go further towards turning animals into things by producing technological animals. But if we accept patenting of animals, that would be a step further in the direction of regarding animals as things.

One may fear that by regarding animals as things, a step is taken on the road towards looking at people as things. Fundamentally it would be unethical to produce a human being for experimental purposes. But the panel said they feared this would be regarded as acceptable in the future.

8.7 IMPACT OF THE CONFERENCE

Most of the 13 consensus conferences arranged by The Danish Board of Technology have directly resulted in decisions taken by the Parliament, the Government or other decision makers. The conference on technological animals has had the effect that a large Danish pharmaceutical industry has changed its policy regarding the use of genetically manipulated animals as bioreactors. Before the conference this firm was reluctant to use bioreactors because of possible negative reactions from the market. The consensus conference was in other words a needed signal about what can and what cannot be accepted regarding the use of bioreactors (S. Riisgaard, Director, Novo Nordisk A/S, personal communication).

For some members of the Danish Parliament the conference emphasized the need for a debate on how aims and means can be balanced. Which kind of regulatory system would be needed, if societal needs were a criterion for approval of gene technology projects? (K.R. Møller, Chairman, Research Committee of the Danish Parliament, personal communication).

In general, consensus conferences get impressive press coverage. Television and radio news programmes usually cover the first and last day of the conference. Newspapers cover the conference and its topic in the weeks before and after. All in all, we seldom get fewer than a hundred press clips with reference to the conference. Most consensus conferences give rise to a boom in press coverage of the topic. The impact of the topic's being thus kept on the public agenda for some time is hard to assess, but it may be the most important effect in the long run.

8.8 OPENNESS PAYS OFF

Danes have a relatively high level of knowledge about biotechnology. As can be seen from the above conclusions from the conference on 'technological animals', they show scepticism about its excellence and environmental safety, but at the same time they give their acceptance to its use. On the one hand, there is no doubt that the Danes are watching the development with critical eyes. On the other hand, industry is left free to act inside the limits of the regulation.

Danish industries often declare that they would prefer to live with stringent regulation rather than not knowing what they are up against – if, of course, the regulation is world-wide. The official policy of the Danish biotechnological industry Novo Nordisk A/S is a rejection of deregulation in the gene technology field.

This peaceful situation, compared with many other countries, has its history. Denmark was one of the first nations – if not *the* first nation – to implement an 'Environment and gene technology act' in 1986. The fact that Danish politicians were serious about the risks and established regulation at an early stage gave returns in the form of acceptance of biotechnology among the population.

Shortly after the establishment of regulation the Danish Parliament (Folketinget) initiated a 'People's education and technology assessment programme' of 23 000 000 DKr (4 000 000 US$) from 1987 to 1992. The programme was managed by The Danish Board of Technology. It consisted of well over 50 projects and produced an immense amount of debating activity. The programme launched 19 conferences, 29 seminars, 600 local debate arrangements, 7 movie/videos, 13 television programmes, 31 reports, 44 debate booklets, 34 books and more than 500 articles in magazines and newspapers.

As mentioned earlier, a characteristic of the development of the Danish debate in the late 1980s is the movement from a standard debate position to a more nuanced dialogue. This must be interpreted as a result of the general gain in knowledge, of biotechnology as well as of the societal impacts of it, in Danish society during that period –

knowledge which has been gained from open debate processes (L. Klüver, http://ingenioren.dk/tekraad).

All things being equal, a society in that situation is in a better position than a society in which dialogue is not present and conflict dominates. The price is regulation and some limitations to the free market. The gain is a more constructive environment and acceptance of the technology inside the prescribed limits.

Note

The author is director of The Danish Board of Technology, the technology assessment institution of the Danish Parliament and Government.

The Danish Board of Technology has a remit of initiating technology assessment, furthering public debate on the impact of technology on society and the individual citizen, and advising the Parliament and the Government.

REFERENCES

Borre, O. (1989) Befolkningens holdning til genteknologi. (Report on the public attitude to biotechnology). *Teknologinævnets rapporter*, 1989/3.

Borre, O. (1990) Befolkningens holdning til genteknologi II. *Teknologinævnets rapporter*, 1990/4.

Eurobarometer (1991) Opinions of Europeans on biotechnology in 1991. *Eurobarometer* 25.1 (INRA Europe. Report to the European Commission, DGXII, 1991).

Eurobarometer (1993) *Eurobarometer* 39.1. European Commission, 1993.

Grundahl, J. (1995) The Danish consensus conference model, in *Public Participation in Science: the Role of Consensus Conferences in Europe* (eds S. Joss and J. Durant), Trustees of the Science Museum, London.

Klüver, L. (1995) Consensus conferences at the Danish Board of Technology, in *Public Participation in Science: the Role of Consensus Conferences in Europe* (eds S. Joss and J. Durant), Trustees of the Science Museum, London.

Klüver, L. Oplysning og teknologivurdering om bioteknologi. (Report on the biotechnology technology assessment activities at the Danish Board of Technology 1987–1995). http://ingenioren.dk/tekraad.

Rienecker, L. and Erichsen, F. (1990) Lægfolk i en konsensuskonference. (Working paper for the Danish Board on Technology on a psychological survey on the development in a lay panel during a consensus conference).

Teknologinævnet (1987) Genteknologi i industri og landbrug (Final document from a consensus conference on Gene Technology in Industry and Agriculture). *Teknologinævnets rapporter*, 1987/4.

Teknologinævnet (1993) Teknologiske dyr (Final document from a consensus conference on Technological Animals). *Teknologinævnets rapporter*, 1993/1.

9

The inevitability of animal biotechnology? Ethics and the scientific attitude

Jeffrey Burkhardt

9.1 INTRODUCTION

Most observers of biotechnology are aware that the main standard critiques of animal biotechnology are based on either animal rights/welfare arguments, ecological-oriented arguments, or socioeconomic consequences arguments. In this chapter, I want to suggest that despite the logic or seeming appropriateness of many of these critiques, they lack *ethical force*. By this I mean that the arguments (and the arguers) are unlikely to actually change the minds of those engaged in biotechnology practices and policy-making (Stevenson, 1944; Olshevsky, 1983). This is because of the orientation or attitude of those entrusted with doing and overseeing biotechnological work with non-human animal species. I will argue that this orientation must change before ethical arguments concerning animal biotechnology, indeed ethics generally (in the philosophical sense as opposed to legalistic or professional courtesy senses), mean anything to the scientific community. Public policy may be one tool to change this orientation but, given the social power of science, the tack more likely to be successful is the moral re-education of scientists and science-policy makers. There is some evidence that sympathy toward ethical concerns is beginning to make its way into the mind-sets of some people in the bioscience establishment. However, given the ease of replies to the standard types of criticisms of animal biotechnology, which I will (partially) catalogue below, there are considerable obstacles to 'ethics in science' in these cases.

There are philosophical reasons, but more important, practical reasons for the proposal in this chapter. Philosophically, while some of the ethical objections to animal biotechnology or to particular biotechnology practices may be justifiable, behind many of them is a misplaced Platonic assumption that the problem with those engaged in animal biotechnology is that they do not know 'the good'. That is, if the scientists or policy makers knew or understood the philosophical objections, then they would stop doing what they are doing. The problem with this assumption is that scientists would have to accept the fundamental criteria for justifiability or reasonableness upon which philosophical argument rests before they would even fathom these criticisms as reasonable ones. This relates to my practical concern.

Practically, the rights/welfare, environment/ecology, and socioeconomic approaches are usually bound to fall on deaf ears. Arguments concerning 'ethics and animal biotechnology' are generally irrelevant, at best, to the actual members of the bioscience community or 'Science Establishment'. Scientists and policy makers may fathom some ethical concerns when their scientific or policy-making 'hats' are off. But to scientists and science-oriented policy makers *qua* scientists and science-oriented policy makers, proponents of animal rights/welfare arguments, environmental/ecological ethics arguments, or social justice arguments, can easily be relegated to the role of 'philosophers crying in the wind' (or howling at the moon). Ethical arguments which do not first assume the *a priori* legitimacy of whatever the scientific enterprise has decided to pursue are bound to be 'external' and 'externalized'. Many people may be hopeful that the science establishment will address those substantive ethical arguments about animal biotechnology or biotechnology in general, and even more significantly, act on ethical stances. This is, however, not likely until such time as the 'inevitability-of-scientific-progress' and 'better living through biotechnology' orientations are themselves subject to ethical questioning. To the extent that those are the dominant, and even growing, orientations of members of that scientific establishment, and indirectly, of the general public, broader ethical concerns likely will remain moot. This again leads me to the conclusion that science either has to be forced to 'be ethical', which may mean giving up particular animal biotechnology practices or animal biotechnology in general, or science will have to awaken to ethics on its own, which (and this is key) *may or may not* entail abandoning animal use or biotechnology. My belief is that we should accept the inevitability of continued animal biotechnology research and development, and hope that the legalistic-type controls now in place in many nations continue to work or work even better. In the meantime, we should also 'sympathetically' impress on scientists the value of ethical reflection on their work.

9.2 WHY THE STANDARD ETHICAL CRITIQUES FAIL

The various kinds and dimensions of (non-human) animal biotechnology are outlined elsewhere in this volume, so I will not explain them here in any detail. Instead, I will only mention that they run the gamut from the actual engineering of animals themselves, e.g. through gene insertion or deletion, to the use of bioengineered techniques to produce products for live animals or for animals used for meat and other consumptive purposes. The only common element among all of these animal biotechnologies, so far as I can discern, is that somehow, non-human animals are involved in either the process or the outcome of the biotechnology activities.

This, coupled with the fact that *biotechnology* is the activity in which non-human animals are used, sets the context for the various rights/welfare and natural kinds sorts of criticisms. In what follows, I simply catalogue some of these critiques, and show the standard science-based replies. My judgement is that many of these replies are correct.

9.2.1 Animal welfare, rights, and natural kinds arguments

Animal rights or animal welfare arguments regarding animal biotechnology arise because in all of these biotechnology activities, non-human animals are *involved*. The strongest argument objects to the use of animals *per se*. The rights argument, articulated so forcefully initially by Regan (1985), maintains that individuality and 'subject-of-a-life'-hood of non-human animals (in particular larger mammals) ethically demands their being treated in a quasi-Kantian manner: as ends in themselves, with appropriate stakes in life, liberty and self-actualization. Genetically altering an individual non-human animal, either before or after conception, *ipso facto* intrudes upon the autonomy of the being. Interestingly, this is true even if the genetic alteration in some way improves the quality of life of the individual, for instance in improving its resistance to disease or some stress-inducing environmental condition. The argument, in its purest form, is that animal biotechnology as *a technology that is used on individual animals* is inherently morally wrong. [An aside: public opinion surveys in the US indicate some support for this position among the general public, although as education levels rise support tends to weaken; see Hoban and Kendall (1992) and Hoban and Burkhardt (1990).]

The philosophical underpinnings to the rights objection to animal biotechnology are easily countered by the scientific community. Most direct genetic engineering of animals (i.e. altering an animal's genetic structure) is performed either so early in the fetal developmental process that a distinct individual animal (in terms of moral autonomy) is indiscernible, or, more often, occurs even before conception takes place. Under any or all of Regan's criterion of the animal's having some rudimentary consciousness, or Singer's criterion of sentience, or Fox's criterion of 'telos-possession' (Fox, 1990), there is no 'subject of a life' (Rachels, 1990) whose inherent value or rights or unique purpose are disrespected through the engineering process. In these cases, opponents of pre-birth or pre-conception animal biotechnology might invoke a quasi-Aristotelian natural kinds argument, or, in more modern garb, suggest some violation of the Darwinian imperative, but that is to move to another kind of critique, not necessarily compatible with the rights/welfare position.

Rights or welfare arguments may be appropriate to an appraisal of some biotechnology techniques, nonetheless. In particular, the use of biotechnologically produced hormones, pharmaceuticals and other agents may be seen as in some way disrespecting animals' rights or may cause a decrease in welfare. Using a chemical which artificially increases milk production in dairy cows, but which also increases incidences of disease (mastitis) and shortens an individual cow's productive life (and by implication, its life), may be bad for the cow in both rights and welfare terms. The issue here is, however, less an animal biotechnology matter than a simple matter of people using cows in dairy production systems. Whatever people do to increase milk production runs the risk of disrespecting the cow's rights or welfare level. In this respect, the ethical culprit is instead the system wherein technology of whatever origin is used for an ethically unacceptable end (from either rights or welfare perspectives or both). The point here is that animal biotechnology *per se* may not be the appropriate object of ethical concern, since biotechnology is used only to produce products for animals, many of which can be produced through non-biotechnological means (even though they are more costly) (Burkhardt, 1992).

One final point on the rights/welfare approaches: if any ethical concession is to be made to the fact that animals are used by humans, then some animal biotechnology may in fact be more ethical than some other research and production practices currently employed. If, as various humane societies have argued, better treatment of non-human animals ought to be our goal, biotechnology might be precisely the means to achieve that goal. This would, of course, depend on exactly what is being done with or to the animals, individually or by species (see, for example, NABC, 1992).

The natural kinds argument is the other main kind of animal-based objection to genetic engineering. Rifkin (1983) argued that the very ideology of genetic engineering – 'algeny' – challenged the naturalness of those species which were either created by God or evolved through natural selection. Independent of the potentially disastrous ecological consequences of tampering with these longstanding kinds, there is a fundamentally immoral audacity in those people who would 'play God' and change the natural order for whatever purposes they intended. It may not be so much that individual animals have rights, but instead that individuals and species of animals (and plants) have some sort of inherent natural worth. As such, they should be left alone to change or evolve at their own natural pace – if at all.

Again, there are science-based replies to natural kinds arguments. Simply stated, humans have used animals for millennia, and in many respects, the very animal species which they used now exist only because of their use. Similarly with plants: Nature may have selected certain plant species to continue to exist in changing environments, but the instrument of Nature was in many respects human activity (conscious or otherwise) (see Busch *et al.*, 1995). Granted, it may now be wrong in some ways to alter animals or intrude on nature, but it is not because of their nature or Nature itself, since their nature and Nature itself are in many respects the products of human activities. Rather, the ethical inappropriateness of some of these activities may simply be due to the negative consequences which result from them. But once again, this is a different kind of argument, which would turn us back either to rights/welfare, or to some other, larger set of consequences.

In terms of ethical force, the appropriate ethical critiques of animal biotechnology are not, at least as animal biotechnology is currently practised, either the animal rights/welfare sorts of argument, or the natural kinds approach advanced by Rifkin. These kinds of criticisms, which I refer to as 'intrinsic' critiques, can generally be met with reasonable points about either the nature of the practices performed on animals, or the extent to which they produce suffering, or the extent to which they are no different in principle from any other animal-using scientific practice. I suggest that other, 'extrinsic' or consequentialist critiques may be more appropriate and forceful challenges to biotechnology, to the extent that there is science-based (and hence, 'reasonable') evidence to support their claims. I will argue, nevertheless, that these criticisms can also fail because they miss major points about what biotechnology, and animal biotechnology in particular, can potentially do.

9.2.2 Consequentialist critiques and irreversibility arguments

Consequentialist-type arguments regarding animal biotechnology usually focus on ecological or socioeconomic cultural ills associated with biotechnology in general, and animal biotechnology in particular. One common theme among these arguments, and perhaps, underlying fear among their proponents, is that these consequences are or may tend to be irreversible: that is, once the technology or its products have been developed, adopted, or widely used or released into the world, severe negative effects will obtain which will be difficult or impossible to stop or reverse (Comstock, 1990).

The ecological arguments are most straightforward, though most originally were advanced with respect to microorganisms and plant species with little thought given to (larger) animal implications. According to this line of argument, a genetically altered individual or species of organism is necessarily different from its natural or wild counterpart. In fact, the reason behind genetic engineering is to design plants, animals or organisms with traits which would allow the organism to cope with the environment in ways different from the non-engineered kin, for instance withstand different and hostile climatic conditions, resist pests, better absorb nutrients from the environment. The goal, of course, especially for agriculturally significant plants and animals, was to increase yields, productivity, and ultimately profits for farmers and ranchers. In other cases, the goal was to solve a different environmental problem, for example to non-chemically control noxious water weeds, or eat crude oil spills in rivers and oceans, or eradicate or control pests to humans and domesticated animals (e.g. flies, ticks). These creatures of bioengineering were intended to perform in their environments in ways to be preferred to those of their natural relatives.

The prime concern of the ecological–ethical critique is that bioengineered organisms, once outside controlled laboratory conditions, might behave in ecologically inappropriate ways. For instance, the engineered species might out-compete its natural relatives to the point of the extinction of the latter; or new predator species might evolve in response to the changes in the original species; or the new species just might grow out of control; or the new species might simply upset the 'biotic community' (Holland, 1990). In each case, purported uncertainty in our knowledge of how ecosystems behave in the face of the new species leads to the conclusion that there is something inherently ecologically dangerous about the bioengineered species. Note that it is again not bioengineering *per se* that is at issue. Rather, it is the *results* of bioengineering. Genetic manipulation through traditional breeding techniques was limited in terms of time and also in terms of what traits could be introduced or eliminated from species. Now, we can radically transform certain species, but no one really knows how they will behave in the larger ecosystem.

There may be sound moral premises behind this critique, such as, 'We morally should not risk ecosystemic disruption because of risks to present or future people or to the ecosystem itself'. Once we have allowed these organisms into the environment, we cannot get them back. Even so, there is again a reasonable reply in this case. We risk ecosystemic disruption all the time, and in fact *cause* ecosystemic disruption through many things much more dangerous than genetically engineered animals or animal products. So, unless the point is that we should leave the ecosystem alone, a practical impossibility, these potential ecosystemic disruptions are not necessarily immoral. Regarding the irreversibility claim, namely, that once we have allowed these new organisms into the environment, we cannot get them back, the reply is equally simple. Because we know the genetic makeup of the engineered species even better than that of the non-engineered ones, we are in fact in a better position to control the new species or even eradicate those individuals who begin to get out of control. The final reply is that all of this concern about risk and uncertainty and irreversibility is a bit misplaced with regard to animal biotechnology, at least beyond the microorganism stage. Microorganisms may be problematical, but most animal biotechnology is either on large animals or for products for large animals, so control is possible; indeed, 'control' is precisely the point.

The ecological critique of animal biotechnology thus also appears to lose much of its ethical force or appropriateness in regard to animal biotechnology, and perhaps even most biotechnology. The moral point about risking people and ecosystems, and the irreversibility issue, remain sound concerns, nevertheless.

Risk and irreversibility are also behind the economic and social consequentialist arguments vis-à-vis biotechnology. The argument here is that, unlike ecosystemic behaviour, we have clear precedents with respect to how new technologies affect social or economic behaviour, relationships, or structures (Burkhardt, 1988). On this basis, it is argued that animal biotechnology is potentially socially disastrous, and hence likely to be immoral in that regard. Indeed, we know that new technologies have negatively affected socially or economically marginal groups of people. Given the nature and probably irreversible consequences of biotechnology, we morally should not allow its development, release, sale, etc. This argument has been most forcefully applied concerning one animal biotechnology product, bovine somatotrophin (BST), also called 'bovine growth hormone'. For at least 10 years, animal activists, farm activists, environmental activists, and philosophers concerned with ethics and biotechnology, have advanced the arguments that BST will put small-time dairy farmers, and indeed, the socioeconomic structure of agriculture, at considerable risk. Accordingly, we must 'stop BST' because of equity or distributive justice or fairness to future generations, or even the health of present consumers (Comstock, 1990).

Despite the force of ethical precedents, the reply to socioeconomic criticisms is straightforward. Recall the reply to the ecological critique:

unless interfering with the environment is inherently morally wrong, then there is nothing uniquely wrong about the release of genetically altered organisms *so long as they do not misbehave*. Indeed, the point of genetic engineering – with all the precision with which organisms can be manipulated associated with these techniques – is to get the organism to behave exactly as planned. Analogously, with respect to the socioeconomic critique, there is nothing uniquely right or wrong about this technology that is not right or wrong about any technology or its products. The objection to biotechnology's socioeconomic consequences is really an objection to technology in general, or perhaps to capitalist socioeconomic arrangements. BST or any other particular product or process might change socioeconomic relations, but that is not the fault of the technology, only the system into which it is introduced (Burkhardt, 1991).

All these replies are, in my judgement, appropriate, and in fact are the kinds of replies scientists and policy makers give to concerned citizens, activists, and the occasional philosopher in their midst (NABC, 1994). There is much misinformation and unfounded concern among many critics of biotechnology. Even when informed and reasoned, however, the critiques all have replies, which are based on either sound science or clear historical precedent. At this point, it may seem as if I am defending scientists or policy-makers involved with biotechnology in general, along with biotechnology in particular. In a sense I am, since both the intrinsic as well as extrinsic critiques focus on particular practices, products or policies, which, as this last example (BST) suggests, are not the best or even proper loci of ethical concern. If there is a problem with animal biotechnology or biotechnology in general, it is a problem with science and technology even more generally (Buttel and Belsky, 1987).

There is one further consequentialist argument to attend to here. This might be called the 'cultural consequence' argument. One of the 'side arguments' made with respect to the BST case was that the continued development, release and adoption of these new technologies will (probably) have disastrous cultural effects. There is an element of this critique in Rifkin's indictment of algeny. The cultural consequences critique goes beyond Rifkin, however, in suggesting that widespread diffusion of biotechnological products (from altered animals and plants to bioengineered chemicals and food products) might open the door for general public acceptance of the ethical appropriateness of engineering *people*. The spectres of eugenics, of Nazi atrocities, and gender-based abortion and infanticide all form parts of this critique. Perhaps this critique rests on a misplaced slippery slope, but the position nevertheless appropriately targets people's attitudes, ethical values, and perceptions of what is 'natural' (Busch, 1984). As more biotechnology becomes the norm, a whole culture might come to accept whatever is bioengineered as even morally preferable to the non-engineered. Given *real* slippery slopes in attitudes, and *real* risks to longstanding human values such as freedom of choice and perhaps diversity among people, biotechnology accordingly is a cultural threat.

Like the potential ecological consequences critique, this last concern plays up the element of uncertainty. Unlike the environmental/ecological position, however, it is less a matter of how bioengineered organisms or ecosystems will behave or be affected than a matter of how people will act and react toward biotechnology. The question is whether there are any reasons for the public or policy makers to be concerned about the standard science-based reply to this position, namely, it will not happen. I will argue in the next section that the answer is predicated on whether there is indeed any reason for us to be concerned that biotechnology will continue to be employed without prior or at least concomitant ethical reflection. There may be little reason to fear biotechnology progressing, but only if biotechnology is either regulated and monitored, or ethics becomes an intrinsic part of the scientific attitude – the fundamental ideological/epistemological basis for science.

9.3 THE BIOTECHNOLOGY CULTURE AND THE SCIENTIFIC ATTITUDE

It has now been more than two decades since what was conceivable in genetics and microbiology became possible, and over a decade since what was possible (in some realms) has become actual. Biotechnology, whether in the generic sense of 'genetic engineering', or in more specialized processes and techniques such as gene insertion, removal, cloning, etc., is, in other words, with us. Nearly every industrialized nation, and all but the poorest developing nations, has some semblance of a biotechnology industry (Busch *et al.*, 1991). Most nations also have some form of national policies regarding biotechnology, whether these are funding policies, regulations, legal protection, or simply national priorities. A realistic appraisal of biotechnology, in general, has to begin with this fact: something in the biotechnology area has occurred, and likely will continue to occur. Philosophers and social analysts have long pointed to the power that science has in modern society. This notion was given contemporary expression and force by Rosenberg (1976) in his notion of 'Scientism' – the ideology of science solving all human problems. Scientism, it is argued, has become another dominant '-ism' of our day. There is little need to reiterate either the philosophical observations or sociological data supporting this idea: I will, in fact, take it as a given that despite at least a century of philosophical critique of the social power of science and technology, Scientism is with us.

The extent to which Scientism undergirds both the biotechnology enterprise as practised as well as public policy regarding biotechnology is astounding. It is this fact which lends credibility to the cultural critique described above. Despite the fact that biotechnology remains relatively new, its processes so esoteric (to the general public and to most policy makers) and its products so few at this point in time, 'advances' in medicine or agricultural research due to genetic engineering or some other biotechnological procedure are routinely touted in the popular press and electronic media. Though the occasional nod might be given to potential ethical questions in an article in, for example, the *Wall Street Journal* or *Newsweek* magazine, by and large the tone of reporting on such advances is precisely that these results are advances (Hornig and Talbert, 1993).

In the public policy arena, moreover, we have witnessed a gradual but steady strengthening of the power of biotechnology or bioscience in general (Busch *et al.*, 1991). Even as funding for some specific kinds of basic research (e.g. AIDS) has been questioned by members of the United States' Congress, the general level of support for biotechnology has grown. In addition, much of the regulatory oversight which grew up in the early years of biotechnology – the late 1970s and early 1980s – has gradually devolved (NABC, 1994). Public policy priorities have shifted from concerns about the potential negative effects of biotechnological research to concerns that the advances in bioscience and especially bioengineered products are not coming fast enough. Indeed, the prospects of a nation being out-competed in the much-touted 'growing global marketplace' are enough to send policy makers concerned with biotechnology (and scientists receiving funding for such research) into a frenzied search for the 'next major breakthrough'. Despite what some scientists have called the 'hype and glitter' (Busch *et al.*, 1991) (and, by implication, necessarily unfulfilled promises) associated with biotechnology, there is little reason to believe that anything short of an environmental catastrophe caused by a bioengineered product or experiment gone awry could actually cause a reduction in the enthusiasm with which biotechnology has been embraced at nearly all levels of research management, oversight or policy-making. One can well imagine that even if an environmental catastrophe were to occur, it would be attributed not to anything intrinsically problematical with biotechnology, but instead to 'human error'.

There are a number of reasons or causes to which the biotechnology craze might be attributed. One might simply be the excitement or enchantment that members of the scientific establishment experience when, as I mentioned above, what was conceivable becomes possible, or what was possible becomes actual. 'Unlocking mysteries' is, after all, part of the (self-described) point of scientific research, and a mystery such as how to make hitherto impossible genetic transfers actually occur certainly caught the attention and imagination of many in the bioscience establishment. As one scientist put it: 'we finally can *do* this!' (J. Burkhardt, L. Busch and W. Lacy, personal interview, 1984; Busch *et al.*, 1991).

There are other, perhaps less noble, reasons for the degree of excitement and commitment to biotechnology in general and agricultural (plant and animal) biotechnology in particular. As mentioned above, a tremendous amount of funding was made available for biotechnology, by federal governments, public and private universities, and major multinational corporations (Buttel *et al.*, 1994). The promise or prospect of profitable new products and processes (to which these corporate actors would have proprietary rights) was not solely the reason, although undoubtedly a large part of the reason. The motivation also appears to have been the time element involved: whereas a new plant variety or pharmaceutical product might take several years or even decades to develop under older research methods, the new biotechnologies offered hope for quicker new products and processes. Again, in an increasingly competitive environment, the quicker the better.

Whatever the reason for the interest in and excitement about biotechnology, there is one additional glaring fact about the scientific attitude concerning biotechnology: ethical considerations such as those discussed above matter little, if at all. This orientation has permeated the science establishment. This lack of concern for deeper ethical matters, as opposed to legalities or professional courtesies, may permeate all society as well (save theologians and philosophers trying to conserve older ways of thinking about what is or is not moral). So long as science continues to deliver or at least forecast new promises – for corporations, for policy makers, and ultimately for the general public – ethics is irrelevant.

9.3.1 The scientific attitude

Biologist Frederick Grinnell, in *The Scientific Attitude* (1987), described what I take to be the underlying reason why 'ethics and science' or 'ethics and biotechnology' have been seen as beyond the pale in terms of attitudes and practices of members of the scientific community. Though Grinnell's purpose was to provide some instruction to beginning scientists (and students) about what the real world of science is about, and hence help them become better professionals, his focus on science as a 'way of being' is particularly appropriate to the thesis I am developing. Grinnell, after some discussion of the nature of science as the pursuit of truth or knowledge (mandatory for texts of this sort), finally hits on the idea that science must become, for scientists, a 'way of seeing' and a 'way of being'.

By 'way of seeing', Grinnell means that the material with which much physical and biological science operates (though perhaps true of social sciences such as economics as well) is only visible once one has come to appreciate and accept the appropriate theoretical or ideological foundation. Making plain the philosophical notion of 'theory-dependent data', it is shown how the basic units of much biological science, 'genes' and 'cells', as well as the processes which go on within and around to those units, are real only if one accepts the operations and methods which attend the study of those units and processes.

The 'way of being' of the scientist is of more direct and critical concern. For, as Grinnell suggests, this means adopting 'the scientific attitude', which in essence is to come to believe in Scientism: Rosenberg's (1976) book was aptly titled *No Other Gods*. And believing in Scientism means always being willing to act on making what is conceivable possible, and what is possible, actual. In other words, to *be* a scientist (in this ideal typology), one must accept the *doctrine* that science defines what is real. Those who do not accept either that reality or its technological ramifications (the tools employed or the products created) are wrong at best, *irrational* at worst. This becomes the crux of the matter.

That this sort of attitude might engender a degree of arrogance or self-righteousness is clear (Feyerabend, 1978). However, not all individual scientists, or even most, need to or do display those personality traits. It is enough that the science establishment – research administrators, policy-setters, leading scientific spokespersons – has the power to define what is or is not real, reasonable or rational. This is power in sociologist Stephen Lukes' (1986) sense of a 'third dimension' of power (the first two being physical force and persuasive ability) – the ability to define the terms in which rational discourse takes place. The work, perceptions, and professional communications of members of the scientific community all take as a given the reality and importance of scientific rationality, and whatever emanates from rational scientific work.

Scientists, including biotechnologists, may in fact be quite humble in the face of new problems, new theories, new frontiers. Nevertheless, there is a sort of moral imperative in the widely shared attitude that 'the work (of science) *must go on*' (J. Burkhardt, L. Busch and W. Lacy, personal interview, 1988). Moreover, significantly, whatever appears to impede or constrain the work of science must be based on some irrational or non-scientific force. As such, lack of funding (to the extent that some of this work lacks funding) is unreasonable. Even more unreasonable, and immoral by this doctrine, are rules, regulations, oversight committees, and reporting requirements. Except as necessary to advance science (for example, biosafety practices), all external interference in the scientist's work is branded illegitimate, counterproductive, un-called-for. In a word, constraints are unreasonable.

This characterization of the scientific attitude and Scientism undoubtedly overstates the case, and may even be questioned for grossly caricaturing science and the scientist. However, the ease with which the science establishment, and many individual scientists, can dismiss criticism or the kinds of objections to biotechnology discussed above is telling. There is not only nothing wrong with biotechnology (that is not wrong with any part of science), but to suggest otherwise is to either fail to understand science or simply be irrational. This refers back to my earlier point: any criticism or ethical concern which does not *a priori* assume the legitimacy of the scientific enterprise and the necessity of using science (including biotechnology) to solve problems must be ignored or, better, rendered impotent. One way to emasculate those criticisms is to fall back on the power that the science establishment has long had in Western society: change the terms of the discourse.

9.3.2 The culture of biotechnology

The case of bovine somatotrophin, one of the first commercial animal-affecting products to emerge from the biotechnology enterprise, is a telling example of the power of the bioscience community to actually change the terms of discourse – to the advantage of the biotechnology enterprise, of course. Bovine somatotrophin is a naturally occurring compound, produced in the pituitary glands of cows, which regulates growth and indirectly affects milk production. Dairy scientists had known for years that administering additional doses of the substance to dairy cattle could increase milk yields without increasing feed intakes significantly. However, obtaining the substance naturally was expensive and time-consuming, and hence, not a practical dairying alternative. Scientists at a number of United States universities, under grants or contracts with Dow Chemical and Monsanto corporations, became able in the early 1980s to produce the compound using recombinant DNA methods. The product could now be produced in greater quantities, and much more cheaply and efficiently.

Bovine somatotrophin was originally named 'bovine growth hormone' (BGH) when scientists and company representatives began touting the chemical for its potential use in animal agriculture. Quite soon afterward, however, representatives of the bioscience establishment began to be met with resistance from consumer advocacy groups, and eventually lawsuits were even filed to prevent the US Food and Drug Administration from permitting the use of BGH in agriculture.

Emphasis was placed on the nature of this compound as a *hormone*, despite scientists' and the industry's assurances that it was a non-steroidal-type hormone, and would not in any adverse way affect consumers of milk from BGH-treated cows. About the same time, however, the term 'BGH' disappeared from scientific publications and company promotions. Bovine somatotrophin became known by its real abbreviation, 'BST'. The resistance and criticisms of the substance did not disappear overnight, but the bioscience establishment managed to diffuse a significant amount of consumer activists' policy-affecting power by simply redirecting the concern away from a 'hormone' to just another productivity-increasing 'treatment' (Browne, 1987).

The whole BGH/BST story is much more complicated and drawn out (Burkhardt, 1992), but just this name change element in the story is sufficient to suggest my point. With nothing more than a semantic sleight of hand, the bioscience establishment was able to effectively control the public forum as well as public policy agenda. There are undoubtedly many, and more glaring cases of science winning a public relations battle or war. The only times the science establishment does not win hands down, it seems, is when it faces an equally formidable foe, for example the tobacco industry in the US, or organized religion, especially the Roman Catholic Church.

We need not explore this last point in any detail, except to note that among the most vocal of the critics of BST were the American Catholic bishops (Comstock, 1990). Further, the bishops, and other religious leaders in the US, have gone on record as being against another biotechnology development, the patenting of (plant and animal) life forms. To date, the scientific community has managed to ignore or diffuse those religion-based objections. However, concerned that the religious objections to particular biotechnology-related activities might turn into larger public and political concern about biotechnology in general, policy makers at the United States Department of Agriculture (USDA) began (in mid-1995) to identify 'reasonable individuals' (their language) (F. Woods, personal communication, May 1995). The USDA sought to find people, including philosophers and scientists, who could articulate the merits of different aspects of biotechnology, and instruct US Congressional staff people and USDA 'higher-ups' in these arguments. The point was similar to the strategy in the BST case: to be able to ward off criticisms pre-emptively should those criticisms begin to reach research-threatening proportions. The sole reason why this ethical think-tank did not come into existence is that the USDA abolished its biotechnology oversight ('ethics') committee, under whose aegis this group was to function. That fact is interesting in itself: among the functions deemed unnecessary for the USDA, given budget constraints, was biotechnology oversight (*Science*, 22 September, 1995).

Critics of the agricultural research establishment have for a number of years pointed to a 'circle the wagons' mentality among people in the science establishment (Busch and Lacy, 1983). Always mindful of potential criticisms – from environmentalists, animal-rightists and animal welfarists, and small-farm and labour activists – the establishment (it was claimed) sought to dismiss or ignore the reasonableness of criticisms. In the case of the new generation of the bioscience/biotechnology community, the strategy seems more intended to pre-empt or co-opt criticisms than to ignore or dismiss them. The result is, nevertheless, that critics become marginalized, unless, again, there is significant political or social power behind them. Given the inherent (and self-defined) 'reasonableness' of the views, activities, and arguments of the scientific community, even formidable social challenges are likely to fail.

These points may suggest nothing more than that the bioscience community, including practitioners of animal biotechnology, probably have little or no reason to fear that their activities will be fundamentally challenged in the actual public arena. Moreover, to the extent that the scientific establishment is becoming more sophisticated about 'science education', the likelihood of even a powerful challenge diminishes greatly. Science writers, science popularizers, and spokespersons for universities and corporations are out in force, promoting the legitimacy and safety of biotechnology. Surveys suggest that as the public becomes more 'informed' and 'educated' about science, concern diminishes significantly. Further, as policy makers become more informed and educated as to the relative benefits and risks of biotechnology (as defined by scientists themselves), strong legislative action is unlikely. Indeed, as mentioned above, the result of all this information and education may well be simply greater levels of funding for the biotechnology enterprise. The only conclusion to be reached is that 'the beast will go on'. The spectre of a broader, social, 'culture of biotechnology', with attendant human genetic engineering, is real.

9.4 CONCLUSION: ETHICS BY REGULATION, COMMITTEE, OR (RE-)EDUCATION

In the early 1970s, a gathering of concerned scientists was held in Asilomar, California, to discuss the risks and benefits of genetic engineering. What emerged from those meetings was a set of biosafety guidelines concerning biotechnology. Many of those guidelines made their way into federal regulations and general governmental oversight. Though fairly stringent at the time, the guidelines and subsequent rules have been gradually weakened. Scientists argue that, as their knowledge about bioengineering has grown, what were reasonable concerns are now known to be unfounded fears. Recall that the USDA abolished its biotechnology oversight committee. Apparently it was thought to be an unnecessary public expenditure. Most universities have in-house biosafety committees, and corporations, it has been argued, exercise extreme caution because of the risk of lawsuits or prosecution under environmental or human safety regulations.

Moreover, most industrialized nations have animal cruelty regulations, and most institutions involved in the use of animals have some kind of animal research rules and guidelines. Although the stringency of all of these varies, all: (i) assume that research involving animals will be performed; and (ii) attempt to reduce pain and suffering as much as is 'reasonable'.

Although citizen involvement is occasionally solicited for participation in biosafety and animal care and use committees, for the most part they are staffed by scientists. Usually, those scientists have backgrounds identical or at least similar to those of the scientists they are overseeing, regulating or constraining. Perhaps it is simply expected that only scientists will understand the nature of the work being monitored. Just as likely, however, is that scientists wish to control the agenda of such oversight or, in some cases, enforcement. Although in many places meetings are open to the public, far too often they are held in difficult-to-access places, at awkward times. And ultimately, critics or concerned citizens have little to gain from attending. The agenda has already been set, both in a literal and figurative sense.

Despite the 'fox guarding the henhouse' potential for self-serving abuse of safety regulations or animal welfare violations, they are, nonetheless, a good bet for at least the beginnings of ethics-based control over some particular kinds of animal biotechnology practices. Although rarely, scientists are occasionally reprimanded or censored for either cruel treatment of animals or unnecessarily replicative biotechnology activities. But with the occasional philosopher or ethically sensitive scientists participating in these committees, some measure of ethics discussion (again, in the philosophical sense of ethics) begins to emerge. And, some scientists appear to listen.

The scientific establishment, nevertheless, continues to use animals, rights or welfare arguments notwithstanding. The scientific establishment continues to engage in genetic engineering practices involving animals. And, technologies will continue to have impacts, some of them negative, on particular socioeconomic groups in society. Barring some sort of major catastrophe, or major gestalt shift, animal biotechnology, biotechnology in general, and even more generally, technological research and development will undoubtedly continue. If ethical considerations are to fit anywhere in this scientific enterprise, it would seem that it would have to be through the current system of oversight and control, or through the force of higher levels of government action.

Given the power of science and Scientism, the latter is unlikely, though not impossible.

One conclusion that can be reached is this: if ethics in a substantive sense is to make its way into the scientific establishment, and the bioscience community in particular, it will have to be at least in part if not exclusively through the moral or ethical re-education of scientists and science policy makers. And the moral or ethical education of young scientists and students would also be a key. Indeed, the ethical force of particular kinds of arguments pertaining to animal biotechnology is dependent on *any* ethical argument having force. And for any ethical argument to have force, fundamental changes in the scientific attitude would be necessary. As Grinnell noted, the way of seeing and way of being of science are *learned* orientations. *Seeing* ethical considerations as inherently part of the scientific enterprise, as well as *being* an ethically aware scientist or policy maker, must also be learned.

Just as there is considerable resistance on the part of the science establishment to external control – to the point of pre-empting *rational* discussion of criticisms – there may also be considerable resistance to the inclusion of ethics as part of the indoctrination into the scientific attitude. Nevertheless, there are enough scientists who do engage in ethical reflection when their scientific 'hats are off' that there is at least some promise for ethics to be a part of the scientific mind-set. Already, there are college and university courses, colloquia, and informal discussion among members of the bioscience community about 'science ethics'. With considerable effort on the part of theologians, philosophers, and social scientists – duly respectful of the ability of science to define the terms of rational discussion – more such inclusion of ethics might continue.

Only when ethics becomes a legitimate – and rational – part of the scientific attitude will concerns about particular aspects of animal biotechnology be taken seriously, or taken at all. Only when ethics is a routine concern among scientists will considerations of whether we should be using animals, or engaging in biotechnology, even be fathomed. I do not believe that we will stop using animals in research (or for food purposes) in the near future. Nor do I believe that the scientific establishment will stop engaging in biotechnology any time soon, if at all. I do believe, however, that any critique which does not first address the need for including discussion of ethics in the very process of 'doing science' is doomed to failure. It does little practical or political good to challenge science from the outside. Rather, rational, informed, science-based discussion of ethical considerations has to be the key to whether continued biotechnological research and development, whether in the animal, plant, or human domains, will simply be inevitable.

Note

A considerable amount of the 'evidence' for the theses in this chapter is based on 'research' performed by the author, a professional philosopher, but whose appointment is in the agricultural science college at a major state university in the US. Although the author wishes to indict no particular scientists or administrators for espousing 'the scientific attitude', or especially indict them for 'ethical insensitivity', both orientations have been found to be extant (though the former far more prevalent) among the physical and biological scientists with whom the author interacts on a daily basis.

REFERENCES

Browne, W. (1987) Bovine Growth Hormone and the Politics of Uncertainty: Fear and Loathing in a Transitional Agriculture. *Agriculture and Human Values*, **4** (1).

Burkhardt, J. (1988) Biotechnology, Ethics, and the Structure of Agriculture. *Agriculture and Human Values*, **4** (2).

Burkhardt, J. (1991) The Value Measure in Public Agricultural Research, in *Beyond the Large Farm* (eds P. Thompson and W. Stout), Westview Press, Boulder, CO.

Burkhardt, J. (1992) On the Ethics of Technical Change: The Case of bST. *Technology & Society*, **14**.

Busch, L. (1984) Science, Technology, and Everyday Life. *Research in Rural Sociology and Development*, **1**.

Busch, L. and Lacy, W. (1983) *Science, Agriculture, and the Politics of Research*, Westview Press, Boulder CO.

Busch, L., Lacy, W., Burkhardt, J and Lacy, L. (1991) *Plants, Power & Profit*. Blackwell, Oxford.

Busch, L., Lacy, W., Burkhardt, J., Hemkin, D., Moraga-Rojel, M., Kaponen, T. and De Souza Silva, J. (1995) *Making Nature, Shaping Culture*, University of Nebraska Press, Lincoln NE.

Buttel, F. and Belsky, J. (1987) Biotechnology, Plant Breeding and Intellectual Property: Social and Ethical Dimensions. *Science, Technology and Human Values*, **12**.

Buttel, F.H., Cowen, J.T., Kenney, M. and Kloppenberg, J., Jr (1994) Biotechnology in Agriculture: The Political Economy of Agribusiness Reorganization and Industry–University Relationships, in *Research in Rural Sociology and Development* (ed. H.K. Schwarzkeller), JAI Press, Greenwich, CT.

Comstock, G. (1990) The Case against bGH, in *Agricultural Bioethics* (eds S. Gendel, A. Kline, D. Warren and F. Yates), Iowa State University Press, Ames.

Feyerabend, P. (1978) *Science in a Free Society*, NLB, London.

Fox, M. (1990) Transgenic Animals: Ethical and Animal Welfare Concerns, in *The Bio-Revolution*, (eds P. Wheale and R. McNally), Pluto Press, London.

Grinnell, F. (1987) *The Scientific Attitude*, Westview Press, Boulder CO.

Hoban T. and Burkhardt, J. (1990) Determinants in Public Acceptance: North America, in *Biotechnology for Control of Growth and Product Quality in Meat Production* (eds P. Van der Wal and F. Van der Wilt), Wageningen Agricultural University, Wageningen.

Hoban T. and Kendall, P.A. (1992) *Consumer Attitudes about the Use of Biotechnology in Agriculture and Food Production.*, Report, USDA Extension Service, North Carolina State University, Raleigh, NC.

Holland, A. (1990) The Biotic Community: A Philosophical Critique of Genetic Engineering, in *The Bio-Revolution* (eds P. Wheale and R. McNally), Pluto Press, London.

Hornig, S. and Talbert, J. (1993) Mass Media and the Ultimate Technological Fix: Newspaper Coverage of Biotechnology, *CBPE Working Paper CBPE 93-4.* College Station TX, Center for Biotechnology Policy and Ethics.

Lukes, S. (1986) *Power*, New York University Press, New York.

NABC (National Agricultural Biotechnology Council, USA) (1992) *Animal Biotechnology: Opportunities and Challenges*, NABC Report 4, Ithaca, NY.

NABC (1994) *Agricultural Biotechnology and the Public Good*, NABC Report 6, Ithaca, NY.

Olshevsky, T. (1983) *Good Reasons and Persuasive Force*, University Presses of America, New York.

Rachels, J. (1990) *Created from Animals*, Oxford University Press, Oxford.

Regan, T. (1985) *The Case for Animal Rights*, University of California Press, Berkeley, CA.

Rifkin, J. (1983) *Algeny*, Viking Press, New York.

Rosenberg, C.E. (1976) *No Other Gods*, Johns Hopkins University Press, Baltimore, MD.

Science (22 September, 1985), **269**, 1659.

Singer, P. (1990) *Animal Liberation*, Random House, New York.

Stevenson, C. (1944) *Ethics and Language*, Yale University Press, New Haven, CT.

10

Needs, fears and fantasies

Andrew Johnson

The debate about animal biotechnology does not take place in a social vacuum, and the ethical stances of opposing protagonists may be more clearly understood in the context of their wider social and political beliefs. These, in turn, can be interpreted in terms of psychological and cultural influences. This chapter will describe in outline some of these factors, which form the background to specifically ethical discussions about new technologies in general, and animal biotechnology in particular.

10.1 NEEDS

An important justification for technological innovation is that it will produce consequent social benefits by satisfying certain needs. This line of argument fits in with a conception of economic rationality in which all production, and indeed all economic behaviour, can be explained in terms of market scarcity. According to such a paradigm, 'needs' include not only the absolute requirements for survival, but also 'market needs' or 'demands' for any goods that are perceived to be scarce (Zadek, 1993). As Ivan Illich (1993) remarks, 'the assumption of scarcity has penetrated all modern institutions. Education is built on the assumption that desirable knowledge is scarce. Medicine assumes the same about health, transportation about time, and unions about work...' This economistic view that all goods are essentially scarce is a product of the 18th century. The Greeks had distinguished clearly between natural needs and unnatural wants, boundless desire for the latter being a form of *hubris*, and this was the basis for a longstanding republican and humanist tradition of associating luxury with corruption. Only with the industrial revolution did stimulation of market demand for fashionable goods become institutionalized as a mainstay of economic progress (Xenos, 1989). Adam Smith perceived the new conflation of needs and luxuries, but while accepting that the pleasures of wealth and greatness were ultimately illusory, he believed that 'it is well that nature imposes on us in this manner. It is this deception which rouses and keeps in continual motion the industry of mankind' (Smith, 1976, p. 183). Of course, there were dissenting voices, such as Rousseau's, but on the whole the idea of continual progress in wealth-creation, spurred on by a constant fear of scarcity, has dominated western society of the 19th and 20th centuries. Only within the past few decades has there been serious questioning of the limits to growth, and a revival of the distinction between natural and artificial needs. If such a distinction is accepted, it becomes pertinent to ask of any new technology whether it satisfies an unfulfilled and real need; whether it satisfies an already fulfilled need more efficiently, thereby releasing more resources for the satisfaction of other market needs; or whether its success depends on the creation of demand for an entirely new kind of product.

Many sceptics of biotechnology accept the need to feed the world's hungry, but argue that traditionally based methods of maintaining soil fertility are the only sustainable solution to this problem and that attempts to push productivity beyond its 'natural' limits will lead to disaster in the longer term. In particular, increased efficiency in the production of meat and other animal-based foods will only encourage more people in developing countries to adopt unnecessary and probably unhealthy western dietary habits – a trend which is already resulting in pressure on the availability of traditional staples as more agricultural land is converted to animal fodder production (Brown, 1995). Thus, the likelihood of animal biotechnology bringing much benefit to the undernourished or starving can plausibly be counted as extremely remote. By contrast, there is a strong possibility that a number of individuals could be kept alive through some of the medical techniques described by Ian Wilmut and Andrew George in Chapters 2 and 3. Moreover, medical demands offer a nice case of needs that are not generally regarded as luxuries, but which, unlike such natural needs as food and shelter, are never fully satisfied: with current medical knowledge, virtually unlimited resources could be applied to prolonging the lives of the sick. It would again be possible to argue that in global terms there are health-care options that are more cost-effective than such applications of biotechnology; it is only in the context of medical budgets in the developed countries that the lifesaving potential of these innovations looks so attractive.

10.2 FEARS

Conflicting attitudes to the new methods of animal biotechnology often reflect their protagonists' wider views about the virtues or otherwise of industrial society, and we should be aware that their arguments may at least in part be a surrogate defence for these wider views. In *Risk and Culture* (1982), Mary Douglas and Aaron Wildavsky developed a cultural theory in which social groups construct 'myths of nature' which depend on the organizational principles of the group as much as on any independent reality. At one extreme, an 'entrepreneurial' vision of nature as benign and resilient to interference means that risks are belittled or ignored; while at the other, a 'sectarian' outlook perceives nature as fragile and ephemeral, to the extent that all action is paralysed for fear of catastrophic results. According to cultural theory, protesters' concerns with issues such as biotechnology or the environment are first and foremost ways of maintaining the purity of the sectarian 'in-group' by blaming perceived dangers on the corrupting influence of outsiders. While Douglas and Wildavsky's work helps to account for the profound differences in perception between different social groups, a theme to which we shall return later in this chapter, it has been criticized for privileging interactions among humans above direct experience of our surroundings (Ingold, 1992). It also has little to say about historical changes in *zeitgeist*, or what, independent of the degree of factionalization or consensus at a particular time, could be described as the 'centre of gravity' of social perceptions and concerns. For a broader perspective on modern beliefs about science and nature it may be helpful to consider how current sociological theories account for the growth in environmentalism, and how they anticipate its future consequences.

Environmental sociologists generally reject the assumption of more or less linear progress through increasing material production which has held sway since its development by Enlightenment thinkers such as Condorcet and Godwin. Evidence that society at large has also lost faith in progress is provided by recent opinion polls in which some 60% of adults expected their children to grow up in a worse world than they were brought up in themselves (Jacobs, 1996). What has gone wrong? For Marxists, the problem is not so much production itself as capitalist control of the means of production, resulting in exploitation of workers and nature alike. A more radical critique of industrialism has been mounted by deep green authors such as Rudolf Bahro (1984), for whom dismantling the means of production as well as the social apparatus for their control is a prerequisite for a sustainable future. Even if such strong medicine has to be swallowed eventually, our present state is probably more accurately represented by the reflexive modernization school of thought (Beck *et al.*, 1994). 'Reflexive modernity' is a new historical phase in which the side-effects of post-Enlightenment modernization and globalization have become central causes of new societal developments. This is particularly so in environmental affairs, where economic 'externalities' such as pollution of air or water, whose costs were previously ignored, have become important political and economic issues. An important implication of reflexive modernization theory is that technology is a process which requires continual efforts to maintain it on track – it is not merely a treadmill, but neither can it be regarded as a one-off 'fix' followed by indefinite free lunches.

In his seminal *Risk Society*, Ulrich Beck argues that the ideal of progress towards material abundance, or the avoidance of real or imagined scarcity, has been displaced by the ideal of security, or the avoidance of real or imagined risk. He believes that a more ecologically sustainable society can be achieved smoothly through science-based institutional restructuring. Some environmentalists are less optimistic, while other commentators, of the 'business as usual' persuasion, deny that any such restructuring is needed (Toffler, 1980; Bauman, 1991). A recent survey of ecological restructuring of the chemical industry tends to confirm Beck's view of what is happening, at least for the time being (Mol, 1995). The shift in focus, from scarcity to risk, has obvious relevance to how enthusiastically we are likely to embrace a new generation of biotechnological developments, though in some ways new technologies are at an advantage: while the risk of smog is readily apparent from its known effects, the risks of genetic engineering are mainly speculative. But in other ways, being so late in the day works against the newcomer, whose advocates have to convince a public who have become increasingly cynical about the promises of scientists, not least because of the extravagant unfulfilled promises and apparent untrustworthiness of the nuclear lobby.

Genetic engineering has certainly been the target of more concerted and intense public resistance than almost any other recent technological innovation. Its supporters have expressed surprise and frustration at the obstacles they have had to surmount, perceiving themselves as the victims of a Kafkaesque situation 'in which the right office to stamp the papers was forever elusive' (Budd, 1993, p. 212). Dorothy Nelkin has commented on the 'paradox' of overt and vehement public opposition to biotechnology, compared with the limited and passive resistance to information technologies which, she suggests, have equally profound and problematic social implications. Nelkin argues that the association of biotechnology with unfamiliar and unknown risks is an important factor in explaining this difference; but even more so, she suggests, is the moral and religious agenda of American fundamentalist Christianity (Nelkin, 1995). Furthermore, in the American mythology, 'the history of agribusiness calls forth quite different associations than Silicon Valley. And helpless animals, like besieged farmers or vulnerable fetuses, become easy lightning rods for social movements'. Nelkin is probably correct in emphasizing the moral dimension of popular reactions against biotechnology; protesters may be motivated partly by prudential concerns about public health and safety but that is not the whole story. She is on shakier ground, however, in relating this moral dimension to America's fundamentalist religious traditions. Resistance to biotechnology has often been tougher in Europe than in the USA, with popular opinion most hostile in Germany and Denmark (Budd, 1993, p. 210). In less-developed countries, regardless of the moral or religious climate, objections in principle to biotechnology would possibly rank lower as matters of public concern than its immediate economic impacts.

Within the industrialized nations, the concern of different social groups about biotechnology issues appears to be correlated with their interest in animal protection and ecological matters. This is confirmed by voting patterns in a 1992 Swiss referendum. Although the original petition for the referendum referred only to biotechnological misuse of human beings, its scope was widened by the federal Assembly to include the protection of animals and plants, and its results showed strong correlation with patterns of voting in favour of the protection of farm animals and for regulation of animal experiments, as well as for various environmental protection issues (Buchmann, 1995, p. 221). The proposed amendment to the constitution was accepted by a majority of 73%, with least support for reform from farmers, those not educated beyond compulsory schooling, and the over-60s. Gender and religious affiliation appeared to have little influence on voting behaviour (Buchmann, 1995, pp. 213–14).

Moral arguments presume a shared viewpoint from which their validity can be judged, even if they cut different amounts of ice with members of different social and economic groups. The extent to which the discourse of philosophical ethics can produce convincing arguments is explored later in this volume. Currently, however, conflict is more evident than consensus, and this may be because proponents and objectors start from positions with radically different world views and are motivated by very different concerns. This polarization fits with a conventional analysis of the dynamics of innovation, expressed in business-speak in terms of 'technology push', 'market pull' and 'consumer resistance', though the 'forces' involved may turn out to be more psychological than economic.

10.3 FANTASIES

Molecular biology is an exciting and highly fashionable research field, which has displaced nuclear and space physics as the acme of scientific progress. It has also, however, inherited from these 'hard' sciences the burden of scientific responsibility. Whatever the skill and the excitement, the unwelcome question of whether what *can* be done *should* be done now presses hardest for an answer in microbiology. Because it is such an exciting field of research, the danger of *hubris* is ever-present: 'a sound Magician is a Demi-god', as Marlowe wrote. 'Creation' is a word in constant use to describe the outcome of genetic manipulation. Of course, for centuries breeders of domestic animals have applied forethought and judgement in selecting which pairs of animals they allow to mate, and the vocabulary they used has inevitably been centred on their personal involvement. The human agent who 'mates' two animals, 'breeds' new strains and 'improves' his stock is the active, driving force; his beasts are as it were the clay with which he works. Such creativity, though, is pretty limited in scope; for an outsider looking in, 'interference' rather than 'creation' may be a more appropriate description of what is going on. Does the choice of words in this context really matter? It may be argued that the consequences of a particular development are independent of the intentions or state of mind of the developers, who should be judged only by the results of their labours. This may be true, but there are still reasons why we should regard carelessly made claims to 'create' new life-forms with caution. There is a tendency in our culture to regard creative genius as an excuse for ethical licence. The great artist is not really expected to conform to social norms, or even to be fully responsible for the mundane consequences of his nonconformity. If it were only a matter of the brilliant professor forgetting to attend faculty meetings, such lenience might well be appropriate; but in a study area where irresponsibility carries grave dangers we should be on our guard against licence of this sort.

The rush to patent transgenic animals also arouses suspicion among the opponents of genetic engineering. Commercially, the need for patenting is justified by the fact that the mere existence of a market for a new product does not guarantee profits for the innovator: imitators who have not spent heavily on research and development are more likely to profit, unless discoveries are protected by patent rights. But it can also be argued that improving the economic incentives for genetic engineering experiments on animals may push welfare concerns more into the background – welfare concerns which are so novel that the whole technology needs the brake, rather than the accelerator, while its ethical implications are worked out. Beyond this lurks the fear that any strengthening of property rights over sentient creatures could encourage cruelty and exploitation. And while patenting may not be the main issue of controversy in such difficult ethical matters as the deliberate production of animals that are destined to develop cancers, a patent may be regarded by objectors as giving social legitimization to something morally offensive, and resisted on these grounds.

There is evidence that the most successful innovations tend to be those for which there is a strong 'market pull' (Roberts, 1981, p. 56). This is not particularly good news for biotechnology, which is plainly characterized by technology push: most of its development has taken place in publicly funded research institutions, in response to advances in basic science and improvements in laboratory technique. Weakness of market demand for animal biotechnology products, particularly for non-medical ones, may encourage their advocates to make exaggerated claims or to downplay their risks or known adverse effects. Those who challenge such claims become classified as biased or irrational. Consumer resistance is an 'obstacle' to be overcome. To date, the disease of 'biotechnophobia' has not been officially recognized, but it could happen: 'cyberphobia' is widely diagnosed in California, and medical intervention is routinely prescribed for citizens who cannot learn to love their computers (Bauer, 1995). A weak economic case may also be bolstered by an exaggerated moral one: while animal biotechnology just might help to feed the hungry, it is probably not the best way to achieve this end. To perceive technology as directed only towards solving problems is to ignore the social constitution of what we mean by a 'problem', as well as the intellectual and economic factors which frequently carry more weight than the moral ones.

Although objectors to animal biotechnology tend to have less direct economic involvement than its more enthusiastic supporters, their moral arguments may also be overstated for a number of reasons. As already mentioned, objections are sometimes based on general distrust of technology, government, or big business. More particularly, interference with the 'nature' of other animals – or of humans – arouses widespread aesthetic revulsion as well as religious condemnation. The extent to which religious arguments can legitimately be invoked is outside the scope of this book, and it is a moot point how far arguments on doctrinal grounds are likely to motivate politicians or consumers. Gut reactions of distaste, backed up by general arguments against 'unnaturalness', seem likely to be more potent, as least in most European countries.

Different species evoke reactions that vary widely between different individuals or cultures, so what chances are there that we can identify certain attitudes as 'natural'? It remains controversial whether the widespread dislike of snakes or spiders, for example, is natural or a product of culture (Davey, 1994). Instead of considering reactions to particular species in isolation, structuralist anthropologists make the claim that we all have a more universal need to sort out and classify; and this activity is threatened by species which appear to breach established boundaries. According to Claude Levi-Strauss, Edmund Leach and their followers, taxonomic classification of animals is a vital tool for understanding and regulating human society. Biblical rules about what flesh it was permitted to eat reflect whether or not the animal it came from could be classified unambiguously according to the Hebrew scheme of creation; so the pig is unclean because 'though he divide the hoof, and be cloven-footed, yet he cheweth the cud'. In *La Pensée Sauvage*, Levi-Strauss argued that the need for such classification is not restricted to primitive cultures; for example, in modern Europe pet birds are much more likely than pet dogs to be given human names (Levi-Strauss, 1996, pp. 204–5). It is precisely because dogs are closer to humans that we need to separate ourselves from them more clearly. The problem of cruelty is likewise most acute with those animals which are closest to us, 'and which therefore present the greatest difficulties for the ordered classification of the world' (Tester, 1991, p. 41).

The structuralist perspective helps to explain the simultaneous fascination and revulsion we tend to feel towards 'monsters' which defy normal classification. This is hardly a new phenomenon, though surprisingly enough, it does appear to be a product of civilization rather than a universal human trait. The first pictorial representations of monstrous hybrids date from around 3000 BC, and it is likely that a much older tradition visualized supernatural beings in the normal shapes of men or other animals, but having such powers as to change shape to that of another species, to make themselves invisible, and so on (Mode, 1975, p. 14). Horace criticized the excessive use by artists of the composite monster motif, as did Bernard of Clairvaux more than a thousand years later. In the Middle Ages, the boundary between humans and other species was a particularly dangerous region. Their 'unnatural' closeness to their horses was a reason to persecute Tartars or gypsies; bestiality was a capital crime; and breaches of the species barrier of one kind or another were among the most frequent charges levelled against those charged with witchcraft (Cohen, 1994). The dividing line between us and other creatures remained a live issue in the 17th century, when the ethical problems of live animal experimentation were neatly 'solved' by distinguishing between humans and other species on the basis of the possession or lack of an immortal soul. More recently, with a decline in religious belief and the general acceptance of Darwin's theory of evolution, this distinction began to look pretty thin, and the past hundred years have witnessed a bewildering variety of new strategies for maintaining the integrity of the human–animal divide according to avowedly scientific criteria.

Even today, portrayals of fabulous beasts, or actual freaks of nature, capture the interest to an extent which suggests the mystery of the species barrier has not been entirely exhausted. While fairground fake mermaids would no longer deceive, and our fascination with freaks is now tempered by greater sympathy than in past ages, worries about demarcation still surface in the context of the animal biotechnology debate. This is not to argue that preserving established distinctions is invariably good: the eradication of 'alien' or 'impure' breeds to preserve the racial purity of others is somewhat reminiscent of racism among humans – a programme few readers of this book would be likely to support (J. Peretti, in press). It is also only fair to remind ourselves that most of the domesticated plants and animal species we accept quite happily are far from natural. Indeed, their breeders were convinced that they were 'improving' on nature to produce fatter, tastier, or more beautiful specimens, even if from an alternative viewpoint they can be seen as enfeebled or in extreme cases actually diseased.

The strength of the 'yuk' factor when the traditional order of things comes under threat has been greatly underestimated and too little studied. Likewise, the 'gee whiz' factor can be a strong, though mostly hidden, spur to achieve what is technically possible rather than what is actually needed. But however powerful these aesthetic sentiments, whatever the social dynamics of progress and resistance, and however deep-seated the human need to classify and categorize, such cultural and psychological factors do not themselves provide a rational justification for either conservatism or change. Rather than a substitute for the moral argument of subsequent chapters, perhaps they are its background – or are they its raw material?

REFERENCES

Bahro, R. (1984) *From Red to Green*, Verso, London.

Bauer, M. (1995) Technophobia: a misleading conception of resistance to new technology, in *Resistance to New Technology: Nuclear Power, Information Technology and Biotechnology*, Cambridge University Press, pp. 97–122.

Bauman, Z. (1991) *Modernity and Ambivalence*, Polity, Cambridge.

Beck, U. (1992) *Risk Society* (trans. M. Ritter), Sage, London.

Beck, U., Giddens, A. and Lash, S. (1994) *Reflexive Modernization*, Polity, Cambridge.

Brown, L. (1995) *Who Will Feed China?* Earthscan, London.

Buchmann, M. (1995) Resistance to biotechnology in Switzerland, in *Resistance to New Technology: Nuclear Power, Information Technology and Biotechnology*, Cambridge University Press, pp. 207–24.

Budd, R. (1993) *The Uses of Life: A History of Biotechnology*, Cambridge University Press.

Cohen, E. (1994) Animals in medieval perception: the image of the ubiquitous other, in *Animals and Human Society: Changing Perspectives* (eds J. Serpell and A. Manning), Routledge, London, pp. 59–80.

Davey, G.C.L. (1994) The 'disgusting' spider: the role of disease and illness in the perpetuation of fear of spiders. *Society and Animals*, **2**(1), 17–25.

Douglas, M. and Wildavsky, A. (1982) *Risk and Culture*, University of California, Berkeley.

Illich, I. (1993) Unpublished lecture, quoted in Zadek, S. *The Economics of Utopia*, Avebury, Aldershot, p. 218.
Ingold, T. (1992) Culture and the perception of the environment, in *Bush Base: Forest Farm; Culture, Environment and Development* (eds E. Croll and D. Parkin), Routledge, London.
Jacobs, M. (1996) *The Politics of the Real World*, Earthscan, London.
Levi-Strauss, C. (1966) *The Savage Mind*, Weidenfeld & Nicholson, London.
Mode, H. (1975) *Fabulous Beasts and Demons*, Phaidon, London.
Mol, A.P. (1995) *The Refinement of Production*, International Books, Utrecht.
Nelkin, D. (1995) Forms of intrusion: comparing resistance to information technology and biotechnology in the USA, in *Resistance to New Technology: Nuclear Power, Information Technology and Biotechnology*, Cambridge University Press, pp. 379–91.
Peretti, J. Nativism and nature: rethinking biological invasion. *Environmental Values* (in press).
Roberts, E.B. (1981) Influences on innovation: extrapolation to biomedical technology, in *Biomedical Innovation* (eds E.B. Roberts *et al.*), M.I.T. Press, Cambridge, MA, pp. 50–74.
Smith, A. (1976) *The Theory of Moral Sentiments* (eds D.D. Raphael and A.L. Macfie), Clarendon Press, Oxford.
Tester, K. (1991) *Animals and Society: The Humanity of Animal Rights*, Routledge, London.
Toffler, A. (1980) *The Third Wave*, Pan, London.
Xenos, N. (1989) *Scarcity and Modernity*, Routledge, London.
Zadek, S. (1993) *The Economics of Utopia*, Avebury, Aldershot.

Part Three

Ethical and conceptual issues

11

Intervention, humility and animal integrity

David E. Cooper

11.1 A NEW WRONG?

Biotechnological techniques, including those of genetic engineering, are being applied to today's animals and will be increasingly applied to tomorrow's. Do such interventions in their lives constitute a new wrong done to animals? There is, of course, a wide range of these techniques and perhaps my question can only be seriously raised about some of them. But even 'benign' ones, like extraction of Factor IX from sheep, presuppose a whole enterprise many of whose activities are clearly less so. To claim that 'it is hard to think of any objection to using genetic engineering to eliminate defects' (Glover, 1984, p. 31) is no more a defence of the enterprise than it is a defence of slavery to find no objection to using slaves to clean hospital wards. Given the existence of slavery or genetic engineering, there are doubtless better and worse uses of it: but maybe neither of them should have been instituted in the first place.

Still, I have mainly in mind techniques which seem to many people far from 'benign', and to which it is not hard to think of some objection. These include the production of animals, like the unfortunate 'oncomouse', programmed to contract diseases; the use of growth hormones to produce creatures barely able, and only at risk to their bones, to carry their weight; and the manufacture of 'transgenic' animals. Further techniques, in a likely future, would be the development, welcomed by one writer, of bulls with 'the bovine equivalent of Down's syndrome', so as to replace today's 'cantankerous animal' with a 'placid, fat, happy' one (Bains, 1990, p. 184); and the engineering, not similarly welcomed by another writer, of 'deaf, blind, legless microcephalic lumps' (Clark, 1994, p. 235) – 'chickens', perhaps, unencumbered by such unprofitable features as claws and intelligence.

If these new techniques constitute a wrong to animals, then is not the answer to my question – whether they commit a 'new wrong' – obvious? But to ask whether a new wrong is committed is not the same as asking if something both new and wrong is being done. Intensive farming of crocodiles is a new venture, but it represents an old wrong, now done to one more creature. By contrast, whoever first trained caged wild animals to perform acts at which audiences would laugh invented a novel kind of wrong, an addition to the already established list. Even in this case, of course, there were precedents, and if our criteria for novelty are sufficiently strict then, indeed, 'there is no new thing under the sun'. But then continuity with the past – similarities with what has gone before – is not the crucial consideration. The straw that broke the camel's back was just like the previous ones in the bale, yet from the camel's point of view it was a very special straw. Sometimes, indeed, we only appreciate something as distinctive and novel by seeing it as the culminating stage – one that reaches a limit – of a continuous process. So the fact that genetic engineering of animals may be continuous with previous practices, such as dog-breeding or force-feeding, does not mean that it is innocent of committing a new wrong. For maybe, with it, 'we reach the absolute limits' of what any morality could tolerate (Linzey, 1994, p. 138).

It is one thing to sense that biotechnology inflicts a new wrong on animals, another to justify and articulate that sense. Proponents of the two best-known approaches in animal ethics – utilitarianism and 'animal rights' theory – find it difficult, in fact, to do these things. On both approaches, biotechnology is, morally measured, just more of the same. Utilitarians, no doubt, can point to new sufferings by animals which biotechnology makes available: but also to the alleviation of some old ones through, for example, manufacturing animal proteins on which to test vaccines in place of actual animals (see Rodd, 1990, pp. 168–9 for further examples). What is unclear, though, is that any new *kinds* of suffering are involved, in the way they might be in the radical genetic engineering of humans. Cloned people might suffer from a hitherto unknown form of 'identity crisis', as might parents of gene-tampered children from the unprecedently large differences, physical and otherwise, between themselves and their offspring (Glover, 1984). Animals, probably, would not encounter these difficulties. Finally, it is hard to see how utilitarians could object at all to some, at least, of the techniques mentioned above, since these do not cause suffering, either of a new or an old kind. 'Bovine Down's syndrome', we are promised, is going to make the bulls placid and happy.

If utilitarians can find nothing distinctively wrong in subjecting animals to biotechnology, nor it seems can those who speak of 'animal rights'. At any rate, there is no right plausibly assigned to animals which these techniques violate that has not already been violated by other means. The 'oncomouse' cannot move about freely, cannot choose a mate or a place to rest: but if a creature's rights are thus violated, they are ones already violated, a billion times, in the case of battery hens. Moreover, it will be difficult for the rights theorist to motivate talk of rights being violated at all in the case of some of the techniques under consideration. The 'microcephalic lumps' envisioned by Stephen Clark do not, as he points out, sound to be creatures with the capacities to count as the 'experiencing subjects', the leaders of lives with 'inherent value', which, on most 'animal rights' approaches, they must be in order to enjoy rights. If it is repulsive to create such objects, it cannot be because *their* rights are violated – and who else's are?

If the sense that biotechnological techniques constitute a new wrong cannot be accommodated by these familiar approaches, then so much the worse for them. That such a sense is both intense and widespread is undeniable, and even if, at the end of the day, it were judged to be misconceived, it is one whose seriousness and *prima facie* authority a moral theory should surely be able to acknowledge. At any rate, a moral theorist, or indeed anyone at all, must be judged insensitive if unable to appreciate that new moral concerns are raised by biotechnology, even if those concerns could be finally laid to rest. Consider, for example, the 'transgenic' animals whose prospect appals many people. Conceivably they could, after long argument, be brought round to accept this prospect of 'ligers and tions' concocted by 'curious officials' in the way they 'breed pomatoes and totatoes' [Ramanujan (1986) 1989, p. 77]. Yet there would remain something chilling in the attitude of the researcher at Beltsville – an institute responsible for breeding arthritic pigs, incidentally – who cannot see what all the fuss is about, since 'much of all genetic material is the same, from worms to humans' (quotation in Linzey, 1994, pp. 150–1). In terms of sensitivity, and logic too, this is like someone wondering why there's all the fuss about Bach and Beethoven, since much of all music's raw material is the same, from popular ditties to oratorios.

11.2 THE PLACE OF EMOTIONS

I want to make a larger claim on behalf of the emotional responses which go with the sense that biotechnology commits new wrongs than that these emotions are data which an adequate moral theory must be able to accommodate and explain. It is by reflecting on these responses that we come to see what is wrong, and distinctively so, with producing 'oncomice', 'transgenic' hybrids, and so on.

This is to attribute to emotions a different and more elevated role than is usually granted them. One writer's views are typical: 'spontaneous human sentiments' should not be 'allowed to enter reflective moral judgements'. In particular, 'the fear and suspicion people may feel towards gene technology do not ... count as a valid objection against it'. As a 'pragmatic consequentialist', this author allows only the following relevance to people's feelings: if they are 'strong enough to outweigh the expected net utility of employing (a) technique', then the technique should not be used (Häyry, 1994, pp. 204, 208). Human sentiments, that is, cannot be any guide to what is right and wrong: it is just that the effects of offending these sentiments must be weighed, along with everything else, on the utilitarian scales. A line almost as dismissive is taken by Jonathan Glover over people's resistance towards crossing species boundaries. 'The temptation to dismiss our resistance as an irrational taboo is one we should mainly yield to' (Glover, 1984, p. 40) – the main qualification being that there might be some 'psychological cost' to people if their 'irrational' resistance is overruled. Presumably these authors, as 'pragmatists' and 'consequentialists', would argue analogously that the only relevance of people's revulsion against paedophilia to the permissibility of that practice is as a datum to be included when calculating the total effects of allowing or prohibiting paedophilia.

These attempts to immunize 'reflective moral judgements' against contamination by sentiment are incoherent. Both the writers discussed defend some genetic engineering on the basis of its contribution to human welfare, particularly the alleviation of human suffering. Their arguments can, therefore, only sway us if people's welfare is something that matters to us and their suffering something that disturbs us. The same would apply had they argued for genetic engineering on some other ground, like that of its being a stage in man's noble pursuit of unlocking the secrets of the universe. Stephen Clark puts the point well: 'If nothing deserves an immediate response of love or horror, then neither does Truth, or Rationality, or Human Welfare. ... If something or other does, why not the very things to which most of us easily react', like manufacturing cancerous animals? (Clark, 1995, p. 26) (my thanks to Stephen Clark for allowing me to see the draft of this article ahead of its publication).

What worries many writers about allowing emotions a guiding moral role is that this raises the spectre of 'subjectivism'. People's feelings about this or that differ, so that if feelings are allowed to arbitrate, no generally acceptable moral verdicts could ever be reached. This is doubtless one reason why Peter Singer, at the beginning of *Animal Liberation*, announces that he will nowhere appeal to his readers' emotions, only their reason (Singer, 1976). Singer's actual strategy, however, like that of any utilitarian, is rather different: it is to appeal to, or take as given and decisive, certain feelings deemed to be universal, notably an aversion to suffering. 'Subjectivism' is then avoided, not by (impossibly) eschewing all reliance on feelings, but by ignoring those feelings which differ from person to person. This strategy is very clear in an 18th century writer like Baron d'Holbach when attacking the 'moral sentiment' school. If, says the Baron, morality is to be 'solid', 'to be at all times the same, for all individuals of the human race', then it cannot be founded on such variable sentiments as sympathy for the woes of other people, but only upon the one indisputably universal feeling, people being 'in love with their happiness' [d'Holbach, (1770) 1990, p. 39].

The strategy, however, is invidious. It may well be that everyone is in love with their *own* happiness, but this is not a feeling that could generate any morality at all, but at best prudential rules, some of which coincide with moral principles. A love of other people's happiness could serve as a basis for a morality, but quite clearly no such love is universal. Some persons are markedly indifferent to the happiness of anyone else, let alone to the greatest happiness of the greatest number. It might be argued that 'there is something wrong' with such people: they are calloused, warped, embittered, or whatever. But, then, why shouldn't the same be argued in the case of other, admittedly variable feelings? Why, for example, should it not be argued that 'there is something wrong' with the Beltsville researcher, mentioned above, in his incapacity even to have a glimmer of the feelings which the prospect of hybrid animals offends? Or with someone who can look forward with enthusiasm to producing zombied bulls and, presumably, to other devices for what Alan Holland calls 'taking a form of life with a given level of capacity ... and reducing that capacity significantly'? (Holland, 1990, p. 172). Or with the biotechnological pioneer, Paul Berg, who proudly announces that he would stop his work 'if there were a sound practical reason, but not if it were an ethical judgement'? (Quotation in Bains, 1990, p. 3).

Certainly, then, there are people without the emotional responses to the genetic engineering of animals that others display. But that can be no reason to leave such responses out of the picture in order that our moral judgements avoid the charge of 'subjectivity'. If 'objective' moral judgements are those based on universal sentiments – ones which everybody as a matter of fact has – then there are no such judgements, since there are no such sentiments. If 'objective' judgements are those based on proper feelings – those had by people with whom 'there is nothing wrong' – then it has yet to be shown that the judgements inspired by aversion to and horror of the bioengineering of animals are not among them.

11.3 ACTIONS, AGENTS AND VIRTUES

One can concede that some feelings should sometimes be left out of the picture. You feel sorry for criminals, however culpable, condemned to spend years in prison: I do not, my sympathies being confined to their victims. But neither of us need think that something is amiss with the other for feeling as he or she does: each understands the other's sentiments, and we could agree that penal policy is not to be settled by appeal to such feelings. Sometimes, though, it is impossible to adopt this relaxed, 'chacun à ses sentiments' attitude. The point cannot be that some matters are too serious – calling for decisions which are bound to accord only with some people's feelings – for this attitude to be viable. Decisions on penal policy, after all, are a serious matter. The point, rather, is that sometimes feelings towards practices are in part directed towards the feelings – or lack of them – evident in those who carry on or condone the practices. If what revolts me is the very coolness of a violent act, its calculated disregard for life and limb, and the accompanying absence of any remorse or compunction, I cannot say, in debate with its perpetrator, 'I won't let my feelings, or your lack of them, sway my judgement on what you did'. For, of course, the 'what you did' which I am judging is partly constituted by the cool, calculated, remorseless character of the act. The wrongness of your act partly consists in your lacking the feelings which, in my case, would either have prevented the violence or given it a totally different character – that of a 'crime passionel', perhaps.

Much more of our moral revulsion is of the kind just illustrated than is usually allowed: that is, it is felt towards agents and practitioners, on account of the feelings or lack of them which their acts or practices betray. This is certainly the case, it seems to me, in the area which concerns us in this book. Much of the vocabulary in which ordinary people express their outrage at current and predicted bioengineering of animals registers that it is the people, the bioengineers, who practise it who are the objects of this outrage. Personalities, not just policies, are part of the issue. This is the import of the familiar charge that bioengineers are 'playing God', or that they are 'Frankensteins', or that they are 'sacrilegious'. In each case, what is being impugned are not simply their techniques (for the suffering they cause, or whatever), but the technicians for being the kind of people they are, lacking 'spontaneous human sentiments' yet equipped, unfortunately, with ones of a very different sort.

Utilitarianism, with its myopic focus on the consequences of actions, blinds us to the centrality, in ordinary moral thought, of judgements on and responses to the people who perform the actions. On rights theories, too, the central issue is never the nature or character of the agent, but whether his or her act violates some right. This emphasis on acts or their consequences, it might be said, is just how it should be. Indeed, it might be charged that I am drifting into paradox: for how can people be judged, how they can they prompt revulsion, except on the basis of what they do and its consequences? Part of the reply is to recall the point: 't 'aint what you do' which is always the paramount consideration, but 'the way that you do it' – or, more accurately, 'the way that you do it', the feelings and attitudes it betrays, are often part of the 'what you do' that is judged.

But there is a more important reply. It is just not true that the traffic between judgements on people and on their deeds is one-way, flowing only from the latter to the former. Of course, unless a person did these things rather than those, we would be in no position to judge the kind of person he or she is. But once such judgements have been made, these can crucially colour our judgements on people's deeds. Actions which 'in themselves', or in terms of their consequences, at first seem relatively neutral take on a very different hue when seen as the actions which only people of a certain sort would perform. In his poem *The Snake*, D.H. Lawrence casually throws a stone at the tail of a disappearing snake: 'And I immediately regretted it./I thought how paltry, how vulgar, what a mean act./I despised myself and the voices of my accursed human education' [Lawrence (1923) 1989, p. 24]. Throwing a stone unlikely to hit or hurt the animal is not 'in itself' overly reprehensible: but it becomes one to regret, to despise oneself for, when one reflects that it is the act of a mean and vulgar person, one corrupted by a certain education. That the wrongness of actions might be a function of the badness of a person, and not vice versa, so far from being paradoxical, was a central tenet of that long tradition of moral thought now referred to as 'virtue ethics'. A conception, like Aristotle's, of a virtuous, flourishing life may begin with certain intuitions about things to do and not to do. But once in place, this conception serves to colour – even dictate – further judgements on right and wrong: for there will be actions, like Lawrence's, which only emerge into a true moral light when this conception is brought to bear. For of all actions, including those about which we have no immediate clear intuitions, the question can be asked 'Is this the kind of thing that a person leading a flourishing human life would do?'

I began the previous section by suggesting that a way to identify what is distinctively wrong with bioengineering of animals is to reflect on the revulsion felt by many ordinary people. I then drew attention to two important, but frequently overlooked, features of moral sentiments: that they are often directed towards other people's feelings or lack of them, and that the judgements on other people which they imply are not simple functions of what these people do, since the moral status of the things people do is often – and rightly – ascertained by reflecting what these things indicate about the character, about the virtues (or lack of them), of the people who do them. I am suggesting, therefore, that what is distinctively wrong with the bioengineering of animals will emerge by reflecting on the revulsion or disquiet – registered by talk of 'playing God', and the like – felt towards those who engage in it, on account (in part) of their apparent lack of the very feelings which their activities offend: feelings which are ingredients in the good, flourishing human life. This suggestion, I think, captures the truth in the familiar thought that those who violate nature violate themselves, in – for example – Richard Adams' remark that 'if we rob the animals of their dignity ... we are, by that act, lowering ourselves' (Adams, 1987, p. 84).

Before trying to articulate what is distinctively wrong with the bioengineering of animals via reflection on what is felt to be wrong with the people who engage in it, I must briefly address the predictable charge that it is 'offensive' to impugn the characters of these bioengineers. I am not sure how much can or should be done to remove the offence. I find it odd, certainly, to find one writer, after a robust attack on genetic engineering and a demand for its total dismantling, saying that he has 'no reason for doubting ... the moral character' of those engaged in it (Linzey, 1994, p. 150). But two 'conciliatory' points can be made. First, I do agree with the same writer that there is no reason for doubting 'the sincerity [and] motivation' of bioengineers. Doubtless most of them are committed to working for human welfare, the search for truth, or something similarly respectable. No one is suggesting they are in it only for the money or for kicks. Second, and more importantly, I do not suppose that bioengineers – not even the makers of 'oncomice', perhaps – are, all things considered, worse than the rest of us. All of us live in glass houses, and we who have thrown stones doubtless deserve some in return. That it is deemed intolerably offensive to criticize people and personalities, not just practices and policies, is a penalty to pay in a society where all of us, from the cradle to the grave, whether by primary school teachers or politicians, are 'buttered up', made to think more of our worth than we should. No one advocates a return to the kind of society where people were brought up with an oppressive sense of sin and guilt. But in a society more candid than our own about people's failure to live good lives, the 'offence' I may be causing would be harder to give.

To return to the main theme: What is it that the sense that 'there is something wrong' with the makers of 'oncomice', 'transgenic' animals, 'Down's syndrome' bulls, and so on, indicates? Put more exactly, what does this sense imply are the virtues – or the sentiments and dispositions which are part of the virtuous life – which these people lack? There are, I think, several such virtues, but I shall focus on just one, to which philosophers have given various names. (Or perhaps it is a tight cluster of virtues, each deserving its own title, or one broad virtue whose various aspects need to be marked.)

11.4 HUMILITY AND INTEGRITY

The best name, perhaps, is the one given it by Iris Murdoch – 'humility': understood not as 'a peculiar habit of self-effacement, rather like having an inaudible voice', but as 'selfless respect for reality' (Murdoch, 1970, p. 95). A related one, conferred by the neglected French thinker, Gabriel Marcel, is *'disponibilité'* ('availability'), understood as the virtue of a person who is not calloused, but open to and ready to heed the calls that creatures and things make upon him (Marcel, 1949). Perhaps Heidegger is speaking of the same virtue when he refers to the capacity, that of 'letting-be', which has been largely lost in a technological world where everything becomes 'standing reserve' or 'raw material', to honour how each thing – 'heron and roe, deer, horse and bull', for example – 'fits into its own being' and is 'in its own way' (Heidegger, 1975, p. 182).

The kind of virtue in question here has an ancient pedigree. It is related to the *apatheia* recommended by the Greek Sceptics, to the reticent non-interventionism admired by the Daoist sages who called it *wu wei,* and to that liberation from 'grasping after' (*tanha*) things preached by the Buddha. For some readers, these references to the ancients give the game away: humility as defined is an 'old-fashioned' virtue. Certainly, it has, until very recently, been out of fashion in a climate where more Promethean virtues – commitment, industry, combating injustice in the world, authenticity and 'self-creation' – have held sway. But it is surely the importance of this 'old-fashioned' virtue which has been brought home to us by critics of environmental degradation, by some feminists, and by writers – some of them quoted in this chapter – who articulate ordinary people's concern at the bioengineering of animals. For it is the lack of humility on the part of bioengineers which has prompted the accusations of being Frankensteins, of playing God, and of sacrilege (the misappropriation for oneself of what belongs outside one's sphere). Indeed, it is because humility has, in their case, shrunk to vanishing point, so that a limit has been reached, that there is the sense of a new wrong being done to animals. Animals have always been used as means but, as Stephen Clark points out, 'old-style pastoralists' had at least to identify the animals' natural ends before working out how to profit from them. 'New-style artificers', instead, 'work out where the profit lies, and mould the ends to suit them' (Clark, 1995, p. 17).

Several reasons may be given why humility, in the relevant sense, is a virtue and why, moreover, it has a special place among the virtues. For it is, first, an exercise of the capacity to 'unself', to look away from one's own concerns, which most of the virtues presuppose. Its exercise requires at a minimum that one takes the ends of other creatures seriously, neither overriding them at will nor, still worse, inventing new ends for creatures who are then made to measure. It requires, therefore, an ability to view things as they are for others, or at any rate not simply in relation to oneself. 'A self-directed enjoyment of nature', writes Iris Murdoch, is 'something forced', and we should 'take a self-forgetful pleasure in the ... independent existence of animals, birds, stones and trees' (Murdoch, 1970, p. 85).

Second, with absence of humility there goes what Jung, reflecting on the vivisections he unwillingly performed as a student, called 'alienation ... from God's world' (Jung, 1967, p. 86). Alienation – a sense of being cut-off from, at odds with, other people, animals and 'nature in this place where we encounter it' (O'Donovan, 1984, p. 5) – cannot be an ingredient in a satisfying, rounded human life, productive as it is of restless dissatisfaction with what is felt to be alien, and of a consequent urge to bend it and assimilate it to the narrow domain in which one feels at home. No one, surely, should want to be like the Austrian novelist Thomas Bernhard, with his perception of nature as merely 'uncanny', 'malignant' and 'always against' him, as something to 'avoid' or at most 'put up with' (Bernhard, 1992, p. 62). This does not mean, incidentally, that we are forced to minimize the differences between ourselves and animals, let alone trees and plants. Too many animal and environmental ethicists, in their zeal to avoid appealing to emotions, are compelled grossly to exaggerate our similarities with other species in order to show that logical consistency alone requires extension of moral concern from humans to these species (Cooper, 1995). What is really required for this extension is a sense of companionship with animals and, perhaps, other species of life: and we can be companions to creatures we recognize as very different and distant from ourselves, though not as alien to us. Sometimes, as Heidegger reminds us, 'nearness preserves farness' (Heidegger, 1975, p. 178).

Critics often accuse bioengineers and others who bend animal and other natural life to human interests of lacking respect for the integrity of non-human life. This criticism is in order, but is best seen as re-describing or bringing out an implication of the bioengineers' lack of humility. The integrity of an animal species is often treated, especially in critical discussions of 'transgenic' animals, as a scientific fact, spelled out in terms of 'real' (versus 'conventional') distinctions among the species between which we discriminate; or as something a species possesses in virtue of some natural *telos*; or even as entailed by God's intention in populating the world with a variety of living creatures.

These suggestions are problematical and we do better, perhaps, to construe talk of integrity as drawing attention to the humility we leave behind when we programme animals with ends to suit ourselves and otherwise bend them to our will. A person abandons humility not *because* animal life has an integrity which he or she dishonours. Rather, integrity is said to be dishonoured when proper humility has been abandoned. If that is the right way round to put it, this vindicates the strategy of discerning what is distinctively wrong with the bioengineering of animals through the sentiments of those who are repelled by it and the lack of a virtue on the part of those who are the targets of this repulsion.

REFERENCES

Adams, R. (1987) Some thoughts on animals in religious imagery, in *Beyond the Bars: The Zoo Dilemma* (eds V. McKenna, W. Travers and J. Wray), Thorsons, Wellingborough, pp. 67–84.

Bains, W. (1990) *Genetic Engineering for Almost Everybody*, Penguin, Harmondsworth.

Bernhard, T. (1992) *Wittgenstein's Nephew*, Vintage, London.

Clark, S.R.L. (1994) Genetic and other engineering. *Journal of Applied Philosophy*, **11**, 233–7.

Clark, S.R.L. (1995) Intrinsic criticisms of biotechnological technique, in *Human Lives* (ed. D. Olderberg), Macmillan, London.

Cooper, D.E. (1995) Other species and moral reason, in *Just Environments* (eds D.E. Cooper and J.A. Palmer), Routledge, London.

d'Holbach, P.T. (1990 [1770]) Système de la nature, in *Moral Philosophy from Montaigne to Kant*, Vol. 2 (ed. J.B. Schneewind), Cambridge University Press.

Glover, J. (1984) *What Sort of People Should There Be?* Penguin, Harmondsworth.

Häyry, M. (1994) Categorical objections to genetic engineering – a critique, in *Ethics and Biotechnology* (eds A. Dyson and J. Harris), Routledge, London, pp. 202–15.

Heidegger, M. (1975) The Thing, in *Poetry, Language, Thought*, Harper & Row, New York, pp. 163–86.

Holland, A. (1990) The biotic community, in *The Bio-Revolution: Cornucopia or Pandora's Box?* (eds P. Wheale and R. McNally), Pluto, London, pp. 161–74.

Jung, C.G. (1967) *Memories, Dreams, Reflections*, Fontana, London.

Lawrence, D.H. (1989 [1923]) The Snake, in *We Animals: Poems of our World* (ed. N. Aisenberg), Sierra Club Books, San Francisco, pp. 22–4.

Linzey, A. (1994) *Animal Theology*, SCM, London.

Marcel, G. (1949) *Being and Having*, Dacre, London.

Murdoch, I. (1970) *The Sovereignty of Good*, Routledge & Kegan Paul, London.

O'Donovan, O. (1984) *Begotten or Made?* Clarendon Press, Oxford.

Ramanujan, A.K. (1989 [1986]) Zoo gardens revisited, in *We Animals: Poems of our World* (ed. N. Aisenberg), Sierra Club Books, San Francisco, pp. 77–8.

Rodd, R. (1990) *Biology, Ethics, and Animals*, Clarendon Press, Oxford.

Singer, P. (1976) *Animal Liberation: A New Ethics for our Treatment of Animals*, Cape, London.

12

On *telos* and genetic engineering

Bernard E. Rollin

12.1 *TELOS*

Aristotle's concept of *telos*[1] lies at the heart of what is very likely the greatest conceptual synthesis ever accomplished, unifying common sense, science, and philosophy. By using this notion as the basis for his analysis of the nature of things, Aristotle was able to reconcile the patent fact of a changing world with the possibility of its systematic knowability. Unlike Plato, whose austere mathematical model of knowledge led inexorably to a denigration of the experienced world, Aristotle saw that world as does a biologist, and found unproblematic the possibility of structural permanence underlying the constant flow of change. Though individual robins come and go, 'robin-ness' endures, making possible the knowledge that humans, in virtue of their own *telos* as knowers, abstract from their encounters with the world. Common sense tells us that only individual existent things are real; reflective deliberation, on the other hand, tells us that only what is repeatable and universal in these things is knowable.

Unlike both earlier and later reductionistic thinkers (the atomists and Descartes respectively), Aristotle saw living things as the paradigm for all things – all things play out their inherent roles in just the way that living things do. To understand the world, then, is not to seek one science of everything, but to uncover patterns which characterize the nature of each thing according to its own kind, just as the biologist attempts to characterize each type of living thing according to the

[1] Major discussions of *telos* in Aristotle occur in: *Physics*, Book II, Chapters 2,3,5,7,8; *Politics*, Book I, Chapter 2; *Metaphysics*, Book I, Chapter 3; *De Anima*, Book II, Chapter 4; *On the Heavens*, Book I, Chapter 4; *Posterior Analytics*, Book II, Chapter 11.

My view of *telos* has been most influenced by J.H. Randall (1960), and by his lectures at Columbia, 1965–68.

unique way it meets the set of characteristics constitutive of all living things – locomotion, reproduction, sensation, nutrition, etc. In this world view, then, biology is the master science, with the concepts of physics derived from the concepts of biology.

For Aristotle, as for common sense, the fact that animals had *tele* was self-evident – the task of the knower was to systematically characterize each relevant *telos*. That knowledge of things, paradigmatically living things, demanded functional categories was patent; why the world happened to be that way was not a sensible question. But with the appropriation of Aristotelianism by Christianity at the hands of Aquinas, an answer was provided, and *telos* as function became *telos* as Divine purpose, thereby indelibly tainting the concept with a supernatural flavour that potentiated its rejection by mechanistic, reductionistic science. (Spinoza's devastating critique of final causes in the Appendix to Part I of the *Ethics* is a clear exemplar of the New Science's derisive dismissal of functional explanations as essentially involving reference to conscious intention.)

Even without the theological overlay that *telos* had acquired in the Middle Ages, the concept would have found no place in a science which sees biology as part of physics rather than vice versa. Whereas both common sense and Aristotle were wedded to the insight that the world is made up of irreducible kinds of things, mechanism and reductionism from Descartes to positivism saw such qualitative differences as secondary and tertiary qualities reducible to quantitative configurations of homogeneous matter in motion. And despite the patently Aristotelian *gestalt* of Darwinism (albeit a dynamic rather than a timeless Aristotelianism), purportedly the root theory of modern biology, it is the molecular which woos most biologists today, so that the study of *telos* is relegated in mainstream science at best to natural history. But despite this reductionistic dismissal, the notion of *telos* has in fact been refined and deepened by the advent of molecular genetics, as a tool for understanding the genetic basis of animals' physical traits and behavioural possibilities. At the same time, the classical notion of *telos* is seen as threatened by genetic engineering, the operational offspring of molecular genetics. For we may now see *telos* neither as eternally fixed, as did Aristotle, nor as a stop action snapshot of a permanently dynamic process, as did Darwin, but rather as something infinitely malleable by human hands.

12.2 CONTEMPORARY AGRICULTURE

Despite the fact that the concept of *telos* has lost its scientific centrality, there are two major and conceptually connected vectors currently thrusting the notion of *telos* into renewed philosophical prominence, both of which are moral in nature. These vectors are social concern about the treatment of animals, and the advent of practicable biotechnology. The former concern reflects our recently acquired ability to use animals without respecting the full range of their *telos*; the latter concern reflects our in-principle ability to drastically modify animal *telos* in unprecedented ways. There obtain significant conceptual connections between the two concerns, but before these are dealt with one must understand the social conditions militating in favour of a revival of the concept of *telos*.

In elucidating the re-emergence of *telos* as a pivotal concept in mid-20th century concerns about animal treatment, I am following Hegel's maxim that at least part of a philosopher's job is to explicitly articulate nascent components of social thought which, while pervasive in society, have not 'become conscious of themselves'. In Plato's apt metaphor, such philosophical articulation should help society 'deliver the ideas it is in the process of birthing'. If the philosophical articulation is off the mark, it will be perceived as irrelevant by society at large. If it is accurate, it can help accelerate the society's 'recollection' of its own ideas, in Plato's other apt metaphor.

The overwhelmingly preponderant use of animals in society since the dawn of civilization has unquestionably been agricultural – animals were kept for food, fibre, locomotion and power. Presupposed by such use was the concept of husbandry; placing the animals in environments congenial to their *telos* – the Biblical image of the shepherd leading his animals to green pastures is a paradigm case – and augmenting their natural abilities by provision of protection from predators, food and water in times of famine and drought, medical and nursing attention, etc. In this ancient contract, humans fared well if and only if their animals fared well, and thus proper treatment of animals was guaranteed by the strongest possible motive – the producer's self-interest. Any attempt to act against the animals' interests as determined by their natures resulted in damage to the producers' interests as well. In this contract, both sides benefited – the animals' ability to live a good life was augmented by human help; humans benefited by 'harvesting' the animals' products, power or lives. One could not selectively accommodate some of the animals' interests to the exclusion of others, but was obliged to respect the *telos* as a whole.

This state of affairs explains why the traditional social ethic for animals in society – the prohibition against cruelty, i.e. overt, wilful, sadistic, unnecessary infliction of suffering or gross negligence – could be so minimalistic yet socially adequate. Only deviant, irrational or psychopathic individuals inflicted significant avoidable suffering on animals; this was guaranteed by the win–win nature of husbandry-based agriculture. To this day, it is axiomatic among western US ranchers – the last significant group of husbandrymen in the US – that 'we take care of the animals, and they take care of us'. Husbandry agriculture was about putting square pegs in square holes, round pegs in round holes, and creating as little friction as possible while doing so.

All of this changed drastically in the mid-20th century with the advent of high-technology agriculture, significantly portended as university departments of animal husbandry underwent a change in nomenclature to departments of 'animal science'. In this new approach to animal agriculture, one no longer needed to accommodate the animal's entire *telos* to be successful. With the aid of 'technological sanders' such as antibiotics, vaccines, hormones and other pharmaceuticals, one could place square pegs in round holes, round pegs in square holes, and limit the friction relevant to human interests. Hence, for example, one can now raise egg-laying hens six to a small cage, with thousands of cages tiered in one building. If this had been attempted a hundred years ago, the animals would rapidly have succumbed to diseases that would have spread like a prairie fire in crowded conditions where natural immunoprophylaxis cannot operate with sufficient alacrity to forestall decimation of the flock. With antibiotics and vaccines, however, one can forestall the disease and create a situation where the animals continue to lay eggs, even though many of the interests dictated by their *telos* – nest-building, dust-bathing, wing-flapping, and locomotion, for example – go totally unsatisfied. Thus, technology has allowed animal producers to divorce productivity from total or near-total satisfaction of *telos*.[2]

High-technology agriculture was not the only mid-20th century force significantly deforming the ancient contract with animals. Large-scale animal use in biomedical research and toxicology is, like intensive agriculture, a creature of the mid-20th century. Like confinement agriculture, too, successful use of animals in biomedicine does not necessitate accommodating the animals' *tele.* Rats, for example, which are of course widely used in biomedicine, are nocturnal, burrowing creatures, yet have traditionally been kept in stainless steel or polycarbonate cages under full-time or significant illumination. Baboons, highly social and intelligent animals possessed of complex behavioural repertoires, were kept in single austere cages, too small for the animal to even stand up. Furthermore, there is none of the contractual reciprocity enjoyed by animals raised under traditional husbandry for animals used in biomedical research. They may be wounded, burned, inflicted with disease, deprived of vital nutrients, frightened, stressed, etc. with

[2] For a detailed account of the problems associated with high-technology agriculture and suggestions for their amelioration, see Rollin (1995a).

no benefit to them, albeit great potential benefit to humans and other animals.[3]

Thus, both the advent of industrialized agriculture and large-scale animal use in science created an unprecedented situation in the mid-20th century by inflicting significant suffering on animals which was nonetheless not a matter of sadism or cruelty. Agriculturalists were trying to produce cheap and plentiful food in a society where only a tiny fraction of the population was engaged in agricultural production; scientists were attempting to cure disease, advance knowledge, and protect society from toxic substances. As society became aware of these new animal uses neither bound by the ancient contract nor conceptually captured by the anti-cruelty ethic, and concerned about the suffering they engendered, it necessarily required an augmentation in its moral vocabulary for dealing with animal treatment. (British society was galvanized into this realization primarily by exposés of industrialized agriculture in the 1960s; US society by exposés of animal research abuses in the 1980s.)

Since, as Plato pointed out, all ethics proceeds from pre-existing ethics, society looked to its moral toolbox for assessing the treatment of humans in search of moral concepts appropriate to the new situations of animal use. Prominently displayed in that toolbox in the mid-20th century was the concept of rights, which exists as a social–ethical solution to the age-old social tension that obtains between the interests of the group or of the society as a whole on the one hand, and the interests of individuals on the other. Whereas totalitarian societies favour the group or the state at the expense of the individuals, and anarchic communes do the opposite, democratic societies strike a mean between these extremes. While making most of their social decisions in a utilitarian way, they nonetheless build deontological fences around human individuals to protect fundamental human interests even from the general welfare. These fundamental interests – freedom of speech, religion, belief, assembly, movement, property ownership, protection from torture – are based upon plausible assumptions about what is fundamental to human nature, in other words, about the human *telos*. Whatever actions society may take for the general good, it is constrained against violating the human *telos* as manifested in each individual. Thus, we may not silence a speaker even if no one wishes to hear him, nor may we torture a terrorist to reveal the location of a bomb, even to save numerous innocent individuals.

It is this notion of rights, based on plausible reading of the human *telos*, which has figured prominently in mid-century concerns about

[3] For a detailed account of animal research and the moral problems it engenders, see Rollin (1992), Part III.

women, minorities, the handicapped and others who were hitherto excluded from full moral concern. It is therefore inevitable that this notion would be exported, *mutatis mutandis*, to the new uses of animals. In essence, society is demanding that if animals are used for human benefits, there must be constraints on that use, equivalent to the natural constraints inherent in husbandry agriculture. These constraints are based in giving moral inviolability to those animal interests which are constitutive of the animals' *telos*. If we are to use animals for food, they should live reasonably happy lives, i.e. lives where they are allowed to fulfil the interests dictated by their *telos*. For the laying hen, this means creating production systems which meet the needs mentioned above – this has been mandated by Swedish law. For the baboon used in biomedicine, this means creating a housing system which, in the words of US law, enhances the animals' 'psychological well-being', i.e. social non-austere containment for these animals that accommodates 'species-specific behaviour' (Rollin, 1989, pp. 177–81). For the zoo animals, it means creating living conditions which allow the animals to express the powers and meet the interests constitutive of its *telos* (Markowitz and Line, 1989).

Thus, *telos* has emerged as a moral norm to guide animal use in the face of technological changes which allow for animal use that does not automatically meet the animals' requirements flowing from their natures. In this way, one can see that the social context for the re-emergence of the notion of *telos* is a pre-eminently moral one: *telos* provides the conceptual underpinnings for articulating social moral concern about new forms of animal suffering. From this moral source emerge epistemological consequences which somewhat work against and mitigate the reductionistic tendencies in science alluded to earlier. For example, it is moral concern for *telos* which is sparking a return of science to studying animal consciousness, animal pain and animal behaviour, areas which had been reduced out of existence by the mechanistic tendencies of the 20th century science that affords pride of place to physicochemistry (Rollin, 1989). In an interesting dialectical shift, moral concern for animals helps revive the notion of *telos* as a fundamental scientific concept, in something of a neo-Aristotelian turn.

12.3 GENETIC ENGINEERING

If our analysis of the moral concerns leading to the resurrection of the notion of *telos* is correct, we can proceed to rationally reconstruct the concept and then assess its relevance to the genetic engineering of animals. By rationally reconstruct, I mean first of all provide an articulated account of *telos* which fills the moral role society expects of it.

Second, I mean to protect it from fallacious accretions which logically do not fit that role but which have attached, or are likely to attach to it for purely emotional, aesthetic, or other morally irrelevant reasons. A simple example of such a conceptual barnacle might be those who would restore the notion of 'Divine purpose' to the concept of *telos*, and then argue that any genetic engineering is wrong simply because it violates that Divine purpose.

What sense can we make out of the notion of *telos* we have offered? In that sense, the *telos* of an animal means 'the set of needs and interests which are genetically based, and environmentally expressed, and which collectively constitute or define the "form of life" or way of living exhibited by that animal, and whose fulfilment or thwarting matter to the animal'. The fulfilment of *telos* matters in a positive way, and leads to well-being or happiness; the thwarting matters in a negative way and leads to suffering (see Rollin, 1992, Part I). Both happiness and suffering in this sense are more adequate notions than merely pleasure and pain, as they implicitly acknowledge qualitative differences among both positive and negative experiences. The negative experience associated with isolating a social animal is quite different from the experience associated with being frightened or physically hurt or deprived of water. Since, as many (but not all) biologists have argued, we tend to see animals in terms of categories roughly equivalent to species, the *telos* of an animal will tend to be a characterization of the basic nature of a species. On the other hand, increased attention to refining the needs and interests of animals may cause us to further refine the notion of *telos* so that it takes cognisance of differences in the needs and interests of animals at the level of sub-species or races, or breeds, as well as of unique variations found in individual animals, though, strictly speaking, as Aristotle points out, individuals do not have natures, even as proper names do not have meaning.

Thus, we may attempt to characterize the general *telos* of the dog as a pack animal requiring social contact, a carnivore requiring a certain sort of diet, etc. At this level we should also characterize gender- and age-specific needs, such as nest-building for sows, or extensive play for puppies and piglets. As we focus greater attention on certain breeds of dogs, we become aware of unique needs or interests determined by their particular natures, which we may call '*sub-tele*'. For example, a shepherd dog bred for herding has an interest in expressing that behaviour; a hunting dog has an interest in expressing that set of behaviours, etc. If we wish, we can look with even greater precision at the needs and interests of individual animals flowing from their unique personalities – Helga wants to carry her red bicycle horn around and toot it, etc. Similarly, we know that some research monkeys are thoroughly entranced by computer games provided to enrich their environment, while others in the same group are totally unresponsive to such attempts at enrichment.

This is perfectly analogous to moral notions we use vis-à-vis humans. Our *ur*-concern is that basic human interests as determined by human nature are globally protected – hence the emphasis on general human rights. We may also concern ourselves with refinement of those interests regarding subgroups of humans, although these subgroups are as much cultural as genetic. Thus, we worry about the specific interests of Inuit hunting cultures in the face of encroaching civilization, not just their general human needs. Finally, we can and do focus attention in certain contexts on the interests of individuals, say on nurturing the musical or artistic interests of a particular gifted child.

Thus, *telos* is a metaphysical (or categorial) concept, serving a moral and thus value-laden function, and is fleshed out in different contexts by both our degree of empirical knowledge of a particular kind of animal and by our specificity and degree of moral concern about the animals in question. For example, the earliest stages of moral concern about the *telos* of laboratory animals focused only on very basic needs: food, ambient temperature, water, etc. As our moral concern grew, it focused on the less obvious aspects of the animals' natures, such as social needs, exercise, etc. As it grew still more, it focused on even less evident aspects. Thus, as mentioned, one of the US federal laws for laboratory animals of 1985 mandates an 'environment for primates which enhances their psychological well-being'. This mandate is noteworthy in that it not only concerned itself with meeting the animals' *telos*, but required that the manner in which it is met is not just at threshold, but that it positively 'enhance well-being'. In this case, the moral concern and the definition of *telos* implicit therein actually went beyond the bounds of our empirical knowledge at the time, at least as far as the scientific community was concerned, and served notice to that community that the demand for that knowledge was non-negotiable, and that they were mandated to acquire it. A wonderful anecdote illustrates this last point beautifully. The story was told at a national meeting of laboratory animal scientists by the chief administrator of the USDA division charged with enforcing the 1985 Animal Welfare Act amendments. He related that the USDA was perplexed at the congressional mandate requiring 'a physical environment ... which promotes the psychological well-being of primates'. Reasonably, he approached the United States psychological community for expert assistance. Not surprisingly, given the ideology permeating psychology that denied consciousness in animals, and saw moral questions as outside of the purview of science, he was told that there was no such thing. 'There will be after January 1987', he tellingly replied, 'whether you people help me or not' (Rollin, 1989, p. 180).

Thus, the notion of *telos* as it is currently operative is going to be a dynamic and dialectical one, not in the Darwinian sense that animal natures evolve but, more interestingly, in the following sense: as moral concern for animals (and for more kinds of animals) increases in society, this will drive the quest for greater knowledge of the animals' natures and interests, which knowledge can in turn drive greater moral concern for and attention to these animals.

It is not difficult to find this notion of *telos* operative internationally in current society. Increasing numbers of people are seeking enriched environments for laboratory animals, and this is even discussed regularly in trade journals for the research community. Indeed, one top official in the US research community has suggested that animals in research probably suffer more from the way we keep them (i.e. not accommodating their natures) than from the invasive research manipulations we perform. The major thrust of international concern about farm animals devolves around the failure of the environments they are raised in to meet their needs and natures, physical and psychological. Rectifying this is the essence of the pioneering Swedish law for farm animals of 1988, which the New York Times appropriately characterized as 'a bill of rights for farm animals' (*New York Times*, 1). Zoos are moving rapidly away from the traditional animal prison paradigm to places which attempt to accommodate animal *telos*; or as one pioneer in this change pointed out, to systems which allow animals to express the 'powers' which constitute their form of life (Markowitz and Line, 1989).

12.4 RESPECT FOR *TELOS* AND THE CONSERVATION OF WELL-BEING

This, then, is a sketch of the concept of *telos* that has re-emerged in society today. Though it is partially metaphysical (in defining a way of looking at the world), and partially empirical (in that it can and will be deepened and refined by increasing empirical knowledge), it is at root a moral notion, both because it is morally motivated and because it contains the notion of what about an animal we *ought* at least to try to respect and accommodate.

What, then, is the relationship between *telos* and genetic engineering? One widespread suggestion that has surfaced is quite seductive (Fox, 1986). The argument proceeds as follows. Given that the social ethic is asserting that our use of animals should respect and not violate the animals' *telos*, it follows that we should not alter the animals' *telos*. Since genetic engineering is precisely the deliberate changing of animal *telos*, it is *ipso facto* morally wrong. I suspect that something like this at least in part underlies the knee-jerk antipathy which many people have to genetic engineering. (The theological notion of *telos* as Divine plan also looms large in this antipathy.)

Seductive though this move may be, I do not believe it will stand up to rational scrutiny, for I believe it rests upon a logical error. What the moral imperative about *telos* says is this:

> Maxim to Respect *telos*:
> If an animal has a set of needs and interests which are constitutive of its nature, then, in our dealings with that animal, we are obliged to not violate and to attempt to accommodate those interests, for violation of and failure to accommodate those interests matters to the animal.

However, it does not follow from that statement that we cannot change the *telos*. The reason we respect *telos*, as we saw, is that the interests comprising the *telos* are plausibly what matters most to the animals. If we alter the *telos* in such a way that different things matter to the animal, or in a way that is irrelevant to the animal, we have not violated the above maxim. In essence, the maxim says that, given a *telos*, we should respect the interests which flow from it. This principle does not logically entail that we cannot modify the *telos* and thereby generate different or alternative interests.

The only way one could deduce an injunction that it is wrong to change *telos* from the Maxim to Respect *Telos* is to make the ancillary Panglossian assumption that an animal's *telos* is the best it can possibly be vis-à-vis the animal's well-being, and that any modification of *telos* will inevitably result in even greater violation of the animal's nature and consequently lead to greater suffering. This ancillary assumption is neither *a priori* true nor empirically true, and can indeed readily be seen to be false.

Consider domestic animals. One can argue that humans have, through artificial selection, changed (or genetically engineered) the *telos* of at least some such animals from their parent stock so that they are more congenial to our husbandry than are the parent stock. I doubt that anyone would argue that, given our decision to have domestic animals, it is better to have left the *telos* alone, and to have created animals for whom domestication involves a state of constant violation of their *telos*.

By the same token, consider the current situation of farm animals mentioned earlier, wherein we keep animals under conditions which patently violate their *telos*, so that they suffer in a variety of modalities yet are kept alive and productive by technological fixes. As a specific example, consider the chickens kept in battery cages for efficient, high-yield, egg production. It is now recognized that such a production system frustrates numerous significant aspects of chicken behaviour under natural conditions, including nesting behaviour (i.e. violates the *telos*), and that frustration of this basic need or drive results in a mode of suffering for the animals (Mench, 1992). Let us suppose that we have identified the gene or genes that code for the drive to nest. In addition, suppose we can ablate that gene or substitute a gene (probably *per impossibile*) that creates a new kind of chicken, one that achieves satisfaction by laying an egg in a cage. Would that be wrong in terms of the ethic I have described?

If we identify an animal's *telos* as being genetically based and environmentally expressed, we have now changed the chicken's *telos* so that the animal that is forced by us to live in a battery cage is satisfying more of its nature than is the animal that still has the gene coding for nesting. Have we done something morally wrong?

I would argue that we have not. Recall that a key feature, perhaps *the* key feature, of the new ethic for animals I have described is concern for preventing animal suffering and augmenting animal happiness, which I have argued involves satisfaction of *telos*. I have also implicitly argued that the primary, pressing concern is the former, the mitigating of suffering at human hands, given the proliferation of suffering that has occurred in the 20th century. I have also argued that suffering can be occasioned in many ways, from infliction of physical pain to prevention of satisfying basic drives. So, when we engineer the new kind of chicken that prefers laying in a cage and we eliminate the nesting urge, we have removed a source of suffering. Given the animal's changed *telos*, the new chicken is now suffering less than its predecessor and is thus closer to being happy, that is, satisfying the dictates of its nature.

This account may appear to be open to a possible objection that is well known in human ethics. As John Stuart Mill queried in his *Utilitarianism*, is it better to be a satisfied pig or a dissatisfied Socrates? His response, famously inconsistent with his emphasis on pleasure and pain as the only morally relevant dimensions of human life, is that it is better to be a dissatisfied Socrates. In other words, we intuitively consider the solution to human suffering offered, for example, in *Brave New World*, where people do not suffer under bad conditions, in part because they are high on drugs, to be morally reprehensible, even though people feel happy and do not experience suffering. Why then, would we consider genetic manipulation of animals to eliminate the need that is being violated by the conditions under which we keep them to be morally acceptable?

This is an interesting and important objection, amenable to a number of different responses. Let us begin with the *Brave New World* case. Our immediate response to that situation is that the repressive society should be changed to fit humans, rather than our doctoring humans (chemically or genetically) to fit the repressive society. It is, after all, more sensible to alter clothes that do not fit than to perform surgery on the body to make it fit the clothes. And it is certainly possible and plausible to do this. So we blame the *Brave New World* situation for not attacking the problem.

This is similarly the case with the chickens. We know that laying chickens lived happily and produced eggs under conditions where they could nest for millennia. It is our greed that has forced them into an unnatural situation and made them suffer – why should we change them, rather than not succumb to greed? This seems to be a simple point of fairness.

A disanalogy between the two cases arises at this point. We do not accept any claim that asserts that human society must be structured so that people are totally miserable unless they are radically altered or their consciousness distorted. Given our historical moral emphasis on reason and autonomy as non-negotiable ultimate goods for humans, we believe in holding on to them, come what may. Efficiency, productivity, wealth – none of these trump reason and autonomy, and thus the *Brave New World* scenario is deemed unacceptable. On the other hand, were Mill not a product of the same historical values but was rather truly consistent in his concern only for pleasure and pain, the *Brave New World* approach or otherwise changing people to make them feel good would be a perfectly reasonable solution.

In the case of animals, however, there are no *ur*-values like freedom and reason lurking in the background. We furthermore have a historical tradition as old as domestication for changing (primarily agricultural) animal *telos* (through artificial selection) to fit animals into human society to serve human needs. We selected for non-aggressive animals, animals that depend on us not only on themselves, animals disinclined or unable to leave our protection, and so on. Our operative concern has always been to fit animals to us with as little friction as possible – as discussed, this assured both success for farmers and good lives for the animals.

If we now consider it essential to raise animals under conditions like battery cages, it is not morally jarring to consider changing their *telos* to fit those conditions in the same way that it jars us to consider changing humans.

Why then does it appear to some people to be *prima facie* somewhat morally problematic to suggest tampering with the animal's *telos* to remove suffering? In large part, I believe, because people are not convinced that we cannot change the conditions rather than the animal. (Most people are not even aware how far confinement agriculture has moved from traditional agriculture. A large East Coast chicken producer for many years ran television ads showing chickens in a barnyard and alleging that he raised 'happy chickens'.) If people in general do become aware of how animals are raised, as occurred in Sweden and as animal activists are working to accomplish elsewhere, they will doubtless demand, just as the Swedes did, first of all a change in raising conditions, not a change in the animals.

On the other hand, suppose the industry manages to convince the public that we cannot possibly change the conditions under which the animals are raised or that such changes would be outrageously costly to the consumer. And let us further suppose, as is very likely, that people still want animal products, rather than choosing a vegetarian lifestyle. There is no reason to believe that people will ignore the suffering of the animals. If changing the animals by genetic engineering is the only way to assure that they do not suffer (the chief concern of the new ethic), people will surely accept that strategy, though doubtless with some reluctance.

From whence would stem such reluctance, and would it be a morally justified reluctance? Some of the reluctance would probably stem from slippery slope concerns – what next? Is the world changing too quickly, slipping out of our grasp? This is a normal human reflexive response to change – people reacted that way to the automobile. The relevant moral dimension is consequentialist; might not such change have results that will cause problems later? Might this not signal other major changes we are not expecting?

Closely related to that is a queasiness that is, at root, aesthetic. The chicken sitting in a nest is a powerful aesthetic image, analogous to cows grazing in green fields. A chicken without that urge jars us. But when people realize that the choice is between a new variety of chicken, one *without* the urge to nest and denied the opportunity to build a nest by how it is raised, and a traditional chicken *with* the urge to nest that is denied the opportunity to build a nest, and the latter is suffering while the former is not, they will accept the removal of the urge, though they are likelier to be reinforced in their demand for changing the system of rearing and, perhaps, in their willingness to pay for reform of battery cages. This leads directly to my final point.

The most significant justified moral reluctance would probably come from a virtue ethic component of morality. Genetically engineering chickens to no longer want to nest could well evoke the following sort of musings: 'Is this the sort of solution we are nurturing in society in our emphasis on economic growth, productivity and efficiency? Are we so unwilling to pay more for things that we do not hesitate to change animals that we have successfully been in a contractual relationship with since the dawn of civilization? Do we really want to encourage a mind-set willing to change venerable and tested aspects of nature at the drop of a hat for the sake of a few pennies? Is tradition of no value?' In the face of this sort of component to moral thought, I suspect that society might well resist the changing of *telos*. But at the same time, people will be forced to take welfare concerns more seriously and to decide whether they are willing to pay for tradition and amelioration of animal suffering, or whether they will accept the 'quick fix' of *telos* alteration. Again, I suspect that such musings will lead to changes in husbandry, rather than changes in chickens.

We have thus argued that it does not follow from the Maxim to Respect *Telos* that we cannot change *telos* (at least in domestic animals) to make for happier animals, though such a prospect is undoubtedly jarring. A similar point can be made in principle about non-domestic animals as well. Insofar as we encroach upon and transgress against the environments of all animals by depositing toxins, limiting forage, etc. and do so too quickly for them to adjust by natural selection, it would surely be better to modify the animals to cope with this new situation so they can be happy and thrive rather than allow them to sicken, suffer, starve and die, though surely, for reasons of uncertainty on how effective we can be alone as well as aesthetic reasons, it is far better to preserve and purify their environment.

In sum, the Maxim to Respect *Telos* does not entail that we cannot change *telos*. What it does entail is that, if we do change *telos* by genetic engineering, we must be clear that the animals will be no worse off than they would have been without the change, and ideally will be better off. Such an unequivocally positive *telos* change from the perspective of the animal can occur when, for example, we eliminate genetic disease or susceptibility to other diseases by genetic engineering, since disease entails suffering. The foregoing maxim which does follow from the Maxim to Respect *Telos*, we may call the Principle of Conservation of Well-being. This principle does of course exclude much of the genetic engineering currently in progress, where the *telos* is changed to benefit humans (e.g. by creating larger meat animals) without regard to its effect on the animal. A major concern in this area which I have discussed elsewhere is the creation of genetically engineered animals to 'model' human genetic disease (Rollin, 1995b, Chapter 3).

There is one final caveat about genetic engineering of animals which is indirectly related to the Maxim to Respect *Telos*, and which has been discussed, albeit in a different context, by biologists. Let us recall that a *telos* is not only genetically based, but is environmentally expressed. Thus, we can modify an animal's *telos* in such a way as to improve the animal's *telos* and quality of life, but at the expense of other animals enmeshed in the ecological/environmental web with the animal in question. For example, suppose we could genetically engineer the members of a prey species to be impervious to predators. While their *telos* would certainly be improved, other animals would very likely be harmed. While these animals would thrive, those who predate them could starve, and other animals who compete with the modified species could be choked out. Thus, we would, in essence, be robbing Peter to pay Paul. Furthermore, while the animals in question would surely be better off in the short run, their descendants may well not be – they might, for example, exceed the available food supply and may also starve, something which would not have occurred but for the putatively beneficial change in the *telos* we undertook. Thus, the price of improving one *telos* of animals in nature may well be to degrade the efficacy of others. In this consequential and environmental sense, we would be wise to be extremely circumspect and conservative in our genetic engineering of non-domestic animals, as the environmental consequences of such modifications are too complex to be even roughly predictable (Rollin, 1995a, Chapter 2).

12.5 CONCLUSION

In conclusion, there is no direct reason to argue that the emerging ethical/metaphysical notion of *telos* and the Maxim to Respect *Telos* logically forbid genetic engineering of animals. In the case of domestic animals solidly under our control, one must look at each proposed genetic modification in terms of the Principle of Conservation of Welfare. In the case of non-domestic animals, there is again no logical corollary of the Maxim of Respect for *Telos* which forestalls genetically modifying their *telos*, and, on occasion, such modification could be salubrious. Given our ignorance, however, of the systemic effects of such modifications, it would be prudent to proceed carefully, as we could initiate ecological catastrophe and indirectly affect the functionality of many other animals' *tele.*

ACKNOWLEDGEMENT

I am grateful to Peter Gillen for his bibliographical assistance and trenchant comments.

REFERENCES

Fox, M.W. (1986) On the genetic engineering of animals: a response to Evelyn Pluhar. *Between the Species*, **2** (1), 51–2.

Markowitz, H. and Line, S. (1989) The need for responsive environments, in *The Experimental Animal in Biomedical Research*, vol. I (eds B.E. Rollin and M.L. Kesel), CRC Press, Boca Raton, Florida, pp. 153–73.

Mench, J.A. (1992) The welfare of poultry in modern production systems. *Critical Reviews in Poultry Biology*, **4**, 107–28.

New York Times, October 25, 1988, p. 1.

Randall, J.H. (1960) *Aristotle*, Columbia University Press, New York.

Rollin, B.E. (1989) *The Unheeded Cry: Animal Consciousness, Animal Pain and Science*, Oxford University Press, Oxford.

Rollin, B.E. (1992) *Animal Rights and Human Morality*, Prometheus Books, Buffalo, NY.

Rollin, B.E. (1995a) *Farm Animal Welfare: Ethical, Social, and Research Issues*, Iowa State University Press, Ames, Iowa.

Rollin, B.E. (1995b) *The Frankenstein Syndrome: Ethical and Social Issues in the Genetic Engineering of Animals*, Cambridge University Press, New York.

13

Intrinsic value and transgenic animals

Robin Attfield

Henk Verhoog has argued that one of the key objections to the genetic engineering of animals lies in the intrinsic value of wild species, which such genetic engineering undermines. Respect for the intrinsic value of animal species supposedly constitutes a deontological ground (a non-consequentialist ground, that is) for a qualified 'no' to genetic engineering, or rather for a 'no, unless'. Here 'unless' is short for 'unless a basic or very serious human or animal interest ... is involved which cannot be met by other means' (Verhoog, 1992a, p. 160). In this chapter I set out to explore what might be meant by the suggestion that animal species have intrinsic value, why philosophers seek to introduce here a deontological or non-consequentialist ground of objection to genetic engineering independent of good or bad states of individual living creatures, and whether what needs to be said can be said in consequentialist terms and on a consequentialist basis. If so, I shall suggest, the objections to genetic engineering can be articulated much more cogently, where they stand up at all.

13.1 VERHOOG'S INTRINSIC VALUE PRINCIPLE

Where the housing systems of food-animals cause them suffering, it is sometimes proposed that, to prevent this suffering, rather than change the housing systems, the animals required to live in those systems be genetically manipulated so as not to suffer. As Verhoog says, this 'goes against the moral intuition of many people' (1992a, p. 155). Yet, as he also says, approaches which locate value in the pleasures and satisfactions of individual animals and treat suffering alone as evil cannot give a satisfactory account of this intuition.

Admittedly not everything need rest on this one intuition, however widespread and persistent. Yet there does seem to be something lacking in theories which, as I have argued elsewhere, imply that there is nothing wrong with breeding or generating animals lacking the capacity to suffer pain so that they may live in otherwise painful conditions, and which imply indeed that no harm whatever can be done to creatures lacking sentience (Attfield, 1995). Thus, Verhoog's search for a better theory so far deserves applause. And the same applies when the position which he is criticizing allows that adjusting animals to their environment might be objectionable, but only on aesthetic grounds; for, while these grounds are often in place, there would also seem, as Verhoog says (1992b), to be a moral issue here as well as an aesthetic one. He is criticizing Rollin (1986, p. 296), who holds that being harmed depends on having conscious experiences.

Verhoog's search for a better theory takes him to the notion of intrinsic value. Now part of his diagnosis of the defective nature of theories such as that of Rollin consists in his view that their locating of intrinsic value in the experiences and states of individual creatures pays insufficient regard to the creatures themselves. Here he is in agreement with Regan (1984), who holds that this approach makes individual animals mere receptacles of value. Indeed, his endorsement of Regan's position extends to his deciding actually to name the view of intrinsic value in which it applies to experiences and states (rather than to individuals) as the 'receptacle view' (Verhoog, 1992b, p. 270). Unlike Regan, however, who writes of the 'inherent value' of individuals, Verhoog, to avoid confusion, decides to use the term 'intrinsic value' for the value borne by individual animals.

Can individual entities be the bearers of intrinsic value? In the classical sense employed by G.E. Moore (1922) and C.I. Lewis (1946), intrinsic value attaches to whatever is valuable or desirable for no reason beyond the intrinsic nature of the thing in question. While the nature of individuals might conceivably underpin their goodness, intrinsic value as intrinsic desirability attaches much more plausibly to experiences or qualities or to the having of such experiences or qualities, rather than to individuals. Individuals as such do not seem to supply reasons for action. Accordingly, the majority of philosophers since Moore have restricted their application of 'intrinsic value' to qualities, experiences and states, as opposed to subjects and individuals. It is important, therefore, to investigate what Verhoog means by 'the intrinsic value of animals'.

Verhoog's position is summarized as follows:

> '...when we say that an animal (or any other entity) has intrinsic value, we mean that (a) the animal deserves to be treated as a goal

in itself, and not merely as a means towards the well-being of others; and (b) the animal has characteristics which make it justified to consider the animal as morally respectable. These characteristics are constitutive of the nature of the animal.'

Verhoog (1992b, p. 270)

(Here 'respectable' almost certainly means 'worthy of respect'.) Shortly, what is meant here by 'goal' and 'nature' will be reviewed; but it should at once be remarked that some kind of Kantian position is implicit here. Thus, it might be suggested that what Verhoog really has in mind, particularly with respect to clause (b), is that animals (or perhaps some animals) have *moral standing,* i.e. have characteristics which make them worthy of respect, and which mean that their interests must be taken into consideration by moral agents. The term 'moral standing' would certainly be clearer in contexts concerning the place of individuals within morality (I suggest) than the language of 'intrinsic value'. But this reading would probably be less than Verhoog means; for it is at least possible that some form of Categorical Imperative is already implicit in his use of 'has intrinsic value'. Certainly at the end of the same paper he proposes the introduction of a deontological element, with respect to the ethics of the treatment of animals, based on their intrinsic value (p. 277). Nor is Verhoog alone in following Kant in such matters; for Alan Holland has written (1995[1]), with at least partial approval, of a Kantian proscription against regarding benefits for one subject as compensation or justification for adverse impacts on another.

Now Holland recognizes that this would involve an extension of Kant's Categorical Imperative, since most animals lack the kind of rationality which Kant required as the basis of respect. In its extended version, the Kantian principle is grounded in a 'common significance ... in every creature's organised form of life', and 'the view, which may be traced back to Aristotle, that every creature is possessed of a *telos* (which literally translated in fact means 'end' or 'goal') which constitutes the fulfilment of its particular form of life'. Holland expounds the Kantian principle as follows: 'According to this view, other things being equal, one ought not willingly and avoidably to subject any creature to conditions which deny it the ability to express its *telos.*' This principle is then ascribed to Michael Fox, who holds that 'transgenic manipulation is wrong because it violates the genetic integrity or *telos* of organisms or species' (Holland, 1995, p. 297; cf. Fox, 1990).

This apparent digression now serves to throw light on what Verhoog

[1] See sections on 'The Kantian principle' and 'Kant revisited'.

probably means by 'the animal deserves to be treated as a goal in itself'; for he wrote in awareness of Rollin's reintroduction of the Aristotelian concept of '*telos*', with a somewhat reduced sense comprising 'the interests and nature of an animal which it possesses as a member of a species' (Rollin, 1986), or 'the set of needs and interests, physical and psychological, genetically encoded and environmentally expressed, which make up the animal's nature'.[2] Though Verhoog quarrels with Rollin's willingness to speak of changing an animal's *telos* by genetic engineering, there is no sign that he dissents from Rollin's sense of the word. Thus, in writing of treating an animal as a goal in itself, Verhoog replaces the Kantian phrase 'an end in itself' with words ('a goal in itself') which probably convey 'the bearer of a *telos*'. And in adding that the animal is not to be treated as a means only, he may well be voicing the Kantian proscription against treating benefits for human beings as compensation or justification for adverse impacts on another – subject, apparently, to the explicit, rather non-Kantian proviso mentioned above, 'unless a basic or very serious human or animal interest ... is involved which cannot be met by other means' (Verhoog, 1992a, p. 160).

Verhoog's basis of respect, however, includes 'the nature of the animal', and about this he writes that 'To have a "nature of its own" is unthinkable without taking into account the species to which the animal belongs' (1992b, p. 273). Here, he is critical of Rollin for not regarding species as legitimate objects of moral concern. In order to employ 'a concept of species which reflects the dynamism and change inherent in the process of evolution' (Verhoog, 1992b, p. 276), Rollin adopts the definition of a species as an 'interbreeding population', but Verhoog points out that the rival phylogenetic theory, in which species are compared to individuals, is equally eligible as far as these same criteria are concerned. The second definition lends itself much more readily to talk of the integrity of the species being disturbed, and thus to the species itself being an object of moral concern. And if so, an animal's species-specific nature, Verhoog may be taken to imply, strengthens the basis for respecting the animal.

Here it is appropriate to express doubts whether, even conceived as an individual, a species has a good of its own, and thus is of itself morally considerable independently of its members. Species can certainly flourish, but they do not go through inherited patterns of development in anything like the ways in which their members do. Nor can it be assumed that their naturalness, as organized forms of life

[2] Verhoog, 1992b, p. 272; Verhoog's text here says '(Rollin 1989: 295)', but should say '(Rollin 1986: 295)'. A closely similar definition appears in Bernard E. Rollin, *The Unheeded Cry: Animal Consciousness, Animal Pain and Science* (Oxford and New York: Oxford University Press, 1989), p. 146.

or as products of evolution, makes it wrong in general to modify them. It is rather their members which have a good of their own (a characteristic which they share with domestic animals and with animals produced by genetic engineering).

Verhoog, however, believes otherwise. He writes of 'a degradation of the intrinsic value' of wild species when domestication takes place, at any rate from an ecocentric point of view (1992a, p. 159), and thus presupposes that at least wild species have moral standing. In addition, he holds that species have intrinsic value, and are thus to be respected, even when they have been eroded or nullified. While this may seem unsatisfactory, what he is in any case able to claim with some plausibility is that, however dynamic and changing, species are sufficiently real at any one time as to confer biologically detectable, species-specific natures on their wild (and sometimes on their domesticated) members; and that these natures are a suitable basis for moral respect. And this conclusion is sufficient to raise questions about genetic engineering, independently of his claims about intrinsic value.

Verhoog does not, however, regard all domestication as wrong. 'When man is viewed as a part of an ecosystem', he writes, 'and a "partner" of other natural entities, then it is not necessarily a bad thing when animals are domesticated for human survival'.[3] 'In the "partner-view", the acceptability of human intervention depends on the "quality" of the new relations arising' (Verhoog, 1992a, p. 159). (As will be seen, this has implications for the acceptability of genetic engineering.)

Meanwhile, there are already problems for domestication relating to the *telos* which has been partially lost. For (a) domestic animals are not to be held to have a changed *telos*; at best they have an imposed, new *telos*, and at worst none at all, except insofar as their wild *telos* remains partially intact. Also, (b) it ceases to be possible to identify psychological suffering, as the nature of this is species-specific and depends on having an intact *telos* (controversial claims these); and (c) the nature of these animals is ceasing to be autonomous and beginning to be artificial. The implications of all this are less than clear; but Holland (1995, p. 300) infers that, for Verhoog, 'the process of domestication is open to critical scrutiny'. This would mean that existing patterns of domestication cannot be regarded as a precedent liable to justify whatever genetic engineering may be shown to be parallel to them. And here, if this is what he is saying, Verhoog is surely right.

When the focus changes to genetic engineering, the moral issues become yet more acute. For there is now no natural *telos* at all, but only

[3] Verhoog, 1992a, p. 159; Holland, 1995: see section on 'The reduction of capacity principle'.

an imposed one. Hence it is no longer possible to discover what the animal's needs and interests are, or when it suffers. (This claim will be contested below.) Again, genetically engineered animals have ceased to be autonomous, or to be involved in relations of partnership with humanity; in these circumstances, it is at least dubious whether they have intrinsic value at all (Verhoog, 1992b, p. 272).

From these implications of his principle of the intrinsic value of species, Verhoog derives his 'no, unless' verdict on genetic engineering. He rejects the reflection that, if we cannot identify the interests of genetically engineered animals, then there are no moral objections any longer, however badly they are treated, though he does see this as a problem for Rollin (Verhoog, 1992b, p. 272). Presumably he rejects this by appealing to what is involved in respecting a natural species and its natural *telos*. This is strong enough to prohibit actions which erode species and *telos*, but not strong enough to do so in all circumstances, since 'a basic or very serious human or animal interest ... which cannot be met by other means' may be involved.

13.2 A CRITIQUE OF THIS PRINCIPLE

One unresolved issue concerns how these precise conclusions can be derived. Thus, the recognition of basic or very serious human and animal interests of other creatures suggests that Verhoog does not rule out a consequentialist appeal to interests or to the possibility of comparing them, and this already clashes with the tacit Kantianism of Verhoog's main principle. Relatedly, his main principle seems specifically to prohibit the sacrificing of the interests of, for example, a transgenic animal for the sake of others' interests, and yet this is just what his overall conclusion permits.

These problems seem to result from an attempt to combine 'the now dominant consequentialist theory of animal ethics' with 'a deontological element based on the intrinsic value of animals' (Verhoog, 1992b, p. 272). But, strictly speaking, it is impossible to combine consequentialist and non-consequentialist theories. For the former make obligations and right and wrong actions turn wholly on consequences (whether the consequences of acts, rules or practices), while the latter do not. And even if this is recognized, and it is acknowledged that would-be mixed theories are, strictly, deontological theories, the problem still arises for any particular such theory of supplying satisfactory criteria for the extent and limits of the consequentialist element in the theory; for it is not plausible to suppose that there is any basis of comparison between the weight of the consequentialist element and the weight of the deontological element.

To tackle these problems, a fairly precise meaning would be required for the principle of the intrinsic value of species; but the terms chosen make this too an intractable problem. For 'intrinsic value' has long functioned as a phrase from the consequentialist vocabulary, and applies to states of affairs which supply reasons for action in virtue of their intrinsic desirability. When Verhoog (to avoid the 'receptacle view') follows Regan and ascribes value to individuals rather than to their states and experiences, there is already a difficulty, since it is not, by and large, individuals which are desirable. Maybe this problem could be cured if 'intrinsic value' were now given a new and unqualifiedly Kantian sense, such as 'never to be treated only as a means to the well-being of anything else'. But Verhoog is too realistic to do this, recognizing that such protection, whether for species, wild creatures, domestic animals or genetically engineered animals, is implausibly strong, and perhaps recognizing too that it would generate large numbers of contradictions, in cases where the interests of more than one protected creature conflict. Hence his sense of 'intrinsic value' turns out to be compatible, as in its consequentialist use, with the subordination of the interests of the intrinsically valuable individual after all.

There is, however, another reason for using 'intrinsic value' in a less-than-Kantian sense. For Verhoog wants to ascribe intrinsic value to species with a natural *telos*; he wants to argue from the respect due to a natural species to conclusions such as the wrongness of genetic engineering which adds to the erosion of a *telos* which was no longer fully present beforehand (having been eroded already through prior processes of domestication). Here there is little or no question of using the species as a means only, since it is supposed to be scarcely present any longer at all. Also there is certainly not supposed to be any question of using its members as means only, as by this stage it is supposed to be impossible to identify their interests; and so it has become unclear what it would be like to treat them as means only. The objection lies rather in failing to respect the natural species which there once was, and at the same time undermining the kinds of relations between species which evolve naturally. But, whether or not there is merit in these objections, there is surely some clearer way of conveying them than through a semi-Kantian use of 'intrinsic value'.

13.3 IS DEONTOLOGY REALLY NECESSARY?

It is time to explore the possibility that Rollin's position can be supplemented, and grounds found for rejecting the generation of desensitized creatures, without a deontological theory grounded on the supposed intrinsic value of species. Rollin's objection to causing or allowing avoidable suffering should be retained, but should surely be extended to include the causing or allowing of harms whether or not they consist in suffering (physical or psychological) of which the suffering animal is aware. This is because harm in the form of unconscious disease or injury constitutes an evil alongside suffering which is noticed and distresses the sufferer. One further extension is also in place; for, as Derek Parfit (1984, Appendix G, pp. 487–90) has argued, a creature can be harmed by bringing it into being if its life will foreseeably be one of misery, or not worth living. In this extended form, though the wrongness of generating desensitised creatures is yet to be covered, consequentialism begins to look suited to the ethics of animal husbandry and of genetic engineering; for it already prohibits generating chickens destined for the conditions of factory farms, where their lives are lives of misery, without in general prohibiting killing animals for food, and it also prohibits engineering pigs (like the Beltsville pigs) destined to live arthritic lives through their genetic endowment, without debarring the generation of, for example, Factor IX sheep.

But before consequentialism can be rehabilitated, including the consequentialist claim that it is states and activities of individuals, rather than those individuals themselves, which are of intrinsic value or disvalue, Verhoog's condemnatory classification of this position as 'the receptacle view' needs to be reviewed. Does this position really encourage or allow lack of respect to individuals?

The first point to make is that, far from failing to ground or underpin respect for individuals, it presupposes that individuals whose lives are capable of intrinsic value and disvalue have moral standing or are morally considerable – that is that they have properties which qualify them for practical respect and that their interests are, morally speaking, to be taken into consideration. That is not to say that their interests will prevail, but it is (as has been seen) to say that they are not to be treated solely as means. Accordingly, non-anthropocentric consequentialism includes provision for respect for individual creatures in its very foundations.

If, further, the objection is that what matters is states of intrinsic value rather than individual creatures, it may be replied that what matters is the having or enjoying of states of intrinsic value by nothing but individual creatures. The relation between pleasure or flourishing and their bearers is not a contingent one, as there can be no pleasure and no flourishing without it being someone's or something's. Free-floating pleasure is a fantasy of consequentialism's deontological opponents.

Sometimes, however, the objection concerns the readiness of consequentialism to aggregate good and bad states of affairs, whoever undergoes them, and to judge by the overall balance of good over evil, rather than safeguarding the interests of each individual concerned. But no one, and certainly not Verhoog, is prepared to reject all comparisons of affected interests; as has been seen, he is prepared for the interests of individual animals to be overridden for the sake of certain interests of human beings. Indeed any theory which made the intactness of every individual's interests an absolute and indefeasible right would rapidly be inundated with contradictory judgements in cases of conflicts of interest, where both action and inaction were forbidden. Since respecting individuals may be supposed not to have these implications, it cannot prohibit comparisons of interests.

But aggregation now becomes the focus of complaint; it might seem that consequentialism is indifferent between diverse distributions of the same good, and disregards, for example, greater needs or stronger entitlements. While an adequate consideration of this issue would take us too far afield, two replies may be given. First, if the satisfaction of basic needs is recognized as of much greater value than other goods, then greater needs will not be disregarded; and it is open to consequentialists to recognize this greater value (and indeed it is open to deontologists to ignore it). Second, consequentialists are free to appraise consequences not as the consequences of single acts but as the consequences of rules or of practices; and if practices embodying entitlements turn out to have a better overall outcome than alternative practices, then consequentialists will favour them. Thus, accusations of unprincipled aggregation and of distributional indifference are wide of the mark.

What remains of the 'receptacle view' objection consists in the impersonal character of consequentialist ethics. It may, by now, be recognized that consequentialism does not authorize the replacement of anyone and everyone by another creature who or which would be happier, but it may still be objected that the willingness of consequentialists to take into account the good of possible creatures which could be brought into being is liable to downgrade the interests of those which are actual. To this, however, there is a forthright reply, stemming, as before, from the work of Derek Parfit. Since most future creatures are at present no more than possible creatures, our responsibilities to future generations would come to nothing if our obligations ran out at obligations with regard to actual rather than possible creatures. Although the numbers and even the identities of future creatures are as yet unknown, we still have responsibilities extending beyond duties to actual, identifiable creatures, to ensure that whichever ones live also lead lives worth living, and not significantly worse than action on our part could have prevented. To put things another way, some spheres of morality are importantly impersonal.

What has been argued is insufficient to develop and defend a nuanced consequentialist theory, but it does suggest that there is nothing amiss with the consequentialist view that reasons for action consist in the good and bad conditions which can befall creatures actual and possible. It is not so much that there is nothing wrong with the 'receptacle view' as that this classification is a misnomer. The concern of the consequentialist is for the well-being of creatures, and not for the dispassionate aggregation of free-floating value; and the fact that this is a more comprehensive form of concern than any deontological position yet encountered does not make it any the less a form of concern, and any more disposed to treat individuals as mere channels, vehicles or receptacles.

But since the ethics of transgenic engineering is in question, and consequentialism has been found wanting in this very connection by Alan Holland, a brief review of his confessedly brief critique is necessary before further progress can be made. (Strictly, Holland was criticizing utilitarianism, the version of consequentialism which seeks to optimize the balance of happiness and unhappiness, but parallel points arise when a consequentialism concerned with value and disvalue in general is in question.)

Holland's first criticism is that pleasures cannot be uncritically counted and balanced, since some pleasures may be wrongfully or shamefully gained. As he forcefully puts matters (1995, p. 296), 'the balance of advantages cannot decide acceptability, since it presupposes it'. Here, the consequentialist can reply that she is not committed to counting or balancing uncritically. There are easily defensible consequentialist rules against robbery (Holland's example), or against shameful practices such as corruption or nepotism; for such practices are enormously destructive of social cohesion and trust, and of the conditions widely necessary for the flourishing of human life.

The second of Holland's criticisms is that 'calculation of cost and benefit cannot ignore the question of the relation between the two. A benefit which is obtained *by* paying a cost cannot be regarded in the same light as one which is not.' Not just any paying of a cost is intended (or this would be an objection to buying from shops). Holland's example is the moral condition which would characterize motorized transport if its advantages could only 'be secured in advance by the sacrifice of a certain number of lives', like the sacrifice of Iphigenia to gain a fair wind from Greece to Troy (Holland, 1995, p. 296). Since Holland acknowledges that currently we appear to think that the advantages of motorized transport outweigh the loss of life involved, a first reply is that this loss of life already includes that among construction workers before roads and roundabouts have even been built. Not that society's acquiescence in this makes all road-building acceptable. Indeed, sometimes the loss of life among construction workers is predictably so great that, at least on a consequentialist basis, construction should not have begun. Further, such are the pollution and accidents resulting from motor vehicles that transport policies in many cities and countries stand in need of drastic change; but this judgement is itself one enjoined by the consequentialist considerations.

Probably, however, Holland envisages cases where death or serious injury is deliberately inflicted on others, as when one country goes to war with another for territorial gain, or, to devise an example closer to motorized transport, where someone is assaulted in order to commandeer their vehicle. These different examples relate to different practices and their infringement, and the consequentialist is as entitled as anyone else to scrutinize distinct practices differently. Thus, she may adhere to a consequentialist version of the just war doctrine, which would plausibly outlaw most wars, including, incidentally, the retaliatory war of the Greeks against the Trojans, and thus the sacrifice of Iphigenia; and she may likewise adhere to laws which (on consequentialist grounds) forbid assault, but which allow commandeering and the use of force in exceptional cases of genuine military 'necessity'. In none of these cases is the consequentialist committed to aggregation unqualified by regard to the distinctive consequences resulting from the practice in question and the diverse circumstances in which it may be invoked; and in all these cases consequentialism plausibly generates defensible principles.

Consequentialism, then, is not a discredited theory, but a defensible one. But it remains to be seen whether it can cope with whatever can be salvaged from Verhoog's talk of the intrinsic value of wild species, and can say what needs to be said about the ethics of generating desensitized creatures. Indeed a consideration of what needs to be said about this is long overdue.

13.4 CONSEQUENTIALISM AND TRANSGENIC ANIMALS

Verhoog is doubtless correct in holding that domestication and its processes cannot be taken as above criticism. Many animals are prematurely killed, and many more mutilated, for example through sterilization; while domesticated breeds are often selected for tractability and submissiveness. While, as Holland points out (1995, p. 293), domestic animals by and large live longer than their wild counterparts, and wild animals lead on average a tougher life, there is considerable room for more compassionate forms of farming, and a strong case for the termination of certain practices.

But protests and objections targeted at the processes of domestication need not be grounded in lack of respect to the species which preceded domestication or to its *telos*. Most obviously (although not, as will be argued, exclusively) they can be grounded in avoidable suffering and avoidable harm, and thus in states of intrinsic disvalue. Certainly Verhoog is right to hold that, to identify some forms of suffering, we need to know the needs and interests of the creature concerned, and that these inherited needs and interests determine whether domesticated creatures are deprived of behaviour patterns natural to their kind. But this does not make the grounds of objection to such deprivation respect for their species or its *telos*. Grounds enough are to be found in concern for the well-being of the individual animals as they now are, intrinsically valuable as this clearly is, and varying as are the forms which well-being may take.

Verhoog's very acceptance of domestication in cases where humanity is viewed as a partner of other natural entities concedes that not all domestication, and thus not all deviation from the *telos* of a wild species, is wrong. As was mentioned above, his criterion of the acceptability of domestication consists in 'the "quality" of the new relations arising'; but this criterion is too vague to serve as such. His further suggestion, that the animals must be autonomous, in that they must not have a humanly imposed nature, seems too strict, as will shortly become apparent; for selective breeding could be advantageous to animals, in virtue, for example, of resistance to disease. One defensible interpretation of the new relations having a suitable quality would consist in the basic needs of the animals (identifiable through the natures of their kind or lineage) being satisfied; certainly nothing less would avert charges of exploitation. This criterion is radical enough to clash with many of the methods of modern farming (as is to be expected of an interpretation of the 'partnership' view). It is also a criterion which can be incorporated within consequentialism, granted a value-theory which attaches a high intrinsic value to the satisfaction of basic needs (or to what they are necessary for), high enough to outweigh the benefits which might accrue to human beings from animals' basic needs being overridden.

Imagine, further, an animal selectively bred to be immune from a painful illness which besets most members of its species, but otherwise leading a flourishing life characteristic of its kind. Since its well-being would seem to be greater than that of most of its kind, generating this animal is hardly open to criticism, or so most people, including consequentialists, would say. Yet if showing respect for wild species as such were what matters, we might have to say that generating this animal is wrong. I am assuming here that the corresponding wild species is not in danger of extinction, for if it were in such danger, there would be more factors to take into account, including the value which could have been present in the lives of species-members which will never now be lived; indeed the loss of this value could account for the inclination to hold that the species is to be respected and preserved as such. But where the animal with the immunity is simply brought into existence alongside the population lacking the immunity, there is hardly a ground for complaint, not even if immune animals are bred in large numbers and become more numerous than members of the traditional species. Here respecting the species seems one deontological principle too far, prohibiting what should not be prohibited.

Holland, however, considers that, if the nature of animals is treated as an end in itself, then there is a Kantian objection even to this benign genetic manipulation. His example concerns building into an animal's endowment 'a resistance to disease and injury' (Holland, 1995, p. 303); and I should acknowledge that this more sweeping measure could be open to objection, as it could have side-effects which made the animal, for example, more timid or more placid (how else has it become resistant to injury?), or generally possessed of reduced capacities. And if fewer fulfilments were open to it, there would plausibly be consequentialist objections to the original manipulation. But as this is not Holland's case, let us imagine the less likely scenario in which there are no such side-effects, and the only difference to the animal's nature consists in its resistance to disease and injury. To convey the Kantian objection, Holland, writing about 'changing an animal's nature for the sake of rendering it less susceptible to disease', claims that 'Essentially, it puts respect for the states of a subject above respect for the subject' (1995, p. 304). The process is, allegedly, disrespectful of the animal's nature, albeit benign to the well-being of the individual animal. Yet all the individuals so treated (plus their offspring if this is an inheritable change) will be healthier and probably more vigorous and longer-living, and *ex hypothesi* there are no side-effects. Besides, even if the animal's 'whole nature' is 'subordinated' to this process (Holland, 1995, p. 304), it must largely remain intact, or the creature's capacities would have been reduced, contrary to our supposition. So, unless changing an animal's nature is wrong in itself, there seems nothing left to object to, even if the process might be held to involve disrespect for the ends implicit in the animal's nature.

Now the ethics of the breeding of the animals with this immunity would be no different whether it resulted from conventional selective breeding or from genetic engineering, if the assumption is retained that the immunity could be genetically engineered without side-effects. (In actual fact there will usually be a risk of such engineering producing unexpected and unintended bad effects, and at least the risk has to enter into appraisals of such processes.) So it turns out that, for some kinds of such manipulation, apart from the consideration of risk, there would be no ethical objection at all, despite the fact that all genetic engineering could be represented as showing disrespect for natural species.

Certainly there is a new difficulty which enters with genetic engineering; for it is much less easy to identify the needs and interests of the animals thus produced, and thus to locate cases where basic needs are being infringed. Since genetic engineering is at least typically intended to benefit human beings, then in view of the desirability of seeking ways of avoiding exploitation of the other party affected (the animals), it would be reasonable to require that people with genetically engineered animals in their charge should patiently seek to discover whether their animals carried the behaviour patterns of their wild or domesticated ancestors, such as rooting in pigs. Where this is the case, the observed behaviour should be treated as an expression of a basic need, at least where the need had been expressed in such behaviour among the ancestors or among other comparable animals. Thus, the 'partner' view has important implications for the ethics of genetic engineering, calling, as it does, for special efforts to identify and satisfy the basic needs of transgenic animals. Yet immunity from a painful disease would still count as a benefit, whatever was discovered about behaviour; for, except in very unusual circumstances, diseases are necessarily disadvantageous. All these considerations can clearly be reconciled with consequentialism.

There are, however, other ethical constraints, I suggest, on genetic engineering. Thus, avoidable harm is done where animals are engineered whose lives are not worth living; as was argued above, a creature is harmed by being made to live out an existence which it would be better spared. Such animals should not be generated at all.

But what of creatures engineered to lack feelings, and thus not to suffer? Philosophers for whom the only evil is suffering cannot regard generating such creatures as wrong, while philosophers who appeal to *telos* are also hard-pressed, as little of a wild *telos* remains, but only individuals whose quality of life is slightly above zero. Again, if such creatures were to occupy the ecological space previously occupied by their sentient ancestors and thus supplant their natural cousins, then the defensible principle would be infringed, that it is wrong to generate creatures with a lesser range of capacities where creatures with a greater range of capacities would have lived instead; and this will be a consequentialist principle if its ground consists in the value of realizing capacities, as plausibly it does. But what is amiss if the alternative would be no animals at all occupying this same space? This is a question put by Holland (1990, 1995, pp. 302–3) with regard to animals whose lives are not lives of suffering, but whose capacities are little above the vegetable.

With regard to animals lacking sentience, it may be replied that the inability to suffer is itself, in animals, a biological disadvantage; thus, animals with this dubious endowment would be disadvantaged as compared with both their contemporaries and their ancestors. Also, generating such animals would often violate what Holland calls the 'reduction of capacity principle' (1995, p. 302), as it is often likely enough that creatures with a greater range of capacities would have existed there instead.

But is it wrong to generate creatures which lead lives more truncated than ones which *could* have been brought into existence instead, and can the generation of creatures lacking sentience be condemned on these grounds? Here Verhoog would probably agree that it is wrong to generate them, but advance the implausible ground that this is wrong because it violates the *telos* which their ancestors once had. For his part, Holland (1995, p. 304) suggests that it is wrong because the zygotes used by genetic engineers to produce the transgenic animals would have grown into an individual with an ampler range of capacities if no genetic engineering had occurred, and thus the very same individual would have been better off, and has actually been harmed. But if this were the only basis of objection, then there would be no objection to the second and subsequent generations of transgenic animals, if we may assume that at least some of the first generation products of genetic engineering are able to reproduce themselves with the genetic changes intact. For individuals of the second and subsequent generations would not have existed with ampler capacities if the original engineering had not occurred; indeed they would not have existed at all.

This returns us to the question concerning the wrongness of generating creatures with reduced capacities (and thus reduced opportunities for fulfilments) where other creatures with traditional capacities could have been generated instead, even if, market economics being what it is, the other creatures just as likely would not have been generated at all. Yet it is quite unacceptable to say that in these circumstances generating creatures of reduced capacities (for example, without sentience) is allowable. For this is like saying the following: it would be wrong to generate creatures of reduced capacities where the alternatives are to generate creatures with traditional capacities and to generate no creatures at all; but it would not be wrong first to decide not to generate creatures with traditional capacities, and then to claim that there is nothing wrong with generating creatures with reduced capacities, since generating no creatures at all is now the only alternative remaining. And what this suggests is that, if action on the part of the agent in question could have generated creatures with traditional capacities, then the likely fact that this alternative would have been rejected out of hand makes no difference to the wrongness of generating creatures with reduced capacities. In other words, the answer to the question 'Is it wrong to generate creatures which lead lives more truncated than ones which *could* have been brought into existence instead?' is 'Yes'.

The fact that the individuals with truncated capacities and the individuals with traditional capacities are not identical thus turns out to make no difference. This finding is in line with the conclusion reached earlier in the context of Parfit's work, to the effect that morality is, in some spheres, impersonal, and does not simply concern whether identifiable individuals are benefited or harmed. Agents are also responsible for the quality of life of whoever or whatever lives, whatever its identity, insofar as this turns on their own action or inaction; and this supplement to the principles of benefiting and of abstention from harm is once again a consequentialist principle. Given also a value-theory according to which an ampler range of realizations of capacities is, other things being equal, of greater value, there is a strong ethical case against bringing into existence creatures with a reduced range of capacities, as creatures lacking sentience would be in comparison with traditional animals.

Holland has, in fact, challenged me to show that reduced capacities are to be regarded as inferior (or to portend an inferior quality of life) (Holland, 1995, p. 302). Yet he himself had earlier written (1990, p. 172), 'I suggest that in general, and within the framework of a biotic community, there should be an objection to any practice which involves taking a form of life with a given level of capacity to exercise options and reducing that capacity significantly.' This view is, I suggest, to be supported because an ampler range of capacities facilitates an ampler range of realizations of capacities, and this is of greater value than a more restricted range. Another way of putting this is, to borrow the phrase which Holland (1995, p. 303) employs, that reducing capacities is 'life-denying' and letting them be is 'life-enhancing'.

He also challenges my (1994) use of 'crippled' to describe a reduced form of life (Holland, 1995, p. 303). This too is a little surprising, as I was taking up Holland's own usage (1990, p. 171): '"Crippling" a virus is, without doubt, harming it, and there is surely something disturbing in the idea of deliberately creating "crippled" life-forms, that is, creating life-forms which are permanently harmed.' Holland's new question concerns 'what is undesirable about a case where the genetic engineer has redesigned what counts as a worthwhile life and in which the condition in question is, precisely, no longer "crippling"' (Holland, 1995, p. 303). Here, granted the form of the reduced capacity principle defended above ('It is wrong to generate creatures which lead lives more truncated than ones which *could* have been brought into existence instead'), it may be replied that Holland's first thoughts were best, and (whether or not we say that the redesigned species is permanently 'harmed') that this principle supplies a sense in which the members of the redesigned species are indeed crippled; for the principle supplies the comparison (with animals of the traditional life-form which could have lived instead) which the analogy demands. This remains true even though the members of the redesigned species have new (reduced) needs and interests, and on this reduced basis may still be capable of leading a (just about) worthwhile life. Imagine a human being redesigned to assemble cars automatically and flawlessly, for just three hours each day. His or her humanity would be reduced, but he or she might still be capable of some kind of worthwhile life, centring on the rest of the day. If this monstrous transformation were possible, condemnation would be in place because of the reduction of capacities involved, in the form of the loss of choice and control over large segments of an otherwise human life.

13.5 CONCLUSION

My theoretical conclusion is that there is no need to appeal to the supposed intrinsic value of species or of their *telos*, or to their nature being an end in itself, to account for the objectionable nature of much possible genetic engineering, for this is much more clearly objectionable on consequentialist grounds, concerning the wrongness of generating lives which are not worth living, of causing suffering or other harm, or of reducing the range of capacities and realizations. Where, however, appeal to the intrinsic value of species or to *telos* is made for cases which are unobjectionable on consequentialist grounds (such as animals genetically manipulated to be immune to disease without the manipulative process having side effects), the appeal is, I suggest, unpersuasive. Respect for naturally evolved life-forms rather than for their members might even be held to involve a new kind of 'receptacle view' of the latter.

At the same time, the practical conclusion is that there are important ethical criteria applicable to genetic engineering and to the husbandry of transgenic animals which extend beyond a concern to prevent avoidable suffering and avoidable harm and to avoid producing them, and which mandate abstention from reducing capacities. Thus, it is wrong to generate creatures which lead lives more truncated than ones which the same agents *could* have brought into existence instead. Further, people in charge of genetically engineered animals should take pains to identify and satisfy the basic needs of these animals. When people become responsible both for the quality of life liveable by animals and for the identities and to some degree the natures of the animals who live, this confers strong obligations, and does not give them *carte blanche* to manipulate as they please. Even creation has its ethics.

REFERENCES

Attfield, R. (1995) Genetic Engineering: Can Unnatural Kinds be Harmed? in *Animal Genetics; Of Pigs, Oncomice and Men* (eds P. Wheale and R. McNally), Pluto Press, London, pp. 201–10.

Evans, J.W. and Hollaender, A. (eds) (1986) *Genetic Engineering of Animals. An Agricultural Perspective*, Plenum Press, New York.

Fox, M. (1990) Transgenic Animals: Ethical and Animal Welfare Concerns, in *The Bio-Revolution* (eds P. Wheale and R. McNally), Pluto Press, London, pp. 31–50.

Holland, A. (1990) The Biotic Community: A Philosophical Critique of Genetic Engineering, in *The Bio-Revolution* (eds P. Wheale and R. McNally), Pluto Press, London, pp. 166–74.

Holland, A. (1995) Artificial Lives: Philosophical Dimensions of Farm Animal Biotechnology, in *Issues in Agricultural Bioethics* (eds T.B. Mepham, G.A. Tucker and J. Wiseman), Nottingham University Press, Nottingham, pp. 293–305.

Lewis, C.I. (1946) *An Analysis of Knowledge and Valuation*, Open Court, La Salle, Illinois.

Moore, G.E. (1922) The Conception of Intrinsic Value, in *Philosophical Studies* (ed. G.E. Moore), Routledge & Kegan Paul, London.

Parfit, D. (1984) *Reasons and Persons*, Clarendon Press, Oxford.

Regan, T. (1984) *The Case for Animal Rights*, Routledge & Kegan Paul, London.

Rollin, B. (1986) The Frankenstein Thing: the Moral Impact of Genetic Engineering of Agricultural Animals on Society and Future Science, in *Genetic Engineering of Animals: An Agricultural Perspective* (eds J.W. Evans and A. Hollaender), Plenum Press, New York, pp. 285–97.

Rollin, B.E. (1989) *The Unheeded Cry: Animal Consciousness, Animal Pain and Science*, Oxford University Press, Oxford, New York.

Verhoog, H. (1992a) The Concept of Intrinsic Value and Transgenic Animals. *Journal of Agricultural and Environmental Ethics*, **2**, 147–60.

Verhoog, H. (1992b) Ethics and Genetic Engineering of Animals, in *Morality, Worldview and Law* (eds A.W. Musschenga *et al.*), Van Gorcum, Assen/Maastricht, pp. 267–78.

14

Organs for transplant: animals, moral standing, and one view of the ethics of xenotransplantation

R. G. Frey

14.1 INTRODUCTION

We commonly use animals as means to certain of our ends. We eat them, use them as experimental subjects, treat them as objects to be observed in zoos, keep them to while away the hours or to forestall loneliness, and so on. It should come as no surprise, therefore, that a case is increasingly being made in the medical research community and among the educated lay public for regarding animals as repositories of organs for transplant into humans. Indeed, this use of animals as spare parts for our bodily renewal will doubtless seem to many a perfectly justified use of them, since the prolongation of human life and the enhancement of its quality will strike many as straightforwardly worthy ends.

Quite apart from my other writings on animal issues, including my writings on the use of animals as experimental subjects generally, I have in two articles (1988a, 1996) presented an argument with respect to xenotransplantation that tries to show how a commonly used defence of the practice exposes certain humans at least to the possibility of having their organs harvested in similar fashion. I still think this argument works, and readers of my 1996 piece will see how I have developed and broadened some of the points in the 1988 piece. (Readers of my other papers on animal issues between 1988 and 1996 will be familiar with the general trend of argument of the 1996 piece.) I now believe there is a much simpler way to make the central point upon which this argument turns, a way, though, which involves a change in my account of the moral status of creatures, whether human or animal. That is, I now see more clearly that in a particular kind of context moral standing is a matter of degree and that this fact plays a central role in certain arguments about the justification of our killing and/or using animals to certain of our ends, including the end of serving as spare parts.

The case for using animals as spare parts is quickly made. At present, in the United States (and most other countries of the world), there are shortages of organs for transplant. This is perhaps ironic, since transplantation of such organs as kidneys, livers, lungs, pancreases, corneas, bone marrow and, indeed, even hearts is increasingly successful in extending the lives of recipients. Campaigns to make the general public aware of the life-saving and life-enhancing results of organ donation have been undertaken, and donation has steadily increased in the United States.

Even so, the number of recipients who could benefit from organ transplantation continues far to outstrip the number of donors. The result is that waiting lists for organs have arisen, as searches for organs and for matches between recipients and donors are undertaken. It often happens, however, that a good many of those on the waiting lists die or suffer further and dramatic deterioration in the condition of their own organs, with consequent effects upon their quality of life. These deaths or impaired lives are held to be unavoidable, at least under the present, voluntary scheme of organ collection.

Of course, the prospect of such deaths and more impaired lives inevitably puts pressure upon our schemes of organ distribution. With these terrible prospects about to become a reality for many of those on the waiting lists, obvious concerns arise about whether some scheme of organ distribution is fair. Do the rich and well-known enjoy advantage, if not in principle, at least in practice? Does queue-jumping in fact occur?

These questions about the fairness of some scheme of organ distribution, however, do not directly address schemes of organ collection, and, therefore, the question of whether the voluntary scheme presently in place in the United States, as enshrined in the Uniform Anatomical Gift Act of 1987, is the most effective way to secure organs for transplant.

There are obvious ways that we might try to increase the number of available organs, quite apart from further and more effective educational campaigns to make the public aware of the seriousness of the shortages and of the life benefits that organ donation can give to recipients. For example, we might try by argument to loosen the grip of some religious or cultural reasons that people may have for rejecting donation out of hand; we might try to quell the squeamishness that some people experience at the thought of allowing a loved one to be cut or mutilated; and we might try more assiduously to keep up to the mark those who indicate that they are prepared to be donors (say, on their driver's license) but who in the end fall prey to backsliding.

A much more radical step, and one that an increasing number of people apparently see as desirable, is to set up a market in body parts, in which people could sell, say, a kidney for money. One would not be coerced to enter such a market, so that it remains a voluntary affair whether or not one does, and it might be that selling one's heart would be proscribed, since that would be tantamount to committing suicide. (Such a charge is or need not be true of the sale of other organs.) Some people reject the notion that such a market would be truly voluntary, since they think that the poor will inevitably be encouraged, if not, at least in some sense, coerced, to sell body parts in order to assist themselves or their families. To others, commodification of body parts can appear controversial in itself, apart from whatever effect such a view has on the practices of the poor. For it may be thought that the human body has a significance that is incompatible with treating its parts as items for sale in a body shop or that such treatment is an affront to human dignity and respect for self. (Though I do not here take a stand on markets in body parts, I do think a reasonably strong case can be made in their favour.)

The result of all these factors that contribute to the problem of organ shortages is to make the case for using animals to solve the problem almost irresistible. Human lives can be saved or enhanced through harvesting organs from animals, and this alone, it may be urged, is justification enough for the practice. A number of additional factors either already add or may in time be thought to add to the attractiveness of this solution.

1. The use of animal parts in humans is not itself a new or alien development; after all, pig heart valves have been transplanted into human hearts for some time now. Nor is there a widespread philosophical worry about whether, if enough animal parts are transplanted into me, I remain a fully human being or person.
2. Since the number of organs available is the problem, we could breed in captivity and for the specific purpose of transplant the number of animals that we require to produce the requisite number of organs. Accordingly, animals in the wild would be unaffected: their number, and so the number of species left in the wild, would not be reduced through harvesting animal organs for transplant.
3. We could stipulate that the organs of household pets would not be harvested, in order to combat any fear that some people might have on this score.
4. Genetic engineering can help to produce, as it were, 'designer' animals. These animals will have been genetically engineered in

order to produce the sorts of organs that best fit the human case, organs that, from the outset, have been 'placed' in animals for our use. These 'designer' animals would be brought into existence in order to harvest their organs; once again, then, species in the wild would be unaffected.

5. It is easy to imagine that, in time, if not already, the costs associated with raising and killing animals for their organs would reduce the overall costs associated with transplants. At present, for example, in the case of bone marrow transplants, costs are considerable; breeding baboons in captivity for this purpose seems very likely in time to be cheaper. (In the case of human kidney transplants from other humans, costs are already cheaper than continual dialysis, if one takes into account the length of time over which a patient will require dialysis.)
6. A major factor that has bedevilled the history of xenotransplantation has been human rejection of transplanted animal organs. Increasingly, however, more and more powerful immunosuppressant drugs are being developed that help deal with this problem, and it seems likely that the problem of rejection will in time disappear. Given this, and given reduced costs in using animal organs, it is unlikely that the history of xenotransplantation to date will impede its future. Certainly, there is a good deal of work in xenotransplantation underway.
7. Though it remains to be determined, for example, whether diseases in baboons can be passed on to humans who, say, receive bone marrow transplants from them, this is an empirical matter that at the moment further information is needed in order to answer. If it transpires that diseases are transmitted, then such transplants could be discontinued; otherwise, they could be continued.

For all these reasons, then, harvesting animal organs for human use can seem attractive. Since we are concerned with the ethics of this practice, I have imagined that xenografts will not founder for practical or empirical reasons; were they to do so, of course, the ethics of the practice might be held to be moot. I do not here prejudge the nature of these empirical findings.

14.2 MORAL CONCERNS

A serious moral worry some people will have about harvesting animal organs is that primates such as baboons have featured prominently in many of the animal to human transplants, and primates are our closest relatives. It is difficult to assess just how insuperable a moral barrier this use of primates actually poses, given that such use can save and/

or enhance human lives. But it is easily imaginable that we shall in time be able to use pigs and other non-primates to secure all the organs we need; certainly, a good deal of genetic engineering of pigs is presently underway with organ procurement in mind. If one's objection to xenograft is that primates are being used, then that practical objection just might in time admit of a practical response, through the development of suitable organs for transplant in non-primates. Here, too, I do not prejudge how successful this development is likely to be in the short term, in order that use of primates may be discontinued.

I stress this matter of primates for the following reason. I think the fundamental question that serious people want to ask morally about xenotransplantation is whether this use of animals in order to save and/or enhance humans lives is justified, *whatever the species* of animal used and *however close* in many respects that species may be to our own. That is, I take the fundamental question to be one that does not turn upon species at all but rather upon what we think justifies us in using animals to save or enhance human lives in the first place. To be sure, there may be further moral worries if primates are used for spare parts, but the central moral worry is surely about what justifies us in harvesting the organs of any animal to save or enhance a human life.

Put this way, the central moral worry about xenograft encompasses the lowly mouse, though at the moment more in the experimental and genetic engineering phases of 'designing' animals than in the harvesting phase. For those who frown upon xenograft, it is important that the mouse be covered; for rodents are main experimental and engineered animals, and without the work done in them it is easily imaginable that the stage of actually 'designing' animals to be spare parts will be impeded to some degree or other. Such experimentation can certainly appear necessary for us to reach the stage where we can breed animals 'designed' to have body parts that can be used by the human body, quite apart from any importance that we might attach to testing immunosuppressant drugs in such animals.

All justifications of xenograft that I am familiar with appeal to the enormous benefits for human health that harvesting animal organs can potentially confer. Let us accept that there are these enormous potential benefits so far as the saving and enhancing of human lives is concerned: if xenograft fails with respect to a wide array of organs for practical or empirical reasons, such benefits will not come to pass, and we need not resort to any ethical reasons for urging the discontinuance of the practice. The central moral issue then becomes whether, in order to obtain these significant benefits for human beings, we may breed, 'design', and use animals as spare parts.

This central moral issue is quickly joined. For it is obvious that we could produce the very same benefits that are supposed to justify xenograft through harvesting human organs, and the success rate we have in transplanting organs from human to human can make this seem a path more certain of success than xenograft. Morality is thought, however, to get in the way. While it would be agreed on all sides that to breed and 'design' human beings to make good the shortfall in organs for transplant is morally impermissible, presumably, if saving and/or enhancing human life is to justify xenograft, then to breed and 'design' animals for this purpose is not morally impermissible. What, then, is the moral difference between the two cases?

Notice the sorts of responses this question does not invite. Our fears about eugenics, that, for example, some race or group will be 'bred' or 'designed' for the ends or purposes of some other race or group, are not really to the point here. Nor is our fear that 'monstrosities' might be produced through mistakes in 'designing' humans. Rather, our question asks why we may not do to humans, any humans, whatever their race or group, what we presently do or propose to do to animals. The focus is upon the moral difference(s) between the human and animal cases.

I am aware that all kinds of differences can be appealed to, in order to try to forge a moral difference between the human and animal cases, and I have in a number of places discussed many of these (Frey, 1980, 1983, 1987a,b, 1988b, 1989). Here, I want to consider again what I take to be the most radical difference one can cite in this regard, namely, the claim that humans but not animals are morally considerable, that humans but not animals possess moral standing or membership in the moral community. Put succinctly, it is morally permissible to do to animals what it is morally impermissible to do to humans because animals do not count morally. They lack moral standing or moral status. (I use these notions interchangeably hereafter.)

So much has been written about the moral status of animals in recent years that I will not bother to rehearse any of the standard positions on it but rather pass at once to the particular points that I want to make with regard to it.

What might be called the usual or standard way of treating moral status has been to link it to the possession of a characteristic that one takes to be a moral-bearing characteristic. If something possesses this characteristic, then it has moral status. This, of course, leaves us with the interesting possibility of something possessing the characteristic but then losing it, so that, if having this characteristic is regarded as a necessary condition for moral status, then the thing in question loses moral status and, presumably, the protections that usually accompany such status. Thus, to give but one example, if the capacity to direct one's life and to make choices within it is lost by a human being, and if that capacity is treated as a necessary condition for moral status, then that human being comes to have no moral status. This view worries a good many people: aware that such a human being would then presumably lose the protections that accompany moral standing, and aware that there are a number of humans who apparently have lost (indeed, have never had) the capacity in question, it is perhaps understandable that a tendency has arisen to turn possession of the characteristic that has been selected as the vehicle of moral status into a sufficient condition for such status. This manoeuvre squeezes all those beings or things which have the characteristic into the protected class but takes no stand on whether those beings or things that lack or have lost the characteristic can be incorporated into the protected class on some other ground.

Bentham's famous link of the capacity to suffer to moral status is of this sort: it is a sufficient condition for the possession of moral standing. It does not follow that beings or things which lack this capacity lack moral status, since they may possess it in virtue of some other characteristic that is regarded as a moral-bearing one; it follows only that having the capacity suffices to confer moral standing on a being or thing. Because so much of the discussion of animal issues following Bentham has made use of his focus on pain and suffering, his view of the moral status of animals has come to be very influential in recent 'animal rights' debates. So far as it goes, I think it correct. But it is by no means the end of the matter. Nor, in other contexts, does it block our using animals to certain of our ends. Just which contexts these are figure below.

What is it about animals that does not count morally? The two obvious candidates that might fill this role are their pains and sufferings and their lives. As for the former, I think something along the Benthamite line is correct. Pain is pain, an evil for any being that can experience it, and certainly the 'higher' animals can experience it. (I am not concerned here with those who deny that the 'higher' animals can feel pain, very much now a minority view, or with those who claim that animals can experience pain but not in a morally significant sense, an even more uncommonly held view. Different arguments address these positions.) Once this is conceded, as I think it must be, the development of the Benthamite line follows a predictable course. For it is not only easy to see how a dog and a human each has moral status, since each possesses the capacity to suffer, but it is also easy to see why, for example, the fact that the human might be able to suffer in more and different ways from the dog (through, say, anticipatory dread) does not affect the issue of moral status. A being who has the capacity to suffer at all obtains moral status on the Benthamite line, obtains it, so to speak, if it has painful experiences. Thus, both dog and human have moral standing, since both have the relevant capacity and experiences appropriate to it.

This leads to a further point. There is nothing inherent in the Benthamite position that says that the human has more moral status than the dog, has a moral standing because he or she is human greater than that of the dog. On Bentham's position, moral status is not a matter of degree but of kind. Beings or things that lack the capacity to suffer are not *per se* morally considerable. (Obviously, various sorts of environmentalists may not accede entirely to this view.) Moreover, there is nothing in the Benthamite position that would have it that, while the pains of dogs and humans count morally, the pains of humans count more. Indeed, because pain is an evil for any being who can experience it, there is usually associated with the Benthamite line of argument the view that the pains of all such beings count the same. This view has a different ground. The horrible pain felt by a cat when it is set alight does not count for less than the horrible pain felt by the child when it is set alight; *who* has the painful experiences does not matter and does not, as a result, confer greater moral weight upon the pains of humans over dogs. Rather, the mere presence of pain suffices, because it is a moral-bearing characteristic for us, to make what happens to a being of interest to us morally, irrespective of who feels the pain. So far as pain is concerned, then, there appears to be no difference between the animal and human cases. If pain is a moral-bearing characteristic, then the pains of any being suffice to confer moral standing upon that being, and if who has the pain is not relevant to the determination of whether pain is an evil, then we have no obvious reason to advantage the human over the animal.

With pain and suffering, this conclusion seems to me to be the truth of the matter. Feel pain, and one has moral standing; feel pain, and one has full-fledged moral standing, as much as any other being that feels pain. There is no matter of degree here: if pain is an evil, and if it is an evil irrespective of who feels it, then the pains of dogs and men are on a par. Thus, in contexts in which pain and suffering are very much to the fore, I see no difference between the moral standing of animals and humans. The pains of both count and count equally. They count because pain is a moral-bearing characteristic for us; they count equally because who feels the pain is irrelevant. Thus, there is nothing speciesist about the position, nothing that inherently discriminates in favour of humans. Of course, strength and intensity of pain enter the picture, but these do not differentiate between dogs and humans.

14.3 THE VALUE OF LIFE

When we turn to the value of lives, however, matters do not appear in a similar light. Here, the vast majority of us think that human life is more valuable than animal life, and I have tried in a number of other places to show why we are right to think this (1993, 1995). One implication of my way of showing this is that, in contexts in which the value of lives is to the fore, and this includes all contexts of killing, moral status is a matter of degree. (Since my views on killing in general and on killing animals in particular are available in a number of other places, I shall pass at once to my central concern here.)

I see no reason to give an account of the value of animal life any different from that which I take to govern the value of human life, namely, that the value of a life is bound up with its quality. Such a view is widely familiar today, and not just in contexts to do with medical ethics. In the way I have developed the view, in, for example, my article 'Medicine, Animal Experimentation, and The Moral Problem of Unfortunate Humans' (1996), the value of a life is a function of its quality, its quality of its richness, and its richness of its capacity or scope for enrichment.

Animals are living creatures with experiential lives and so are things with a welfare and a quality of life. Their welfare and quality of life can be increased or diminished by the quality of their experiences, including those which they have as the result of our treatment of them. Their lives can go well or badly. It is not just humans of whom these things are true. Richness of content determines a life's quality, whether human or animal, and a creature's capacities or scope for enrichment determine richness of content. It is clear, therefore, that a quality of life view requires that we have access to the subjective experiences or inner lives of animals, that we be able, in terms appropriate to the species of animal in question, to determine to a reasonable degree how well or how badly their lives are going. That we do this all the time, at least at some level, is clear; that is why we play fetch with our dogs. But if quality of life judgements can sometimes be difficult in the human case, even though they often seem unproblematic in broad measure, it stands to reason that they can be difficult in the animal case.

I accept that animals – certainly, the 'higher' animals – have subjective experiences, that those experiences determine their quality of life, and that the quality of their lives determines the value of those lives. This is exactly what I accept in the human case. With animals, we use behaviour and behavioural studies to give us access to their inner lives, and what is rough and ready in all this we can come to use with more confidence as empirical studies of animals yield more information about them. Perhaps I can never know exactly what it is like to be a baboon, but I can come to know more and more in this regard, as we learn more about them and their responses to their environment.

Now most of us do not think that animal life is as valuable as human life, and it is plain that a quality of life view of the value of a life can explain why. The richness of normal adult human life vastly exceeds that of animals: our capacities for enrichment, in all their variety, extent, and depth vastly exceed anything that we associate with dogs or even chimps. That dogs have a more acute sense of hearing than we do does not make up for this difference in the variety, extent, and depth of capacities; for that to happen we should have to think that the dog's more acute hearing confers on its life a quality that approximates the quality that all of our capacities for enrichment, along multi-dimensions that appear unavailable to the dog, confer on our lives.

14.4 THE THESES OF LIFE

Now I do not need to go further here into a discussion of richness of content to make the point that, quite apart from the question of whether pain and suffering are moral-bearing characteristics for us, animals are members of the moral community because they are experiential creatures with a welfare or well-being that can be affected by what we do to them. But they do not have the same moral standing as normal adult humans, since the value of these human lives far exceeds anything that we associate with the lives of animals. The truth is that not all creatures who have moral standing have the same moral standing. Alas, however, this truth applies to the differences between human lives as well as to the differences between normal adult human life and animal life.

Two theses I draw out of this discussion of how lives have value flow readily out of quality of life views of the value of a life. I concede, of course, that with quality of life views of the value of a life the devil is in the details, but here I want only to show how two theses that emerge from my discussion about humans affect the case for xenograft. These two theses I call the *greater value thesis* and the *equal value thesis.*

According to the greater value thesis, normal adult human life is more valuable than animal life and for non-speciesist reasons. I have argued for the non-speciesist view that, since value is a function of richness and richness of the capacity or scope for enrichment, then normal adult human life is more valuable than animal life because of the extent, variety, and depth of our capacities for enrichment. It is all these different capacities for enrichment that we refer to when we say of a person after death that they led a 'rich, full life,' and what we refer to by this expression in the normal adult human case extends far beyond anything we refer to were we to speak of such a life in the case of a dog. In turn, this discussion enables me to explain why it is worse to kill a man than a dog: it amounts to the destruction of something of greater value. Thus, taking a life and saving a life are different sides of the same coin: if we can save a life, then all other things being equal we save the life of greater value; if we take a life, then we take the life of lesser value. It will be obvious, however, that this outcome accounts for the greater value of normal adult human life over animal life but not in any way that represents a barrier to the lives of some animals having greater value than the lives of some humans. We are all familiar with tragic human lives, lives whose quality has fallen to such an extent that even those living those lives no longer wish to do so; hence, the increasing attraction of physician-assisted suicide. This leads to the equal value thesis.

According to the equal value thesis, all human lives have the same value. I think that this thesis is plainly false and that the truth of the greater value thesis shows why. There are some human lives of such an appalling quality that no one would wish to live them, and where, in the past, religion maintained that all human lives were equal in value in the eyes of God, such a line rings very much more hollow today, as those who live such lives seek release from them. It seems bizarre in the extreme to insist that, though all of us know there are human lives that no one would pick to live, those lives are as valuable as normal adult human life. Quality of life views of the value of a life make such a view indefensible today. Indefensible, that is, unless one has recourse to metaphysical abstraction, such as some distinction between the value of lives and their inherent worth.[1] In fact, the burden of all such distinctions is simply to replicate the former religious claim that all human lives, of whatever quality, however low, are equal in value to all other human lives.

In my view, then, the greater value thesis is true, and for non-speciesist reasons, but the equal value thesis is false. The conjunction of the two leads straightforwardly to the possibility of using humans as we use animals. If we need to perform experiments on retinas in order to enhance human health, and if we still must use living models in order to do so,[2] then we morally must use beings of a lower quality of life to beings of a higher quality of life. Normal adult humans will have a higher quality of life than rabbits, but anencephalic infants, the brain dead or those in a permanently vegetative state, etc., will not. If a

[1] Something along the lines of this distinction is made by a good many writers on animal issues. The aim is to distinguish between the quality of a life as determined by its content or experiences and the worth of a life as determined by something else, which in turn allows us to speak of 'respecting' a life or 'according' it proper dignity, whatever its quality or content. All this kind of talk presupposes that we can distinguish value from inherent worth, which I doubt. One of the best discussions of the matter is to be found in Regan, 1995, pp. 13–26. I do not think Regan successfully makes out the distinction (see, for example, Frey, 1987a.)

[2] I take no position here on whether we shall in the immediate future be able to achieve the same information we currently receive from animal testing in non-animal models. Replacement of animals in experiments is a goal I share, provided that it can suffice to give us the information we need for the use of whatever in the human case.

living model has to be used, what then is the case for still using the rabbit? So far as I can see, what we need is something that always ensures, in each and every case without exception, that a human life of any quality whatever, however low, is more valuable than an animal life of any quality whatever, however high. I have dealt in several places with attempts to provide humans with this magical ingredient, from appeals to religion and culture to outright appeals to partiality and affinity; all I have found argumentatively unconvincing (Frey, 1980, 1983, 1987a,b, 1988b, 1989, 1996).

It is easy then to see the argument structure of my position, say, on using animals in medicine: human benefit drives the search for medical improvements to save and/or enhance human lives; those benefits may be obtained through using animals or using humans; while normal adult human life is of higher quality than animal lives, not all human lives are of the quality of normal adult human life; some human lives are of a quality so low as to be exceeded by the quality of life of perfectly healthy animals; therefore, if living creatures are to be used at all, that creature with the lower quality of life is to be used in preference to the creature with the higher quality of life. This argument exposes certain unfortunate humans - admittedly, the weakest among us - to a possible fate of the sort that we presently impose on some animals. Whether we could ever bring ourselves to use such humans as we presently use animals is obviously a doubtful affair; but how then do we defend our continued use of animals of a high quality of life for our experimental purposes? This question about the continued use of animals is the crucial one and, as we shall see, fits perfectly into the debate about the morality of xenotransplantation.

The point I want to stress is that, in contexts in which the value of a life is to the fore, moral status is a matter of degree. Some lives have more moral status than others. Normal adult human life, with a very high quality, has moral status in the highest degree. As human lives begin to fall in quality, however, their value and moral status begin to fall; when these lives fall to a disastrously low quality as, for example, in the case of the brain dead or anencephalic infants, their moral status has reached a very low ebb indeed. In fact, on a quality of life view of the value of a life, some lives can fall to a quality so low as to bring into question their continued value at all. No one familiar with discussions of killing in medical ethics today will be surprised by this result; indeed, in discussions of the morality of euthanasia, physician-assisted suicide, treatment in futile cases, withdrawing/withholding nutrition and hydration, and so on, such a view of human lives is part of what fuels the discussion and is very much of contemporary concern.

I quite appreciate, of course, that much fuller discussions of the greater value and equal value theses are required, in order for me to flesh out my views about the comparative value of human and animal life. But the comparative value question is not what is at issue here; here, the concern is that, in different sorts of contexts, moral status can appear to be either a matter of kind or a matter of degree. And this can pose a difficulty – one that I think utilitarians in particular have often overlooked.

By emphasizing pain and suffering and conferring moral standing through these, utilitarians invite opponents by way of response to focus upon, say, some treatment of animals that does not involve the infliction of pain and suffering or upon a magical ingredient that could dispel any pain or suffering that animals might feel. If the objection to what is being done is the pain and suffering inflicted, what is being done will no longer be objectionable on that score. Then, if what is done is to be objected to, some other ground must be found for the objection. *This* kind of focus, however, does not really get at what seems a deeper problem here. For what one *wants* to make the object of moral concern here is not so much whether what one is doing to animals involves the infliction of pain and suffering, however powerful the case may be for the view that these are moral-bearing characteristics for us, but rather how what we are doing to animals involves using up valuable lives. Even if their lives are not of equal value to normal adult human life, their lives have some value; using up their lives involves the destruction of these things of value, and this remains true even if we had the power completely to eliminate any infliction of pain or suffering. Put differently, using up valuable lives requires justification, and justification is not provided merely through observing that what we did to the animals in question did not involve pain and suffering.

Suppose, then, the argument from benefit now reasserts itself: what justifies using up animal lives in the course of what we do to them, including using them as spare parts, is human benefit. The argument from benefit, however, does not tell us which lives to use up; it needs to be supplemented. The point of my earlier discussion of the greater value and equal value theses should now be apparent: using a quality of life account of the value of life, whereby we find the former thesis to be true but the latter one to be false, it turns out that lives of higher quality have greater value than lives of lower quality and that taking a life of higher quality in preference to a life of lower quality is worse. *This* line of argument thus tells us *which* lives to use up, assuming we have to use up some lives in the first place, in order to realize the human benefits in question. We should use lives of lower rather than higher quality. Typically, lives of lower quality will be animal lives, but we simply cannot guarantee that we shall find this always to be true; in some cases, it seems quite clear that healthy animals will have a higher quality of life than some unfortunate humans. Then, the argument demands that these human lives be used up to produce the benefits in question, if, that is, any lives have to be used up at all, in order to produce the benefits.

14.5 XENOTRANSPLANTATION

Xenotransplantation would appear to fit this argument line exactly. That is, the argument will certainly justify, *if* we employ the argument from benefit, using some animals as spare parts; but it will also justify using some humans, e.g. anencephalic infants, as spare parts. There is nothing strange or peculiar in it so justifying this. For if we can save a life, we save the life of greater value to the life of lesser value; thus, it would be very odd indeed to save a person who was going to die anyway *in preference* to a person who was not going so to die or to save the life of a person in the final agonizing moments of senile dementia *in preference to* the life of a healthy person. The taking of life is simply the reverse of this: we take the life of lesser value in preference to the life of greater value. Most of the time, this will involve us in taking animal lives over human lives; but there will be times when this is not the case, times when there are humans whose lives fall below the quality of lives of the animals involved. Thus, xenotransplantation, to be justified at all, must involve us in using lives of lower rather than higher quality, and I know of nothing whatever that ensures always, in each and every case, that animal lives turn out to be of lower quality than human lives.

While nothing demands that one rely upon the argument from benefit in order to justify xenotransplantation, I have never found any attempt to justify it that did not. Thus, if one appeals to this argument in order to justify xenotransplantation, and if one uses quality of life views of the value of a life, views which are very widespread today, then there are going to be cases in which using certain humans in preference to animals will be indicated. Almost certainly, we will not use these humans. Thus, it will be interesting in the extreme to examine what is cited in order to ensure that all human beings, including those in permanently vegetative states, have a higher quality of life than any animal. Without this, the argument from benefit will only justify xenotransplantation if one is prepared to use humans of a very low quality of life in preference to animals of a higher quality of life; since it is most unlikely that we would ever agree to use these humans, I conclude that the case for anti-vivisectionism is far stronger than is usually believed by those who, like myself, favour continued use of, say, non-primates in medical research. Where killing is concerned, then, the case for using animals to save or enhance human lives is, I believe, harder to make out than is usually thought, by those who appeal to human benefit, who rely upon quality of life views of the value of a life, and who seek to deny animals moral status.

If I am right so far, then I think we can now see the step that defenders of xenotransplantation who rely upon the argument from benefit are going to have to take, a step, it is clear, that places them at odds with what both they and others may well want to maintain in other important cases in medical ethics. For it seems almost required, in order to block the case for using certain humans, that defenders of xenograft give up quality of life views of the value of a life. If the value of human life can be traced to some other source, then perhaps one can argue that a human life, no matter what its condition, no matter how poor its quality, no matter how strongly the human living that life desires to be relieved of it, is always more valuable than an animal life. Clearly, those who think some cases of physician-assisted suicide right, based either wholly or partially on the deterioration of the quality of the life being lived by the human in question, use quality of life views of the value of a life in their arguments. If defenders of xenotransplantation do this, however, using the organs of certain humans must be envisaged. Therefore, these defenders require some other kind of account of the value of human life, a kind of account that guarantees that the brain dead, anencephalic infants, and those in a permanently vegetative state all have more valuable lives than the life of any animal, however high its quality of life may be. In the past, religion gave us this comfortable assurance; increasingly, however, it seems that fewer and fewer people find in religion either a source of comforting assurance or a source of the value of life. Indeed, even many religious people appear increasingly to wonder whether those living lives so dreadful that they seek release from them should be forced to live out what their religion otherwise maintains are lives every bit as valuable as normal adult human lives.

In addition to religion, one might appeal to the tradition of which we are a part, in which the use of humans as spare parts, indeed, the use of humans in any of the ways we use animals is forbidden. But this kind of appeal is often nothing more than a disguised religious one, since those who appeal to tradition often see that as a euphemism for the moral, social, and political views that they think their religion underwrites. In Anglo-American medical ethics contexts, this often takes the form of appeal to a Judaic–Christian ethic. So unless one can go on to demarcate one's tradition in a way that separates it from the underlying religion that underwrites the particular moral, social, and political views in question, this appeal is not different from the religious one. And if one tries to go down the path of appealing to cultural traditions of the kind actually practised in a particular locale, without their being seen through the prism of some religion or other, then I really cannot see any reason *per se* to take those practices as definitive of anything moral or good. The cultural traditions into which I was born, and the practices in the locale in which I lived, were those of the American South and were heavily racist.

What the defender of xenotransplantation requires, then, is an equivalent of what used to be established by appeal to religion behind the Judaic–Christian ethic. As I indicated earlier, one might try in this regard to distinguish the value of a human life from, say, its inherent worth. It might be maintained that, while human lives of different quality may have different value, absolutely all human lives, whatever their quality and so their value, have the same inherent worth. Notice that even if we were to accept some such distinction, it would not, in and of itself, give the defender of xenotransplantation what is desired. For if the lives of animals had inherent worth, then we might be forced to choose between the levels of inherent worth of human and animal lives, in order to decide which creatures' organs to use in a particular case.

So, if defenders of xenotransplantation are going to use talk of the inherent worth of human lives to block arguments that might expose some unfortunate humans to being used in medicine in the way we use animals, then they seem likely to have to maintain either that animal lives have no inherent worth or that the inherent worth of human lives is always greater than the inherent worth of animal lives (irrespective of quality in both cases). But why should we think either of these things? Why should we think that some living things have inherent worth and other living things do not? Indeed, why should we think that *only* living things have inherent worth? Why could not a volcano have such worth? And the claim that human lives have more inherent worth than animal lives seems self-serving, in order both to answer the problem of using humans instead of animals in certain cases of transplant and to make sure that our lives are, as it were, always and inevitably beyond compare with the animal case. These assumptions look too convenient for the human case, even if we were to concede, what seems to me to be highly problematic, that we actually knew what inherent worth was and could state the criteria in terms of which we identified its presence. All criteria I have come across in this regard, such as our exhibiting wonder and awe, fail miserably to suit; for these are things many of us exhibit not only in the case of volcanoes and other parts of nature but also in the case of cars, computers and other human artefacts.

There is, moreover, a further problem with using some distinction between the value and inherent worth of a life, in order to block the kind of argument I have been concerned with above. Suppose for the sake of argument that: (i) we allow that we can make sense of what is meant by the inherent worth of a life; (ii) we know what the criteria are by which to recognize the inherent worth of lives; and (iii) we do not contest the claim that human lives of absolutely abysmal quality are of equal inherent worth with the lives of normal adult humans: nothing whatever about the distinction shows that moral status is to be allied with the inherent worth as opposed to the value of a human life. This, I think, is a crucial point; for if the defender of xenotransplantation hopes to use animals but not humans as sources of organs in part through arguing that all animals lack but all humans have moral status, then it is by no means clear that this is shown by arguing that all humans have inherent moral worth. Nothing as yet ties moral status to inherent worth as opposed to value of the life lived. And it is far from clear that anything *will* tie moral status to inherent worth, since, if the lives of animals have inherent worth (even if less inherent worth than the lives of humans) but the defender of xenotransplantation argues that they lack moral status, then clearly moral status is not turning upon the possession of inherent worth.

With this the case, however, the defender of xenotransplantation seems required to argue that the lives of animals lack inherent worth. But now the criteria for identifying the presence of inherent worth in a being or thing become crucial: what are the criteria for having inherent worth, such that the brain dead, those in a permanently vegetative state, anencephalic infants, those in the very final stages of senile dementia, etc. have it but that all healthy animals with a high quality of life lack it? What could confer inherent worth on an anencephalic infant but not a healthy dog? The mere human appearance? The mere bodily shell? Very far down *this* road, I suspect, and we shall encounter something like earlier religious stories that invest the human body with special religious significance.

Finally, on quality of life views of the value of a life, the content of a life determines the value of it, and most people today concede not only that the content of human lives can vary but also that it can vary massively in a negative and tragic direction. Such lives turn out to be less valuable than normal adult human life, and lives of less value have, in the way indicated, less moral standing. To be sure, one can move to block this kind of account by insisting, through, say, some distinction between the value and inherent worth of a life, that the moral standing a creature has is tied to that creature's inherent worth, not the value of its life. But now another puzzle arises: the value of a life is determined by its content, by the experiences the life contains; unless inherent worth is to reduce to the same thing, it must be determined by something other than the content of the life and be such that a human life, even if devoid of content altogether (or all but the barest content), nevertheless has such worth. What could this worth consist in? In the absence of some story of a religious kind, what could make it the case that a human life of disastrously low quality, a quality so low that we would not wish that life on anyone and would move heaven and earth to avoid such a life for ourselves and our loved ones, had equal inherent worth with the lives of normal adult humans? If we concede that the value of the lives are different, but hold that the inherent worth of them is the same, what could this inherent worth consist in? In the earlier stories, these different lives were equal in that they were equal in the eyes of God; in what way are they equal apart from this sense? The puzzle is this: if moral standing turns upon the value of a being's life, and the value of that life turns upon its quality, and if its quality turns upon the richness of content of the life, then moral standing is going to turn out to be a matter of degree. What degree will be determined by the content of the life actually being lived, and this content will reflect the differences in capacities and scope for enrichment present in different lives in question. This view makes the moral standing and value of a life turn upon the actual content of the life as lived, not upon some metaphysical abstraction that is devoid of any connection with the content of the life lived. It seems odd in the extreme to proffer an account of the value of a life that has nothing to do with the actualities of that life as lived, and that is exactly what some claim of inherent worth would appear to do.

14.6 CONCLUSION

In sum, in contexts in which pain and suffering are to the fore, moral standing is a matter of kind and not of degree; in contexts in which killing is to the fore, moral standing is a matter of degree and not of kind. By focusing upon pain and suffering exclusively, as so many contemporary discussions of animal issues do, including those by utilitarians, we can overlook claims about the value of the lives that the practice of xenotransplantation uses up. Once we focus upon this question of the value of the lives involved, however, we see that xenotransplantation can be justified but only at a human cost, a cost that most people will be unprepared to pay, and to proceed further without moral justification is unthinkable for morally serious people.

REFERENCES

Frey, R. (1980) *Interests and Rights: The Case Against Animals*, Clarendon Press, Oxford.

Frey, R. (1983) *Rights, Killing, and Suffering*, Blackwell, Oxford.

Frey, R. (1987a) Autonomy and the Value of Animal Life. *The Monist*, **70**, 50–66.

Frey, R. (1987b) The Significance of Agency and Marginal Cases. *Philosophica*, **39**, 39–46.

Frey, R. (1988a) Animal Parts, Human Wholes: On the Use of Animals as a Source of Organs for Human Transplant, in *Biomedical Ethics Reviews 1987* (eds J.M. Humber and R.F. Almeder), Humana Press, Clifton, NJ, pp. 89–101.

Frey, R. (1988b) Moral Standing, the Value of Lives, and Speciesism, in *Between the Species*, **4**, pp. 191–201

Frey, R. (1989) Vivisection, Morals and Medicine, in *Animal Rights and Human Obligations* (eds T. Regan and P. Singer), Prentice-Hall, Englewood Cliffs, NJ, pp. 223–36.

Frey, R. (1993) The Ethics of the Search for Benefits: Animal Experimentation in Medicine, in *Principles of Health Care Ethics* (ed. R. Gillon), John Wiley, New York.

Frey, R. (1995) The Ethics of Using Animals for Human Benefit, in *Issues in Agricultural Bioethics* (eds T.B. Mepham, G.A. Tucker and J. Wiseman), University of Nottingham Press, Nottingham, pp. 335–44.

Frey, R. (1996) Medicine, Animal Experimentation, and The Moral Problem of Unfortunate Humans. *Social Philosophy and Policy*, **13**, 181–210.

Regan, T. (1995) The Case for Animal Rights', in *In Defense of Animals*, Blackwell, Oxford, pp. 13–26.

15

Making up animals: the view from science fiction

Stephen R.L. Clark

15.1 THE REAL AND THE FANTASTIC

There are stories, and invented creatures, which bring us pleasure largely because we know they are not real.

> 'When Sir Arthur Conan Doyle claimed to have photographed a fairy, I did not, in fact, believe it: but the mere making of the claim, the approach of the fairy to within even that hailing distance of actuality – revealed to me at once that if the claim had succeeded it would have chilled rather than satisfied the desire which fairy literature had hitherto aroused.'
>
> Lewis (1943a, p. 9)[1]

On the other hand, there are stories, and newly discovered creatures, which bring us pleasure largely because we know that they are true. Finding that they are not true after all is a colossal disappointment. It seems obvious that these are contrary emotions.

Some of the pleasure that we take in the Unreal is merely voyeuristic or sensational. People fantasize about horrendous monsters, even about rape and murder, but do not therefore wish these things to happen. Similarly, some of us (apparently) enjoy rollercoasters: we would not enjoy a ride we thought was *really* dangerous, really out of control. We enjoy (perhaps) the fantasy precisely because it only affects our feelings, not our real beliefs: the real thing has unforeseen, and undesired, corollaries. Conversely, the pleasure that we take in Truth is often merely utilitarian (in the vulgar sense). We enjoy it because we

[1] See also Clark (1990) pp. 247–65.

plan to make some use of it: being told you've won the Lottery gives lasting pleasure only if (if at all) it's true.

But Fantasy and Fact can both give pleasure of a less sensational or vulgarly utilitarian sort. We sometimes delight in the real presence of a living creature, or a mountain range, or an ancient monument irrespective of any use we plan to make of them: we're simply glad they're there, and consequently glad to know they are. We sometimes delight in the fantastic, not as any sort of wish-fulfilment dream, but because it opens up some wider, wilder possibilities than our present, actual lives afford. Realists may instruct us that such fantasies are bound, in fact, to be less interesting, really, than the real, and that we are excited by them only because we're very ignorant. Surrealists may tell us that the world of fact is bound to be more boring than the products of a rich imagination,and that we are excited by mere truths only because we're very uninspired. The rest of us may plausibly insist that we like both, and perhaps for similar reasons. Fantasy and Fact alike may give us a sense of Otherness: we like them because they are not much like us, because they are not under our control. We might not like real out-of-control descents from mountains; we certainly like real out-of-control mountains.

Our recognition of the genuinely Other is a source of awed delight, and also of morals.[2] There are purely self-regarding virtues, and right choices about our wholly private lives, but the greater part of morals rests on there being, and our knowing of, the Other. Levinas has been criticized, not quite unjustly, for equating Otherness with the Eternal Female. Men easily forget that women have a life, even or especially when they address Woman respectfully as Other than Man. But it may actually be true that many of us do first become aware of something genuinely Other that makes demands on us, and excites our admiration, in our experience of the other sex (whichever). It is not wholly implausible – though politically incorrect – to wonder if homosexuality is a refusal to encounter otherness on its own terms. But (sex apart) we also encounter it in our experience of vulnerable and defenceless creatures, who control us by being, exactly, vulnerable and defenceless. We may also find it in a sudden shock: the mountainside is *real*, and can kill us; or the sea-bird flying up aloft is real, and utterly indifferent to us; or the dragon is unreal, yet powerful.

The right kind of fantasies are spoilt by being turned into accessible facts. They would cease to have that flavour if they turned out to be tameable. Paul Veyne remarks that he finds himself afraid of ghosts *because* he doesn't believe in them (Veyne, 1988, p. 87). If they were merely ectoplasmic fauna we could call in the pest controllers. Because

[2] See Murdoch (1970) p. 85; Levinas (1989) pp. 37–58.

we know they're 'only' workings of our minds (which are perhaps the very things that we have least chance of ever understanding or controlling), we cannot deal with them. Fear, in this case, is unpleasant: but it may create a similarly welcome sense of Otherness, with respect to our own workings. The right kinds of facts are spoilt by being rewritten as a sentimental story, freed from danger, strangeness, Otherness. Both sorts of error are displayed in Disneyworld (and Disney's films), which simultaneously vulgarize our fables, and sentimentalize our histories. Disneyworld is what our world will be if the world-transformers have their way (Berleant, 1994). No doubt in practice, Otherness will sneak back in: the rides will break down, the actors miss their cues and the facades themselves will begin to gain a character. Since even our selves are strange to us, we can be fairly sure that everything else is too. The problem is that we forget it.

15.2 THE MORAL STATUS OF ARTEFACTS

Everyone agrees that artefacts are different from living creatures. If a champagne bottle were tricked out with wires, an electric light bulb and a fancy shade it would be recognizably a table lamp. If a living cat were heavily disguised to look just like a rabbit, it would still be a cat, even if it were conditioned into behaving like a rabbit. Living creatures are not identified by the functions they perform for us, but by the kind of thing they are. Peoples from all over the world, and speaking many different languages, make the same distinction (Pinker, 1994, pp. 424–7). An artefact is identified, and judged, by the function it performs, and how well it does it. Tool users may have strong opinions about the right way to treat their tools, but rarely (except in fantasy) reckon that their tools have any interests other than their function. Living creatures are not identified only by their usefulness to us. Further evidence lies in the standard responses to Aristotle's *ergon* argument (which I defended in my first published work, 1975, section 2.1). A knife has an *ergon*, Aristotle says, and so does a cobbler: how can a human being not have one? By my own account this means that there is something human beings do, by virtue of their being human beings, strictly so called. The life that human beings live is one of deliberate action: what we do is *choose*, and the good life, for us, is one where we make good choices. There is, in other words, no fallacy of the kind that is often still alleged, that Aristotle (and Plato before him) illicitly assume that because a knife has been designed to do a job, and a cobbler has been trained to do one, therefore all human beings have been allocated such a special task. The truth is, so the objectors say, that human beings have to choose what they should do (which is, of course, what Aristotle says). However confused as a reply to Aristotle, the objection may be important. Once again, there is a difference between artefacts, and things that have a life. A knife has a function, but not the metal from which it is made. A war-horse has a function: it does not follow that all horses do. We know what a good war-horse is, but maybe not what a good horse is (because, perhaps, we have no insight into what life it is that horses naturally lead). But in all such cases we know what a thing is by what it characteristically does.

Everyone makes the distinction, but of course it is often subverted: what is a good dog but a dog that does exactly what we want (Clark, 1985, pp. 41–51)? To treat a living creature as a tool is to reckon that it is good or bad, well-treated or ill-treated, according to its usefulness. Artefacts offer no moral barrier to any appropriate use: 'Can the pot speak to the potter, and say "Why did you make me like this?" Surely, the potter can do what he likes with the clay' (*Romans*, 9.21)? 'A combined physical and linguistic subterfuge first cripples an experimental animal to the point where it is barely alive and then calls this wretched being "a preparation" – which implies that it has been turned from a being into a "thing" no longer deserving compassion. That this reflects a complete parting of the ways between the "scientist" and reality hardly needs to be stressed' (Devereux, 1967, p. 236). People who wish to eat animals may think of them as so many pounds of steak, or as devices for transforming grass to milk and meat. They had better be kept 'healthy', as a knife must be kept sharp, to do their jobs, but it makes no sense (if they are artefacts) to ask if the job is one they want to do. Thus, Callicott (1989) suggests that a concern for the welfare of farm animals must be misplaced: such creatures, he says or said, are artificial forms, with no identity outside their function.[3] The claim may be ridiculous, at present: farm animals are not much different in kind from their wild cousins, and many are plainly capable of lives outside the farm. But Callicott's argument, such as it is, does draw attention to the issue that I am addressing: what moral status has an artefact, or something made into or considered as an artefact? Consider a world invented by Terry Pratchett (long before Discworld), whose barrenness has been alleviated by careful imitations. 'A small mechanical fly blundered into the spider web. There was a minute blue flash.... The spider slowly dismantled the protesting fly with two spanner like legs'

[3] 'There is something profoundly incoherent (and insensitive as well) in the complaint of some animal liberationists that the 'natural behavior' of chickens and bobby calves is cruelly frustrated on factory farms. It would make almost as much sense to speak of the natural behavior of tables and chairs'. Callicott does later admit that factory farming is 'clearly' in violation of 'a kind of evolved and unspoken social contract between man and beast' (p. 56): the two claims do not seem to me to be compatible.

(Pratchett, 1978, p. 138). This might be horrific as an episode of natural history: as an artefact (for 'fly' and 'spider' are just two halves of a toy) it is only clever.

O'Donovan (1984) has raised the question with respect to artefacts constructed from a human stock: say someone bred to be an easy-going slave, or even a source of transplants. What complaint can any such genetically engineered hominid have against her makers? She has not been made to have a *worse* life than she might have done: the only life that *she* could have is one within the conditions that the engineers laid down. Anyone without that genetic makeup would not, obviously, be her. Obviously, in any case, she won't complain (or else the engineering failed), but if she did, she could not have a case. Consider, for example, C. J. Cherryh's azi. Cherryh has been constructing, for some years, a future history (the stories of Merchanters' Universe)[4] in which human beings are reared from carefully selected, genetically modified sperm and egg, in artificial wombs, and deep-taught, with the aid of drugs and tapes, from their earliest infancy, to fulfil certain roles. Since their controllers are humane, according to the standards we apply to the treatment of expensive animals, such azi are content, and can have no complaint against their makers. In wartime they are regularly reprogrammed; even in peace they are not citizens, and may be 'put down', humanely, if they outstay their welcome, or begin to suffer psychological distress. They are in fact the natural slaves, the living tools, that Aristotle feared or hoped were real.

So it seems there is some willingness to think that artefacts can have no reasonable complaint against their maker. Any complaint they make, any show of disobedience, must be a symptom of disease, a reason to remodel or destroy them. Asimov's robots are expensive tools, even if they talk like people, and even the roboticist whom Asimov (1950) pretends loves robots (as being 'better' than born-men) is willing to destroy them, and their minds.[5] But perhaps the issue is not quite the one I have identified. An artificial, or pretend-person, may not have the rights (or duties) of real people, because it *is* an artefact, and does not really think or feel the things that a naive observer may imagine. It no more makes sense to say that we should respect a robot, than that we should respect an IBM PC, except, again, as an expensive tool. The reason that some people treat non-humans badly is that they suppose, however absurdly, that non-human creatures do not think or feel, that they cannot suffer pain, loneliness or boredom. If we really could imagine an artificial person, a real robot, wouldn't we, and

[4] A great many of Cherryh's stories are concerned with the moral value and effect of unreasoning loyalty, even on the part of born-men.

[5] Asimov's treatment of the roboticist, Susan Calvin, is as sexist as 'her' treatment of the robots is, effectively, racist.

shouldn't we, respect it and its views? What difference could its origin make, if it were really what it seems like? 'Robot Rights' would then not be ridiculous at all, any more than Animal Rights are.

Perhaps this is so, and our hesitation, Asimov's hesitation, about allowing robots to be morally considerable rests simply on our strong suspicion that they are sophisticated fakes. The 19th century manikin, Maillardet's Automaton, 'was a musical lady, who was advertised, rather alarmingly, to perform most of the functions of animal life, and to play sixteen airs upon an organized pianoforte, by the actual pressure of the fingers' (Heyer, 1965, p. 113). Modern automata may perform more functions, but are no more living, sentient and intelligent creatures.

But I am not sure that the argument is so easy. Suppose, what is not altogether absurd, that some of our tools do wake up. Suppose they really are what they profess to be: they perceive, and feel, desire, intend and plan, or do whatever it is that we are doing when we think we do those things. Suppose what is quite likely, that we can no longer quite predict their actions. After all, it is already true that we cannot predict what complicated computer programs do (which is why so many of them have so many bugs). Suppose that we can no longer tell, no matter how long or complex is the verbal test, whether we are dealing with a born-man or a robot. Is it obvious that people would at once agree that robots had as many rights as they did? Or would we think that evidence of thought and feeling was as bad as evidence of bugs? A tool that really 'thought for itself' and had its own goals, notions and associations would be a very dangerous (therefore a bad) tool. Asimov's Three Laws were devised to avoid the immediate destruction of a tool that reasoned for itself: without the assurance that it could not overstep the bounds we set, we simply would not permit it to continue. The Three Laws are wholly absurd as models of a humane morality, *pace* Asimov[6], but they are also absurd as measures to prevent revolt. Asimov's own stories show that robots notionally 'obedient' to the First Law ('a robot may not injure a human being, or, through inaction allow a human being to come to harm') must either 'go catatonic' or conclude that they may injure any number of human or putatively human beings, as long as they have good cause. So, lacking any way to restrain a really thinking thing forever, we would certainly not create it in the first place, or allow it long to live.

In other words, we probably would not acknowledge the new thinking creature's rights: we would extinguish it. After all, we made it, and its contrary ways are only a design fault, as they would be in the case of Cherryh's azi.

[6] See Clark (1988b).

15.3 MODERN ENGINEERING PROJECTS

Living creatures, it is said, are not defined or explained by their usefulness to us: they have, in that sense, no functions. But it could easily be argued, as by Callicott, that domesticated animals are bred and conditioned just to serve our purpose. So, for that matter, are human beings themselves. We have, our ancestors have, quite seriously, bred us, all of us, to serve our purposes. 'Human being', like 'warhorse' or 'hunting dog', is after all a functional expression, and not just a biological one.[7] Our characters are explained by the breeding programme that our ancestors initiated: a 'good' King Charles's Spaniel or 'good' Friesian is identified by the breeders' rules, and so is a 'good' human. We have been bred for special sorts of cleverness, obedience, and occasional kindnesses. More obviously, and easily, we have bred animals and plants to serve our purposes, selecting useful differences from the wild and propagating them, under our care. No doubt those domestic creatures could revert (unless we have made sure that they cannot breed at all without our help: like turkeys, or chihuahuas or hybrid roses). No doubt they still give evidence of features that we do not like. The most carefully bred animal, under laboratory conditions, when faced by a carefully controlled stimulus, does exactly what it pleases. To that extent, they show us that they are not wholly artefacts. Are we about to make a larger change than any since the Neolithic?

There are already tomatoes engineered to have a longer-lasting flavour. An American firm has won an American patent over all varieties of cotton (though it will be seriously challenged as far too extensive). Oncomice are patented: that is, mice with a genetic disposition to get cancer. There are sheep (in Edinburgh) which secrete human proteins essential to a healthy life for humans. Genetic engineering offers us the hope that we could do quickly what only generations of selective breeding could produce. Maybe we can even transfer characters from one species, or one phylum, to another, in ways that no amount of selective breeding (which can only work on characters already present in the gene-line) could ensure.

Of course, such practices are far more difficult than popular imagination reckons. The new piece of DNA is inserted into the genome, not with 'surgical accuracy', but by showering the genome with copies of that DNA: sometimes some copies stick. The piece of DNA which had, perhaps, a particular effect in its original place within its original genome, may have a different, or additional, effect within the new. Only a very few of the manipulated genomes will produce a creature,

[7] See Clark (1995).

plant or animal, that expresses the character the engineer was seeking. Only a few of them breed true. It is hardly easier to cut out a piece of DNA that, it is said, produces (usually) a phenotypic character we do not want. What else have we removed? What other stretches of DNA take up the task? The truth is that it is the whole genome, positioned in its biochemical, environmental, and social context that (at best) produces the effects we see in the adult organism. We may believe that it does so through a complex process of biochemical interactions – but we have no hope at all of calculating what those interactions *would* produce within a given context (and so no way at all to confirm that only those interactions have an effect). For all we actually know the embryo is guided to its adult form by flower fairies, or a ghost in the machine. Even if the biochemistry is all, it is far too complex for us to predict or even retrodict its workings. Genetic engineering is far more miss than hit.

Popular imagination is still ready to believe in resurrected dinosaurs, man–lion hybrids, talking beasts and useful symbiotes created by the new magicians. If once we know how things are put together, and what changes need to be made to get at least some visible effects, what limit can there be? The present fact is that there are real limits. There are no successful, major changes. Even minor ones (like making pigs who produce a leaner pork) may have unforeseen, and awful, consequences: the Beltsville pigs turned out to be seriously arthritic. Genetically engineered cereals may be more susceptible to pests, not less, and need more active intervention to keep them growing. Every sensible breeding programme always keeps a mongrel, wild population, so that there is a reservoir of genes to correct the errors of the programme. This is as true of engineering as of selective breeding programmes. So any attempt to think about more extreme cases relies on science fiction, not successes.

One of the earliest science fiction inventions was Pohl and Kornbluth's Chicken Little, an ever-living slab of chicken flesh, from which any number of slices can be cut: in essence, an edible tumour. I am not quite sure whether this, or various forms of bacterial exudations (aka pus), or Van Vogt's cockroaches, wins the accolade of most-unappetizing menu. In every case, such dishes are the staple diet of people at the very bottom of the socioeconomic ladder. Probably they prefer it to starvation. But what exactly is wrong with Chicken Little? Such an edible tumour would not suffer pain, nor be deprived of any decent life by being eaten. Douglas Adams' 'Dish-of-the-Day', by contrast, is deranged: a creature ludicrously eager to be killed and eaten. People who wish to dine on chickenflesh will do less harm by eating Chicken Little, than by financing factory farms, or even more traditional farms. Chicken Little, in fact, is just what factory farmers want: a chicken that does not peck its neighbours, nor pluck out its feathers, nor get its feet entangled in the mesh, nor excrete tons of chemically polluted guano, nor excite 'misplaced' compassion in the British public. What could be wrong, or morally wrong, with eating it? Would it not be positively right to breed, or engineer, it as a deaf, blind, featherless, legless, anencephalic lump that we could not injure further? We could do the same with pigs, and thereby realize Chrysippus' dream as narrated by Porphyry: pigs (or porkers) that are really living meals, with merely vegetative souls instead of salt to keep them fresh.

Utilitarian calculation (assuming that there will not be many devolving generations on the way to Chicken Little, or Pure Pork) seems almost to demand that we initiate the project. However aesthetically unpleasing it may be, the result will be cheap meat, without an unacceptably high cost to living animals. Nor is it easy to suppose that moralists who speak of 'rights' can much object. An edible tumour is not the sort of thing that has 'a life', nor is it 'the subject of a life', in the sense Tom Regan means. It is living, but no more so than tubers. Its rights are therefore not being violated when slices are carved off, or its nutrients varied to produce a fashionable taste. With luck, even its ancestors were treated fairly well: they contributed no more than sperms and ova (as of course they do already, with some discomfort and some loss of dignity), and those cells were fiddled with until they began to grow into an undifferentiated mass of muscle tissue. No embryo was ever allowed to grow up to a point where it might be the subject-of-a-life: if any trace of a complex nervous system, or a brain, was seen, the embryo was hurriedly aborted, and that gene-line halted.

But it does not altogether follow that this programme is OK. Both utilitarians and rights-theorists assume, without much argument, that concepts like 'a travesty', 'a cruel joke', 'obscene', 'an insult to the integrity of nature' are no more than expressions of conservative emotion: what is sometimes called 'the yuk factor'. Properly conditioned decision-makers, utilitarian or Kantian, despise the actual emotions of the people that they rule. Perhaps that is one reason why the people so often despise decision-makers. Maybe engineering or compelling dairy cattle to give birth only to female cattle would be the best available solution to one contemporary scandal (granted that our tastes are too corrupt to give up our dependencies). Engineering anencephalic lumps comes close to blasphemy. Even of a fish it is blasphemy to say it is *only* a fish, or of a flower that it is 'only a growth like any other' (Chesterton, 1962, pp. 54, 58, 68).

Consider other science fiction artefacts: Frank Herbert's float-homes, bedogs and chair-dogs. All these items are living creatures bred from natural stock and conditioned to serve as attractive houses, or as comfortable furniture (Herbert, 1972, pp. 15, 17, 116, 1977, pp. 17, 24,

291). 'The chair-dog is flesh that is ecstatic in its work.' Herbert's universe also contains vat culture food, descended from cattle cells, like Chicken Little. The other living artefacts satisfy slightly different urges, for unquestioning service, warmth and comfort. Bedogs and chair-dogs enjoy holding, massaging, and responding to their masters. If vat flesh has no feelings, furniture flesh has feelings that compel it to cooperate with what its masters wish. John McLoughlin's beasts inhabit a future when Earth, and industrial manufacture, has been abandoned: our human descendants, up aloft, inhabit and make use of genetically engineered organisms, 'synes' (McLoughlin, 1986). The consequence is a manipulative culture, hypocritically dedicated to 'the forces of selection'. All living creatures, including the notionally human, are artefacts in the power of the rulers. A similar suggestion is made in Poul Anderson's *After Doomsday*: a culture dedicated to the biological rather than the physical sciences is less respectful of the organisms that it understands, and can remake. So also Cherryh's dream of Union, whose scientist–politicians manage the lives they engineer: maybe they are 'humane', but what can that mean when they control their victims' goals and characters?

15.4 VISIONARY PROJECTS: SUPERMEN AND TALKING BEASTS

Animals and plants that have been bred and engineered to serve our usual purposes are not the only projects being considered. Nothing, so far, has been bred or engineered outside its kind: transgenic sheep are sheep, and oncomice are mice. Even Chicken Little is probably a sort of chicken, if it is any sort of living animal at all. But there are wilder dreams. Species differences are often not as great as we suppose: a species, after all, is simply a set of interbreeding populations, and the natural barriers to successful interbreeding may develop slowly. Genetic information sometimes flows between populations that are, in the abstract, different species: one sort of gull can breed with another sort, and that with a third, and so on round the Pole, until two sorts of gull, which do not breed directly with each other, can be linked by several stages (Clark, 1988a, pp. 17–34, 1993, pp. 113–25). In other cases, animals of distinct species can nonetheless be bred together, even though the offspring is usually as infertile as a mule. Lions and tigers, sheep and goats are close enough to hybridize – and so, it has been suggested, are human beings and apes. When the first successful hybrid will be bred – and what will be her fate – is quite uncertain.

The thought that creatures could be bred 'with human genes' (but not be human) is one that clearly alarms some commentators. The Edinburgh sheep I mentioned before do, technically, have some human genes, or have some genes that copy pieces of human DNA. Most of our DNA, in fact, is shared with other creatures, and it would be quite wrong to say even that the DNA we do not share is somehow quintessentially human. It is the whole genome that, suitably positioned, makes one sort of creature rather than another. Genetic engineers mean by 'chimeras' only creatures whose DNA has been composed from several distinct sources, not creatures with phenotypic characters appropriate to different species. Once again, we have to rely on fiction for examples. Dr Moreau's victims have been surgically transformed to mimic human character, and told to despise their 'animal' or 'subhuman' ancestry. Right-conduct, we are constantly informed, requires us to transcend our 'animal' nature, though we could as easily insist that animal affections are what move most of us to treat each other well. That marks, in fact, the major division between fantasists. The Wellsian version suggests that beast–human hybrids must struggle even more strenuously than we must to sustain a reasonable, 'humane' life against their animal impulses to rape and kill. Stapledon's Sirius (a naturally born dog with the power of human speech) has no real place in rational society, reverts to a pre-domestic wildness, and is shot. Any creature that steps out of its place is 'mad', and dangerous. Alternatively, there is a tradition taking its start from Plutarch's *Gryllus.* Gryllus is, or was, a human being transformed by Circe into a pig's shape, who much prefers this state to being 'that vain animal who is so proud of being rational'. Beast–human hybrids turn out to be wise, compassionate, and eventually victors: witness John Crowley's lion–men, in *Beasts*, or Cordwainer Smith's (appallingly sentimental) underpeople. Reason – or fashionable reason – on this account divorces us from our natural, loving ancestry, our place. That is to say, again, that anything that steps out of its place is mad, and dangerous, and that's what we have done.

A further twist in the story: beasts who incorporate a human or a sort-of-human mind are either very bad or very good; human beings who incorporate some animal power or attitude are either worse-than-beasts or gods. Maybe we do not always see much difference between the two. In Williamson's dark fantasy (*Darker than You Think*), conspirators breed back into existence a lost race or sub-species of wizards and shapeshifters, the prehistoric origin of all our tales of vampires, werewolves and gods. The fantasy was printed in the late 1940s, and has clear Cold War elements (indeed, clear racist elements). Likewise other stories: consider Pournelle's Saurons, genetically engineered 'supermen' who lack all human (or mammalian) empathy. Even less overtly unsympathetic 'supermen', like Stapledon's Odd John, are imagined as homicidal, even genocidal, thugs. All these stories can perhaps be read as warnings against the Humean fallacy, the idea that life would be improved if only we had additional powers or somewhat different motives. Every attempt, at least in fantasy, to create such creatures, ends in war and tyranny, just as every merely technical improvement leaves us with new problems, and with our morals untouched. Trans-humanist fantasies of life aloft, with power to dictate exactly what we wish, end – usually – in disaster. One imagined piece of engineering that does not aim at 'supermen' occurs in Brunner's *Stand on Zanzibar*. The one phenotypic trait needed in a grossly overcrowded, maddened culture not so far from ours, is the power to transform aggressive emotion into mere amusement. A really successful mutation, replicable in principle, gifts a West African tribe with peace-making pheromones. Unfortunately, in the story, the one scientist capable of working out the transference gets killed. But it is not entirely clear that success would have solved our problems. Imperialistic attitudes are not wholly founded on bellicose emotions: we do not go to war, nor even civil war, simply because we are irritated by our immediate neighbours. That story, nonetheless, does draw attention to the real need: a moral transformation that cannot, in its nature, be gene-engineered. There are no single genes for moral decency, or saintliness. We have bred ourselves to be obedient, and sometimes kindly: it is just those traits that sometimes lead to wickedness. Improved mathematical ability, greater stamina, resistance to identified disease, and peace-making pheromones may all be goods. They are not always good for us, and none of them is a virtue.

15.5 THE ABOLITION OF NATURE

The rationalizing demand that everything be made or remade so as to fit our plans for it is almost certainly to be disappointed. Contemporary fantasies about genetic or molecular engineering seem deeply implausible, if only because we know how inefficient ordinary engineering is, and how dependent on chaotically unpredictable accidents. They seem implausible also because we know how easily new desires emerge to make us discontented with our lot. No-one can make anyone else happy, and we can help ourselves to be happy only by submitting to the Way. But the dream persists. We are, apparently, affronted by the thought of accidents, irrelevancies, Otherness.

> 'There is a world for you, no?' said Filostrato [pointing at the moon]. 'There is cleanness, purity. Thousands of square miles of polished rock with not one blade of grass, not one fibre of lichen, not one grain of dust. ... There is life there. ... Intelligent life. ... An inspiration. A *pure* race. They have cleaned their world, broken free (almost) from the organic. ... They do not need to be born and breed and die; only their common people, their *canaglia* do that.

The Masters live on. ... They are almost free of Nature, attached to her only by the thinnest, finest cord.'

Lewis (1945), p. 213 ff)[8]

What Lewis imagined, pejoratively, back in the 1940s has been acclaimed since then. Where there was Id, shall Ego be. What was once Matter, will at last be Mind, *our* Mind. The more honest bricoleurs will add that it will not be a world where every presently human being will be a master. Most of us will be *canaglia* or cattle. 'Man's power over Nature means only the power of some men over other men with Nature as the instrument. There is no such thing as Man – it is a word. There are only men' (Lewis, 1945, p. 217).[9]

Filostrato, and his deluded colleagues,[10] may seem extreme. But 'if there is one notion that virtually every successful politician on earth – socialist or fascist or capitalist – agrees on, it is that "economic growth" is good, necessary, the proper end of organised human activity. But where does economic growth end? It ends at – or at least it runs straight through – the genetically engineered dead world that the optimists envision' (McKibben, 1990, p. 159). It is that death of nature that we should fear, and therefore suspect what leads to it. Nature probably cannot in fact be killed: what can die is our own life in nature. What would be lost with nature's death is just that Otherness, that strangeness, which we desire and serve. The wish to kill it is the magician's dream, the Oedipal desire to reign in Hell rather than serve in Heaven.

One of the philosophemes that has had most success in scientific and in broadly intellectual circles is the notion that Facts and Values are distinct, that 'natural' is not the same as normative. The extraordinary confusions implicit in this argument lie beyond my present brief. It is of course quite true that what happens in fact is not necessarily what ought to happen. Moral and political judgement positively demands that we can compare What Ought to Be, and What Sometimes or Often or Always Is. At the same time moral and political judgement demands that the two realms have some connection. If What Ought to Be has no effective influence on What Is, then we cannot even guess what ought to be: it would lie beyond all human knowledge. Common reactions to

[8] See also Lewis, 1945, p. 337.

[9] Lewis examined the point more academically in his best, and wrongly neglected book, *The Abolition of Man* (1943b).

[10] Frost, for example, who supposes that 'the body and its movements are the only reality' and that 'the self' is a nonentity (p. 444; see also p. 317); or Wither, for whom 'the indicative mood corresponded to no thought that his mind could entertain' (p. 438); or Straik, who imagines that 'man – or a being made by man – 'will finally ascend the throne of the universe, and rule forever' (p. 218). Do these characters seem familiar?

Chicken Little, and to the projected Death of Nature, suggest that at least one aspect of our moral consciousness is the recognition of an Other that demands, deserves respect. The natural, after all, is sometimes normative. One theoretical route to understanding this is Platonism – and to that I briefly turn.

To be is to be something. To be something is to embody, though perhaps ineptly, some one form of the many forms which shape, or are shaped in, the mosaic of the divine intellect. Every individual thing, that is, is a more or less distorted embodiment or reflection of that intellect. 'The realities we see are like shadows of all that is God. The reality we see is as unreal compared to the reality in God as a coloured photograph compared to what it represents. ... This whole world is full of shadows' (Cardenal, 1974, pp. 83–99). For everything, the something that it ought to be, by which it is measured, is what it is, in a sense, already. To disregard that pattern, that presence, so as to remake things in our image, apart from the one image, is indeed to insult 'the integrity of nature'.

A complete elaboration of Platonic theory demands a distinction between substances and properties, relationships, events. On the one hand, a thing's being is shown in what it does; on the other, the perfection of its real being need not be matched by any perfection in its actions. What happens need not be what should happen; what really is, should be. Defects and discrepancies and failures are all relative to what pattern it is that seeks embodiment at one place or another. It need not always be obvious what that pattern is: we do not necessarily know here-now what, for example, lions really are, what Form it is that lions-here resemble. If we did, we might be absurdly tempted to encourage the transformation of lions-here into some better version of the one true lion. But as Dante records of eagles:

> Their limner needs no model; His own best
> Model is He; we know Him, implicit
> As power and form in very sparrow's nest
>
> *Paradiso*, 18.109ff.[11]

Dante, of course, was speaking within the tradition of Christian Platonism, which identified the divine intellect with Christ. Other brands of Platonism found its embodiment in the Torah, the Koran, or simply in the cosmos of which we have so partial, so fragmentary a view. It might even be possible to envisage an atheistical Platonism, although the difference between a causally active Good and God Himself is hardly clear. In none of its various brands is Platonism vulnerable to the argument that what exists in the world is merely a chance event. Beauty

[11] Dante, *Paradise* 18, 109ff. Tr. Sayers and Reynolds, p.217.

is not exhausted by the presently beautiful, and neither do we need to suppose that what comes into being is prepared for by what came before. Our history and our present state is only one of many possible: new creatures and new worlds may come to view, and what now exists need not be preserved in time for ever. But what does exist deserves respect, and should not be subjected to our eager will to interfere. The mere fact that something is really there, that it is real, gives it a moral edge. God's Plan, so to speak, not ours, should rule.

So, like it or not, it will. The impulse to control our world, by turning it into a fantasy or by realizing fantasies, is bound to fail. Genetic and molecular engineers alike, like breeders and domesticators, can only trade on what comes naturally. Even artists can only uncover what is, in a sense, already there – a view which is not peculiar, as Harré supposes (1984), to the Kwakiutl.[12] All decent artistry respects the material of its art – because that material is already informed, or on the edge of form. Artists respect their material. Those who come after must respect the artists' work as well. For the very division between artefact and living, real thing that I endorsed at the beginning is readily subverted. Even Disneyworld, in time, would take on itself a being of its own, and future developers would be asked to preserve it for a more distant future. Even artefacts, if they go on existing, exist not by our will, but God's. Much of the damage we now do to nature will, one day, be part of nature, and defended from a future artificer.[13] It does not follow that we should not defend nature here and now. On the contrary, even artificers need their work defended: how much more the world that sustains us all.

REFERENCES

Asimov, I. (1950) *I. Robot*, Grafton, London.
Berleant, A. (1994) The critical aesthetics of Disneyworld, *Journal of Applied Philosophy*, **11**, 171–80.
Callicott, J.B. (1989) *In Defence of the Land Ethic: Essays in Environmental Philosophy*, SUNY Press, Albany.
Cardenal, E. (1974) *Love*, tr. D. Livingstone, Search Press, London.
Cherryh, C.J. (1983) *Downbelow Station*, Methuen, London.
Cherryh, C.J. (1986) *Forty Thousand in Gehenna*, Methuen, London.
Cherryh, C.J. (1989) *Cyteen*, New English Library, London.
Chesterton, G.K. (1962: first published 1929) *The Poet and the Lunatics*, Darwen Finlayson.
Clark, S.R.L. (1975) *Aristotle's Man*, Clarendon Press, Oxford.
Clark, S.R.L. (1985) Good dogs and other animals, in *Defence of Animals* (ed. P. Singer), Blackwell, Oxford, pp. 41–51.

[12] Compare pseudo-Dionysius tr. Rolt (1971), p. 194f.
[13] See Clark, 1987, pp.7–12.

Clark, S.R.L. (1987) The land we live by, in *The Ecological Conscience*, BANC Occasional Papers, Gloucester.
Clark, S.R.L. (1988a) Is humanity a natural kind?, in *What is an Animal?* (ed. T. Ingold), Unwin Hyman, London.
Clark, S.R.L. (1988b) Robotic morals, *Cogito*, **2**.
Clark, S.R.L. (1990) On wishing there were unicorns, *Proceedings of the Aristotelian Society*, pp. 247–65.
Clark, S.R.L. (1993) Apes and the idea of kindred, in *The Great Ape Project* (eds P. Singer and P. Cavalieri), Fourth Estate, London.
Clark, S.R.L. (1995) Herds of free bipeds, in *Reading the Statesman* (ed. C. Rowe), Academia Verlag, Sankt Augustin, pp. 236–52.
Dante *Paradise* tr. D. L. Sayers and B. Reynolds (1962) *The Divine Comedy*, Penguin, Harmondsworth.
Devereux, G. (1967) *From Anxiety to Method in the Behavioral Sciences*, Mouton & Co, The Hague.
Harré, R. (1984) *Personal Being*, Harvard University Press, Boston.
Herbert, F. (1972) *Whipping Star*, NEL, London.
Herbert, F. (1977) *The Dosadi Experiment*, Berkley Medallion, New York.
Heyer, G. (1965) *Frederika*, Bodley Head, London.
Levinas, E. (1989) Time and the other, in *The Levinas Reader* (ed. S. Hand), Blackwell, Oxford.
Lewis, C.S. (1943a) *The Pilgrim's Regress*, 3rd edn, Bles, London.
Lewis, C.S. (1943b) *The Abolition of Man*, Bles, London.
Lewis, C.S. (1945) *That Hideous Strength*, Bodley Head, London.
McKibben, B. (1990) *The End of Nature*, Penguin, Harmondsworth.
McLoughlin, J.C. (1986) *The Helix and the Sword*, Futura, London.
Murdoch, I. (1970) *The Sovereignty of Good*, Routledge & Kegan Paul, London.
O'Donovan, O. (1984) *Begotten or Made?* Clarendon Press, Oxford.
Pinker, S. (1994) *The Language Instinct*, Allen Lane, London.
Pohl, F. and Kornbluth, C.M. (1953) *The Space Merchants*, Ballantyne, New York.
Porphyry, tr. T. Taylor, (1965) *On Abstinence from Animal Flesh* (ed. J. Wynne-Tyson), Centaur Press, Fontwell.
Pratchett, T. (1978, first published 1976) *The Dark Side of the Sun*, NEL, London.
Pseudo-Dionysius tr. C. E. Rolt (1971) *Writings*, Paulist Press, New York.
Veyne, P. (1988) *Did the Greeks Believe in their Myths?*, tr. P. Wissing, University of Chicago Press, Chicago.

16

Species are dead. Long live genes!

Alan Holland

16.1 INTRODUCTION

In a little book called *The Right to Life*, Norman St John-Stevas wrote that: 'One reason for the rise of Hitler was precisely the number of people who could command no metaphysical view of human nature' (p. 15). This remark is, of course, open to the retort that another reason for Hitler's rise to power was the existence of a sufficient number of people who *did* command a certain view of human nature. Either way I simply want to take from the exchange the point that a 'metaphysical view' is not an unimportant matter. I wish more particularly to show how a 'metaphysical view' is unavoidably in question when it comes to determining how, as a society, we should address the difficult and largely uncharted ethical problems which arise in the wake of biotechnology. The argument of the chapter will be that it is not so much the technology as the accompanying genetic theory which presents the greatest challenge.

St John-Stevas wrote of the treatment by humans of their fellow-humans; but the point holds equally for human treatment of their fellow-animals. In recent times, one of the most common objections levelled by philosophers[1] against the treatment generally meted out by humans to other animals has been the charge of 'speciesism' – the charge that humans unjustifiably favour the interests of members of their own species. It is usually portrayed as an undesirable proclivity, although an opposing 'humanistic' view might see it, quite to the contrary, as an obligation. On the other hand, an objection which is increasingly voiced against animal biotechnology in particular is the charge that it constitutes an invasion of the animal's species-specific

[1] Peter Singer, Tom Regan and Richard Routley, to name but a few.

nature, or *telos*. In the case of both objections, ethical positions are advanced which make certain assumptions about the nature of species – certain 'metaphysical' assumptions.

Although they both defend a similar ethical standpoint, in that they both express concern for animal welfare, there seems, on the face of it, to be some tension between these two lines of objection. For it seems that what in the one case humans are criticized for defending, in the other case they are criticized for violating. More precisely, the former objection appears to carry with it the suggestion that boundaries between species are or should be of no account, morally speaking. The latter objection, on the other hand, appears to carry with it the suggestion that species boundaries do or should count a great deal. Thus, not only do these two lines of objection make certain metaphysical assumptions, but they seem to make conflicting ones. I shall first suggest that this tension may be more apparent than real.

16.2 THE ATTACK ON SPECIESISM

In its most radical form, speciesism is the view that in the moral domain, human beings are the only creatures who count. In its less radical form, it is the view that human beings count for more than any other creature, simply by virtue of being human. This latter position can be further refined. If we make a rough and ready distinction between the vital and the non-vital interests of any creature, one version of this position would hold that both the vital and the non-vital interests of humans take precedence over any interests of non-humans. Another version would hold that only the vital interests of humans take precedence over any interests of non-humans. Clearly, which version of this position is held may depend on which non-humans are being considered.

In the account of speciesism so far, however, we have not touched on the crucial factor which makes these positions speciesist, namely, that they are prepared to rest entitlement to moral consideration solely on an appeal to species membership. Opponents of speciesism hold that species difference alone is a morally irrelevant consideration, and that therefore treating creatures differently for this reason amounts to unjustifiable discrimination. In reply, many attempts have been made to show that the species boundary correlates with criteria which *are* morally relevant, such as intelligence or self-consciousness; but these have met with doubtful success. The well-rehearsed problem is that, whatever criterion is chosen, some non-humans will be found to possess it, or some humans will be found to lack it. In other words, the objection is not so much that there are no species boundaries, but that the criteria which define species boundaries do not coincide with those which qualify a creature for a certain level of moral consideration.

Now it is certainly a misunderstanding of those who voice this objection to suppose they necessarily hold that members of all species should be treated equally – so-called 'species egalitarianism'. Far more commonly, their position is that we should not assign greater weight to an interest simply because it happens to belong to a member of a particular species. This is quite compatible with, and is usually held along with, the view that there are quite considerable morally relevant differences between members of different species which depend on their different capacities and susceptibilities. But more disturbingly, it is also quite compatible with the view that we may assign *lesser* weight to an interest even though it happens to belong to a member of the same species who, for whatever reason, has a reduced level of capacities or susceptibilities. Thus, what works to some extent as a defence of non-humans can also work to the detriment of humans with incapacities.[2] The attack on speciesism is not without cost.

16.3 THE DEFENCE OF *TELOS*

One of the most distinctive defences of animal interests in the face of animal biotechnology is the charge that it constitutes an invasion of the animal's species-specific nature.[3] Bernard Rollin, in particular, has revived a version of the Aristotelian concept of *telos* as a way of articulating this objection. By *telos,* he understands the interests and nature of an animal which it possesses as a member of a species.[4] It is important to this defence that the individual creature is seen as belonging to a kind. For it is by reference to the kind to which an individual belongs that concepts of health and welfare are usually thought to be established,[5] and therefore the criteria of what it may or may not be permissible to do. So far as the individual is concerned, if there is no norm established by the kind to which it belongs, it might seem as if there could be no criterion of what may or may not be permissible.

Some advocates of this defence[6] suppose that respect for *telos* would lead to a proscription of animal biotechnology. But Rollin, for one, denies this. Violation is distinct from modification; hence respect for *telos* would not rule out modifications which show due regard for welfare.

[2] I have explored this concern elsewhere: in 'On behalf of moderate speciesism', *Journal of Applied Philosophy* 1, 1984: 281–291.
[3] For example, Jeremy Rifkin, Michael Fox and Bernard Rollin.
[4] Rollin, B. E., 'On *telos* and genetic manipulation', *Between the Species* 2, 1986: 88–89. A much fuller account is given in ch. 12 of this volume.
[5] So, for example, L. Reznek (1987) who includes in his normative account of a 'pathological condition' that it 'harms standard members of A's species in standard circumstances' (p. 167).
[6] For example, Michael Fox (1990).

Moreover, it is not clear that the defence of *telos* constitutes any defence of domesticated species, but rather might seem to be an abandonment of that whole class of animals inasmuch as they no longer exemplify the pure, natural kind. Their *telos* has already been modified. Either it has been changed to some new artificial form in the process, or it has been damaged. If the former, then the introduction of further artificialities through biotechnology scarcely seems to matter. If the latter, then biotechnology might in some cases offer the only prospect for repair or amelioration. Hence, the defence of *telos*, too, is not without cost.

16.4 METABIOLOGY – THEORIES OF SPECIES

We have seen how both the attack on speciesism and the defence of *telos* do not come without cost. But the more fundamental problem with both positions is that the 'metaphysical' (or perhaps we might say 'metabiological') view of species which they in fact share – namely, the view of species as groups of organisms united by a set of common characteristics – is scarcely any longer defensible. Although this 'phenetic' characterization of species in terms of a set of properties still has adherents, and will continue to function as a guide to identification, it is much more common now to find species characterized in relational terms (Ridley, 1985, Chapter 6). And if philosophical reflection ignores this fact, it will cease to be taken seriously.[7] According to the so-called 'biological species concept', species are communities of naturally interbreeding organisms (Mayr, 1987). According to the 'cladistic species concept', species are lineages between speciation events (Ridley, 1989). In a word, species are no longer understood as 'kinds', not, at any rate, if this is taken to mean a group of individuals united by virtue of the fact that they share common characteristics. According to the alternative relational conception of species, nothing *constitutes* being a harebell, a hedgehog or a human, other than being related in a certain way. Of course, members of the same species will tend to exhibit common properties, but these are the de facto consequence of the relations which hold them together, they are not defining properties.

A relational view of species puts both the attack on speciesism and the defence of *telos* in a different light. But it does not mean that the concerns which both express are denied articulation. Arguably, indeed,

[7] Notwithstanding the critique which is to follow, the view taken here is that scientific description, because of the standards to which it adheres, carries exemplary weight. But
i) this does not mean it has a monopoly over 'truth';
ii) it is compatible with recognizing that it is not wholly neutral (because no description is), that it is provisional, and that it can be to varying degrees culturally loaded; and
iii) for that very reason we need to be particularly vigilant about how it is deployed.

it permits their articulation without the accompanying costs. Given a relational view of species, for example, there is no clear reason why a broadly humanist perspective should entail the exclusive stance of speciesism or the essentialist stance which marginalizes even some humans. Further, to acknowledge the permeability of the species boundary still appears to leave room for a notion of integrity founded in the organism and for an ethic based upon relations rather than upon intrinsic properties which would, in addition, embrace domestic as well as wild creatures.

16.5 METABIOLOGY – THE THEORY OF GENIC SELECTION

As the profile and even the clarity of the concept of species has appeared to diminish, so has that of the gene correspondingly risen, being championed especially in a series of books by Richard Dawkins. There is no doubting that the genic re-reading of Darwin offers a powerful new perspective on the theory of natural selection. It also presents a more radical challenge than do the new understandings of species to the project of finding clear ethical ground-rules with which to approach the new genetic technology. For it appears to undermine not simply the centrality and integrity of species, but the centrality and integrity of organisms also. Moreover, as Evelyn Fox Keller has perceptively observed: 'with life relocated in genes, and redefined in terms of their informational content, the project of refashioning life, of redirecting the future course of evolution, is recast as a manageable and doable project' (Keller, 1995, p. 63). This remark both underlines the crucial fact that there has accrued around the new technology a 'metaphysical narrative' which drives its progress and greatly facilitates its acceptance, and pinpoints the conception of genes as units of information as a key element in this narrative.

In answer to this challenge, and chiefly with reference to Dawkins' work, it will be argued in what follows that the metaphysical narrative which accompanies his re-reading of Darwin, and in particular the accretion of the discourse of information technology, is profoundly misguided. In particular it will be argued that it:

- lacks coherence, and
- is unfaithful to its Darwinian paradigm.

The gist of Dawkins' position may be gathered from John Cornwell's succinct summary in a review of his most recent book: 'organisms are merely vehicles for genes, which compete to leave more copies for the next generation. In other words, it is information contained in the genes that is of supreme consequence in the story of life on this planet' (Cornwell, 1996, p. 46).

Two preliminary points of clarification are worth making. Although what Dawkins calls the 'gene's eye view' of natural selection assigns a more fundamental role to genes than to individual organisms in the theory of evolution, it should not be inferred (*pace* Cornwell) that his position is reductionist. Organisms are not *merely* vehicles; they are also 'survival machines'. If Dawkins' position were reductionist, then genes would provide both an exclusive and a comprehensive explanation of natural selection. But they do neither. Not comprehensive because, in the business of living, the organism is described as 'on its own', while the genes 'can only sit passively inside' (Dawkins, 1989, p. 52). Not exclusive because gene and organism are not rivals but 'candidates for different, and complementary, roles in the story' (Dawkins, 1989, p. 254).[8] The second point is that even if it were true that 'organisms are merely vehicles for genes' it certainly would not follow that it is what happens at the level of genes which is of 'supreme consequence'. If genes were in turn 'vehicles for electrons' this would not make electrons 'of supreme consequence' – at any rate in the sense which seems to be intended. What is 'of consequence' presumably needs to be related to human-sized units of significance.

But let us now move on to specifics.

16.6 GENES AND INFORMATION TECHNOLOGY

As a sample of what we have termed the 'metaphysical narrative' we might fairly turn to Chapter 5 of *The Blind Watchmaker*, where Dawkins reflects in engaging fashion on the seeds of a willow tree which are being cast on the waters of the canal in front of him: 'These fluffy specks' he observes, 'are, literally, spreading instructions for making themselves' (p. 111). There is no mistaking that we are here in the grip of the information technology model. 'If you want to understand life', he urges, 'don't think about vibrant, throbbing gels and oozes, think about information technology' (p. 112). This accretion of the language of information technology is quite pervasive and deserves, if anything does, to be cast as the 'commanding metaphysical view of genes'. It is assumed, for example, by the scientific contributors to this volume, and is common in recent textbooks. In one such book we read that 'A human cell contains a DNA library with instructions on how to

[8] The Preface to the 1989 edition of *The Selfish Gene* certainly indicates that he has departed from the modest position of the original, which represented the 'gene's eye view' of natural selection, compared with the view which has the individual organism in the central role, as simply a different view 'of the same truth' (p. ix). But, although p. 254 of the same work reaffirms the replicator/vehicle distinction, it is equally clear that the individual organism (the 'vehicle') plays a far from passive role in the processes which constitute natural selection.

produce and maintain humans ... the genetic information of a species is encoded into the structure of the DNA of the cells of that species ... to get the information out of DNA for use by the cell, the DNA strands, or tapes, are transcribed, or rewritten, into a closely related molecule, RNA'.[9] Moreover, this conception of the gene also informs public policy, as evidenced in the UK's recent 'Clothier Report': 'Genes are the essence of life; they carry the coded messages that are stored in every living cell, telling it how to function and multiply and when to do so. Until recently a genetic message could be altered only by accident or chance. Now human ingenuity makes it possible to manipulate these messages deliberately and with increasing precision.'[10]

The picture of the gene inside the cell telling it how and when to operate is, of course, a throwback to the homunculus theory of the human mind – a little person inside each of us telling our bodies how and when to operate – and as such, is harmless enough. But the same cannot be said of the residual concept of the 'message'. Here, we get our first inkling of how the metaphysical narrative has its effect. For, once genes are construed as carrying coded messages, then of course it will seem as if the only 'novelty' introduced by genetic engineering consists in whether the messages are sent 'deliberately' or 'by chance'. The notion that no novelty is involved is secured by pretending – for that is what it is – that what is there 'by chance' is the kind of thing – a message – which human agency might have wanted to put there.

Returning to Dawkins' text, it appears that we cannot in fact take his remark quite 'literally', since it is 'literally' incoherent to speak of something as spreading instructions for making *itself*: *it* already exists. The formulation in *The Selfish Gene* is more careful, when it describes the (original) Replicator as having 'the extraordinary property of being able to create copies of itself' (p. 15). Accordingly we can construe the 'fluffy specks' as spreading instructions for making – not themselves, but – copies of themselves.

But now there are three kinds of conceptual difficulty which arise from this account of the gene.

1. The first concerns the concept of copying. Essentially, the difficulty is that being a replicator is different from being an ancestor, and that being a replicator is the wrong notion. The relation between ancestor and descendant is a purely causal one. The copying relation, on the other hand, is both causal and formal – there must be something in the copy *answering to* something in the original.

[9] J. W. Brookbank, *The Biology of Aging* (New York, Harper & Row, 1990): pp. 65–66.
[10] Clothier, 1992, Introduction, p. 1.

Often some kind of similarity is involved, but this is not necessary (consider a young child's attempts to copy). Nor is it sufficient. Even exact similarity between two tokens of a given gene, coupled with the fact that one has causally contributed to producing the other, does not make the second a copy of the first. A further, 'teleological', relation of purpose, or intention, needs to obtain, which will be discussed in the next section. What proto-life discovered was not the art of self-replication but the art of being an ancestor. Indeed, this is precisely what Dawkins himself says about whole organisms in the opening few pages of *River out of Eden*. But the objection being advanced here is that it is this relation, rather than the copying relation, that must be seen as going all the way down to the gene.

A symptom of the fact that 'copy' and 'replica' are the wrong concepts surfaces in the unease over cases where replication supposedly 'fails'. For on the one hand, when this occurs, the exponents of Darwin feel obliged to talk of 'mistake' and 'error', yet at the same time, they acknowledge that 'errors' of this kind are what drives evolution. 'On rare occasions, such errors are harbingers of new evolutionary directions' writes S. J. Gould (1983, p. 133), while Richard Dawkins advises us to 'banish' from our minds 'all pejorative associations' of the word 'error', admitting that 'it is possible for an error to result in an improvement' (1988, p. 130).[11] Thus, such 'errors' become the 'norm' for subsequent generations. In addition to their reservations about using the term 'error', the authors cited fail to take the one further step of admitting that it is not simply the pejorative associations of 'error' which are unwarranted, but the very concept itself, together with the notion of replication on which it rests.

2. A second conceptual difficulty lurks in the account of the gene which appears in *The Selfish Gene* – where it is defined as 'any portion of chromosomal material that potentially lasts for enough generations to serve as a unit of natural selection'. What is evident here is the confusion of type and token. What the fluffy specks can do is spread instruction for making the *type*, of which they themselves are *tokens*. Whereas for Dawkins, the gene seems to be the type or structure that strands of DNA have in common, David Hull,

[11] Dawkins 1989, p. 17 is even stronger: 'it is ultimately these mistakes which make evolution possible'. The point being made in the text is not the one anticipated by Dawkins (ibid, pp. 17–18). Commenting on the paradox of supposing that an error might bring about an improvement, he responds that although evolution might seem a 'good thing', nothing actually 'wants' to evolve. The point here is that nothing 'wants' to stay the same either, and it is therefore misleading to represent a symmetrical situation with an asymmetrical distinction between what is an 'error' and what is (presumably) 'correct'.

on the other hand, seems to understand the gene to be the token exemplifying the type which 'passes on its structure largely intact in successive replications' (see Harms, 1996, p. 360). In fact, the criteria of identity for genes are opaque.

The confusion of type and token is reminiscent of nothing so much as the mistake which Plato was alleged to have made in his doctrine of 'forms'. The confusion even extends to the fact that Dawkins, like Plato before him, attributes immortality to the hybrid which results from the confusion.[12] The truth is that insofar as genes are types or patterns of DNA, they cannot be said to be immortal; only spatio-temporal individuals are (logically) capable of being immortal. The objection to treating types as tokens, which Plato himself anticipated (*Parmenides*, 132c–133a), is that it generates an infinite regress. Formulated in what became known as the 'third man argument', the objection ran as follows. What the individual 'men' Tom, Dick and Harry have in common is that they all exemplify the form or pattern of 'man'. But if this form or pattern is in turn taken to be an individual, capable – say – of immortality or selfishness, then there must be a third 'man' which is the form or pattern exemplified by Tom, Dick, Harry *and* the form or pattern which they exemplify. And so on, ad infinitum. Just so, what 'genes' as individual tokens of a DNA sequence – call them T, D and H – have in common is that they all exemplify the form or pattern of 'gene' M; but if this form or pattern is construed as an individual capable, say, of selfishness or immortality, then there must be a third 'gene', M1, which is the form or pattern exemplified by T, D, H and the gene M which they exemplify, and so on to infinity. The parallel is so complete that we may christen this the 'third gene argument'.[13]

3. A third difficulty arises in any attempt to apply the concept of a 'copy' in the absence of a copier. Essentially, the difficulty is that there is no criterion for establishing how faithful the 'copy' has to be to the 'original', nor for establishing how large a departure will make it no longer a copy. Nor is it clear that complete fidelity to the original is enough to make something a copy if each occurs in a different context. To take a linguistic analogy, it is unclear whether we should count 'cat' as it appears in 'catapult' as a copy of 'cat' on its own.

[12] 'The genes are the immortals', Dawkins 1989, p. 34.

[13] This criticism might seem unfair in view of Dawkins' acknowledgement that 'genes, like diamonds, are forever, but not quite in the same way' (1989, p. 35). However, the view being urged here is that genes are not 'forever' in *any* way. Further, to say that a gene is 'forever' 'in the form of many duplicate copies' (ibid. p. 35) is to assume the validity of the 'copy' view, which is here being challenged.

16.7 GENES AND TELEOLOGY

However useful the information model of the gene may be, it is indelibly teleological. In a recent article, Kim Sterelny and others make the point particularly clearly (1996, pp. 384–9). They correctly observe that talk of 'information' is predicated on talk of function. To suppose that genes are units of information is to suppose that they are 'for something'. An acorn genome can be said to carry information about the oak tree to the extent that its function can be understood to be to contribute to the production of the oak tree. Other things contribute too, such as light and moisture, but it is not necessarily their function to do so. As a much earlier text has it, 'Zeus does not send the rain in order to make the corn grow'.[14] The acorn genome is different because it is doing 'what it is supposed to do' – that is, what its predecessors did that ensured their replication and the present acorn genome's existence.[15] Sterelny and colleagues are correct to imply that the concept at the heart of the 'gene's eye view' – replication – is also indelibly teleological. Replication is a form of copying and 'copying is a teleological notion' (p. 396). For such a notion implies that iteration has as its function or purpose the production of a 'copy' of some 'original'. It also grounds the corresponding notions of 'error' and 'mistake' in cases when iteration fails.[16]

The reason why iteration is such a pervasive phenomenon, like the reason why organisms cluster around species norms rather than being strung out like beads on a string, can itself presumably be explained by reference to the mechanism of natural selection. But the fact that iteration is the norm should be recognized for what it is, a *de facto*

[14] Aristotle, *Physics* Bk. II, ch. 8.

[15] But notice that what its predecessors did, and what ensured the present acorn genome's existence, comprised a mix of perfect and *imperfect* iteration.

[16] In fact, Sterelny and colleagues are keen to accept, and indeed argue for, such a teleological understanding, and take themselves to be disagreeing with Dawkins in so doing. Dawkins, they say, has argued that "because genes are the *beneficiaries* of adaptation, they are not *for anything*". They also draw the correct inference that "if that were right, a mutation would not be a mistake, only a change" (p.388). Whether Dawkins holds steadfastly to the view ascribed to him is unclear. But if he does, and if Sterelny and colleagues are correct about its implications, and about the teleology implied in the copying notion and in the language of information technology, then the view being argued for here follows – that in presenting the 'gene's eye view' Dawkins ought to jettison the whole information technology narrative together with the concept of the replicator.

It is instructive to observe that the functional notion of copying would, on the other hand, be quite at home in the Aristotelian scheme of things in which every living thing seeks to "partake in the divine [and also therefore everlasting] in so far as they can"; but because no naturally perishable thing can persist forever, "what persists is not the thing itself but something like itself, not one in number but one in species" (*De Anima* 415a25–27). Hence, "the end is to generate something like oneself" (416b23), and "it is a natural function of living things ... to produce another thing like themselves".

contingency resulting from the processes of natural selection, rather than an inherent feature of such processes constituting what they are *for*. Thus, my accusation against the present day Darwinian evangelists is not the common one, that they are too Darwinian, but the uncommon one, that they are not Darwinian enough.[17] For all the modernity of the concept of the gene, the prevailing metaphysical narrative which accompanies gene technology is not simply pre-Darwinian, but positively Platonic in both its teleological and essentialist implications. According to the analysis given here, the teleological and essentialist connotations embedded in the older concept of species have simply been displaced onto the gene.

16.8 AN ALTERNATIVE NARRATIVE

Darwin never sought to explain why there is anything at all (or why there is life). Nor, more importantly, did he seek to explain why what there is persists (or why life persists). What he sought to explain is why life as we know it is so diverse – more precisely why it exemplifies the particular kinds of diversity which it does – those to be found, for example, on the 'tangled bank' referred to in the final paragraph of *The Origin*. Changes in the distribution of alleles (the alternative forms of a given gene) are determined by changes in population size and mating patterns, migration and natural selection, but mutation is responsible for new alleles (Berry, 1990, pp. 49, 60). A crucial part of the explanation of diversity, therefore, is provided by deviance or 'error', and the survival of deviance is explained either by the absence of competition or by the fact that what is there already is not (fully) fit – if it were, the deviant would not gain a foothold.[18] Evolution has taken the particular course that it has because the 'copying' process is not perfect. If it had been perfect, evolution would not have occurred at all since natural selection would have had nothing to operate on. In fact it has occurred without there being any copying process at all. Iteration (as distinct from copying) has been the norm at the genetic level because, in

[17] Neither, of course, is Darwin himself, who often betrays teleological or perfectionist tendencies. The end of the penultimate paragraph of *The Origin*, for example, speaks of 'progress towards perfection'.

[18] One might well say, therefore, that what makes diversity possible is a combination of deviance and unfitness. Picking up an intriguing observation by Elizabeth Hay that regenerative phenomena 'are closely related to the devices for asexual reproduction' (1966, p. 1), one might add to this deflationary account the speculation that reproductive life itself results from a failed regenerative process. 'Darwinism works only because – apart from discrete mutations, which natural selection either weeds out or preserves – the copying process is perfect' (Dawkins, 1995, p. 19). This seems just wrong unless, at any rate, the preservation of discrete mutations is given more than mere parenthetical status.

circumstances which change rather slowly, natural selection will tend to favour offspring equipped in much the same way as their parents, and therefore also will tend to favour the mechanisms which produce such offspring.

16.9 SOME IMPLICATIONS OF THE ALTERNATIVE NARRATIVE

An overriding aim of the alternative narrative is to emphasize how evolution is nothing but a contingent historical process. The Darwinian theory of natural selection is construed not as the triumphal story of how the fit have survived, but the story of how a series of absolutely humdrum events – of chance escapes and encounters, strivings and resignations, strokes of luck and exercises of judgement – have conspired to produce life on earth as we know it.[19] It would be understandable if one were to think that this alternative prospectus might encourage an 'anything goes' kind of ethic, rather than offer critical purchase for assessing animal biotechnology. This seems to be very much the view of Brian Goodwin (1996): '... in Darwinism, [species] don't have a nature, because they're historical individuals, which arise as a result of accidents. ... The Darwinian theory makes it legitimate to shunt genes around from any one species to any other species' (p. 7). Goodwin is undoubtedly correct regarding the biological implications of Darwinism, but we need not follow what he says regarding the alleged normative implications. We shall briefly review three sets of considerations.

1. Consider, first, the nature of species. Goodwin argues that an 'anything goes' mentality stems from the denial that species have natures. We have seen, however, that Rollin, for one, certainly seems to believe that species have natures, yet regards this as compatible with the biotechnological project of creating new natures. It seems, therefore, that species natures do not offer much defence against animal biotechnology. What this suggests is that we should not be seeking to reinstitute the notion of species kind, but rather to exorcise genic essentialism. On the view of species we have endorsed, species membership, even if this is still thought of as 'belonging to a kind', is not a matter of satisfying some check-list of characteristics. It is more like, indeed exactly like, being a member of a family. What this in turn suggests is that a more appropriate

[19] Thus, on the view taken here, fitness is not an explanatory concept, not even in the minimal sense of being well designed 'by an engineer's criterion' (Gould, 1980, p. 42). Darwin's genius was not to provide some super-explanation for evolution – fitness – but to show how everything that needs explaining is explained in perfectly humdrum ways, and how (in general) no further explanation is required.

ethical approach is to focus not so much on 'qualifications' for moral consideration, but more on relationships.

Once we can learn to put behind us the tedious question of *whether* any morally significant things are going on in animal minds,[20] and focus on the challenging and interesting question of *what* morally interesting things are going on, we might begin to learn more about the way animals relate to one another. Those who associate with non-human animals and trouble to become familiar with their forms of social relationship will testify, for example, to the significance of family and of parenting among many species. Consider the female rat who dares all when her nest of young is discovered beneath the henhouse: a refusal to acknowledge this as a form of parental devotion is hard to sustain. It is just as hard to sustain a position which makes family values a cornerstone of human morality while at the same time treating as morally invisible the parental and other claims of non-humans. For this reason, the implications of dairy farming and the separation of cow and calf should no doubt be taken far more seriously than it usually is. The greeting of cat and dog who have been brought up together and meet after a period of separation signals the significance of individual relationships among some animals.

The particular grounds for concern, then, which receive a higher profile under the alternative narrative is the effect that gene manipulation may have, not so much on a creature's capacities as on its form of life. The claim that this is a matter of concern need not rest on the contentious thesis that animals are members of the human community, but on the thesis that animals are members of animal communities sufficiently analogous to human communities to make certain forms of consideration appropriate, and certain forms of treatment oppressive.

2. Along with Plato's theory of forms went the idea that individuals were simply the 'meeting place' of forms, and had no independent metaphysical status of their own. For Dawkins, correspondingly, the individual organism is the meeting place of genes – a sort of street corner where they momentarily gather, or a taxi ('vehicle') in which they take a short ride. This conception is bound to pose a challenge to conventional ethics built around notions of the integrity and inviolability of the individual organism. Such a view might well be thought to legitimate 'shunting genes around'. It also facilitates a sometimes disturbing demonization of specific genes such as we find, for example, in Bodmer and McKie's description of the search

[20] Even philosophers have not thought the problem of other minds sufficient reason not to be concerned how they treat one another.

for the cystic fibrosis gene: 'The hunters were closing in for the kill' (1995, p. 121); 'the honour of the final kill went to Lap-Chee Tsui' (p. 122). But it was a view which we have challenged.

The alternative narrative encourages a somewhat different account of the relation between the individual and their genes. Genes play a quite special role in the constitution and the identity of the individual organism. As Saul Kripke has convincingly argued (1980, pp. 110–15, 140–2), an individual could not have originated from anyone other than their actual parents. If they did, they would be a different individual. Likewise, they could not have originated from anything other than the token DNA from which they actually originated. Accordingly, insofar as gene technology manipulates DNA, it challenges individual integrity, not at the level of some species nature but at the level of the individual's origins and identity. Perhaps it is only humans who can be said to have a sense of their origins and a sense of their birthright. But presumably humans are concerned about such things because they are thought to matter independently; nor is it clear why they should matter only in the case of humans.

3. Emphasis on the historical contingency of evolution might certainly seem to license an open-ended approach to what we do from here on, and therefore to license a biotechnological future as much as any other. But there are reasons to draw a very different response:
   (a) What is precious does not have to be enduring. It is often ephemeral. Therefore there are grounds for defending fragile happenstance against the managerial ambitions of a very small section of a very small and unrepresentative sample of a precocious and recently arrived species. The ambition seems all the more vaulting if this should turn out to be the only experiment in life that there ever was to be in the whole history of the universe.
   (b) Much criticism of gene technology has focused on the risks involved if it should go wrong, and there should be accidents, escapes, and unforeseen effects. The greater risk is if it should succeed. For this betokens the closing down of options; in the human appropriation of the living world, something fundamental would be lost. We might be assured that genetic technology will never succeed in designing out 'useful' genes in view of the plasticity and variability of the material. But a reassurance based on the claim that a technology will fail is none too reassuring. We might also reflect, for example, that if Ian Wilmut's speculation is right (Chapter 2, p. 18) that the effectiveness of genetic manipulation depends on a combination of the recipient DNA's susceptibility to rupture and inadvertent assimilation of the foreign DNA, then the more the world is populated by geneti-

cally engineered organisms and their descendants,[21] the more it will be populated by organisms whose DNA has a susceptibility to rupture. In a word, there will be selection in favour of the fragile genome.

16.10 CONCLUSION

I have tried to indicate the importance of a 'metaphysical view' and its implications both for thought and action. I have also suggested that the 'metaphysical view' which lies behind the current practice of biotechnology in general and animal biotechnology in particular is incoherent, and that to persist in the practice without engaging in a great deal more reflection is tantamount to fumbling in the dark.

Given the Human Genome Project, and given the huge significance attributed to the gene under the new genetic re-interpretation of Darwin, it is disingenuous to claim, and people are right not to accept, that genetic technology raises no new issues.[22] The excising of an ingrowing toenail, or implanting of a hip, simply fail to raise the metaphysical issues awakened by the excising or implanting of a piece of DNA; no-one has (yet) propounded the doctrine of 'hip selectionism' or proposed a billion dollar human toenail mapping project. For similar reasons, it is disingenuous to claim, and people are right not to accept, that genetic technology 'in itself' is ethically innocent. No technology ever can be isolated from a given cultural context in the way that this implies. It is sobering to reflect that if genetic technology had been available in the time of the Roman emperor Julian, when superstition and a belief in signs and portents was in vogue, then it would have been put to very different uses from those which we now contemplate.

Sections of the scientific and 'managerial' community (I include governments) have been fond of portraying concern over genetic technology as based on shallow, ignorant and irrational fear. The tenor of the argument sketched in this chapter has been to suggest, to the contrary, that it is based on a deep and well-founded 'metaphysical fear' centring on concerns over the implications of this technology for conceptions of individual identity, integrity and origin which are foundational to our world view. Animal biotechnology is fully implicated. For a combination of the view that 'organisms are merely the vehicles

[21] 'Ancestors are rare, descendants are common', Dawkins, 1995, p. 1.

[22] In contrast to a judgement contained in the foreword to the Clothier Report: 'We should perhaps emphasise that nowhere in our deliberations did we discern the existence of, or the need for, some new ethical principle. Rather we found ourselves considering the application of familiar and established principles to a new and promising field.' Report of the Committee on the Ethics of Gene Therapy (Chair, Sir Cecil Clothier). London, HMSO, Cm 1788, 1992.

for genes' with the realization that species boundaries are fully permeable, brings it home that we should do well to ponder long over our treatment of non-human animals lest we should come to treat our fellow humans likewise.

REFERENCES

Aristotle (1968) Aristotle's *De Anima* Bks II and III (translated, with introduction and notes, by D.W. Hamlyn) Clarendon Press, Oxford.
Berry, R.J. (1990) *Inheritance and Natural History*, Bloomsbury Books, London.
Bodmer, W. and McKie, R. (1995) *The Book of Man*, Abacus, London.
Clothier, Sir C. (1992) Report of the Committee on the Ethics of Gene Therapy – 'The Clothier Report'. Cm 1788, HMSO, London.
Cornwell, J. (1996) Chance is a fine thing (Review of Richard Dawkins' *Climbing Mount Improbable*). *New Scientist*, **150**, no. 2027, 27 April, pp. 46–7.
Darwin, C. (1968 [1859]) *The Origin of Species*, Penguin, London.
Darwin, C. (1899 [1868]) *The Variation of Animals and Plants under Domestication*, John Murray, London.
Dawkins, R. (1989) *The Selfish Gene* (new edn), Oxford University Press, Oxford.
Dawkins, R. (1988) *The Blind Watchmaker*, Penguin, London
Dawkins, R. (1995) *River out of Eden*, Weidenfeld & Nicholson, London.
Fox, M. (1990) Transgenic animals: ethical and animal welfare concerns, in *The Bio-Revolution* (eds P. Wheale and R. McNally), Pluto Press, London, pp. 31–45.
Goodwin, B. (1996) How the leopard changed its spots – conversation with David King. Gen*Ethics* News, **11**, March/April, pp. 6–8.
Gould, S.J. (1980) *Ever Since Darwin*, Penguin, London.
Gould, S.J. (1983) *The Panda's Thumb*, Penguin, London.
Harms, W. (1996) Cultural evolution and the variable phenotype. *Biology and Philosophy*, **11**, 357–75.
Hay, E. (1966) *Regeneration*, Holt, Rinehart & Winston, New York.
Keller, E.F. (1995) Fractured images of science, language and power: a post-modern optic, or just bad eyesight? in *Biopolitics* (eds V. Shiva and I. Moser), Zed Books Ltd, London, pp. 52–68.
Kripke, S. (1980) *Naming and Necessity*, Blackwell, Oxford.
Mayr, E. (1987) The ontological status of species: scientific progress and philosophical terminology. *Biology and Philosophy*, **2**, 145–66.
Plato (1926) *Parmenides* (tr. H.N., Fowler), Heinemann, London.
Reznek, L. (1987) *The Nature of Disease*, Routledge and Kegan Paul, London.
Ridley, M. (1985) *The Problems of Evolution*, Oxford University Press, Oxford.
Ridley, M. (1989) The cladistic solution to the species problem. *Biology and Philosophy*, **4**, 1–16.
Shiva, V. and Moser, I. (eds) (1995) *Biopolitics*, Zed Books Ltd, London.
St John-Stevas, N. (1963) *The Right to Life*, Hodder & Stoughton, London.
Sterelny, K., Smith, K.C. and Dickison, M. (1996) The extended replicator. *Biology and Philosophy*, **11**, 377–403.

Part Four

Policy and regulation

17

Biotechnology policy: four ethical problems and three political solutions

Paul B. Thompson

Policies to govern and regulate animal biotechnology are plagued by ethical problems in four areas: food safety, animal welfare, environmental impact, and social consequences. Although the problems of animal welfare and animal rights are probably of most interest to the readers of this book, the interaction between each of these four areas has a significant effect on the way welfare and animal rights policies are negotiated and administered. Each of these regulatory areas is typically controlled by distinct divisions of government, and each raises somewhat distinct ethical concerns. I will argue that consensus solutions are available for three of the four problem areas. Only social consequences appear to involve philosophical differences that pose insurmountable obstacles to political compromise. Yet the lack of consensus on social consequences precludes social progress in the three areas where policy solutions appear possible.

17.1 ETHICS AND PUBLIC POLICY

Ethical issues are inevitably associated with regulation and public policy. Regulations and public policies are always fashioned in accordance with background assumptions or politically negotiated consensus about moral questions. What is the public good? Who should make decisions and control events? What are the fundamental purposes of government and society? Even when tradition or consensus rules that individuals are to have complete discretion in exercising moral judgement (as is the case with respect to religious duties in most parts of the world today), the choice to make such judgements 'personal' and beyond the scope of public policy is itself a public policy. Ethical principles entail answers to the fundamental questions that establish the limits of public authority and the principles on which public laws and procedures are based.

The influence of ethics on public policy does not end with this straightforward transition from principle to practice, however. Public policies themselves may have implicit normative dimensions. They may reflect traditional or customary practices that have no legal status. Customary beliefs govern social practices for the use of animals in many ways that are not legislated. The determination of which animals may be used as food in a given society is an obvious example. Yet even when statutes govern a particular area of conduct, the interpretation and administration of those statutes may be determined by informal consensus or by the public arguments of speakers who have no statutory authority to establish administrative criteria. Given these multiple patterns for linking ethics and public policy, we may say that policies present ethical problems when

- the ethical consensus – what Bernard Rollin calls the informal social contract – provides inadequate or viciously ambiguous guidance for public conduct, legislation and the administration of existing legal codes, or
- public policies themselves fail to enforce an existing consensus.

Of course, there is seldom complete unanimity of opinion in a consensus, and even the most influential arguments are opposed by dissenting voices. To say that a public policy is either ethically problematic or is potentially solvable is therefore to make a contestable statement. Most of the chapters in this book discuss the ethical issues raised by animal biotechnology, with particular emphasis on animal welfare and animal rights. Clearly many of these authors have described positions that are so deeply inconsistent with each other that the idea of an ethical consensus seems ridiculous. Yet it is possible to have relatively orderly procedures for providing ethical guidance on public conduct, legislation and the administration of regulatory codes even when opinion and argument is quite diverse. Sometimes such policy solutions to ethical problems involve compromise. It may be possible to stipulate public practices that are not fully consistent with the ethical goals and arguments of any single constituency, but that are preferred by all constituencies to the alternative of contentiousness and conflict. At other times, policy solutions stipulate procedures for establishing which of several incompatible ethical principles will prevail. Many interpretative problems in regulatory policy and administrative law are settled by presidential elections in the United States, for the

President appoints the administrators who will make the key decisions. In parliamentary systems, a similar pattern of appointments is made by the ruling party or coalition government. As long as political appointees are themselves clear about the ethical principles that should guide policy, this procedure is sufficiently well established to permit adequate determination of public policy, even when the public remains divided.

To say that there are consensus solutions to ethical problems in animal biotechnology policy is therefore *not* to say that it is possible to resolve the philosophical disputes documented in the other chapters of this book with a deft policy move. For the purposes of this chapter, an ethical problem is solved whenever custom, public pronouncement or deliberative judgement can be brought to bear on policy choice. This means that philosophical or moral disagreement does not paralyse the process of setting policy for science and technology. Hence the next question: What policies must be made regarding animal biotechnology?

17.2 ANIMAL BIOTECHNOLOGY POLICY

In its broadest terms, biotechnology policy includes intellectual property rules and budgetary decision making, as well as regulation of research and products utilizing recombinant DNA techniques. This chapter will focus exclusively on the regulatory issues. The term 'animal biotechnology' is itself vague. Genetic engineering performed on animal genomes represents the most controversial application of animal biotechnology. However, genetic engineering is already being used for animal drugs. The international controversy over recombinant bovine somatotrophin (rBST) provides an excellent case study for evaluating public policy problems for animal biotechnology, but rBST is an animal drug. The genetic engineering that made rBST possible was performed on a microbe, which in turn produces rBST for use on dairy cattle. Although regulatory issues for genetically engineered food and research animals will certainly differ from issues associated with genetically engineered animal drugs, it will prove more useful to take a broad view of animal biotechnology, to include genetic engineering of plants or microbes intended for diagnostic or therapeutic use on animals.

Given these parameters, there are four areas of animal biotechnology policy to consider. First, *food safety policy* regulates disease and injury risk associated with consumption of all foods, including meats and animal products, and assures standards of quality that may have an aesthetic or cultural basis. Animal products that utilize genetic technologies may pose unique risk issues, and clearly challenge cultural definitions of wholesome foods. The primary philosophical conflict in food safety stems from the tension between consequential approaches to regulation, which seek efficient achievement of health objectives, and approaches that stress informed consent, relying on mechanisms such as labelling and certification.

Animal welfare policy regulates the impact of research, husbandry and other handling practices on the well-being of animals themselves. The manipulation of animal genomes can affect an animal's welfare in several respects. Modifications can create dysfunctional animals that endure lives of discomfort, pain and hardship, but they can also protect animals from diseases and produce animals that are better adapted to the environmental conditions of domestication. In this area, there are philosophical controversies over the appropriate standards of well-being, as well as between advocates of the status quo and those who would dramatically reduce human use of animals.

Environmental policy regulates the impact of human activities on ecosystem processes. Some ecosystem processes affect human health directly. Toxic substances found in animal waste, for example, accumulate in soil, air or water, and humans who are exposed to them face risks of disease and injury. The human health considerations in environmental policy share many features with food safety. However, other ecosystem processes are crucial to the habitat and hence the reproduction of flora and fauna, yet have little visible or immediate effect on humans. Recent environmental philosophy has proposed several approaches to understanding the ethical significance of such 'non-human' impacts.

Although each of these three policy areas pose philosophical problems, each is also characterized by a political structure such that opponents can secure partial fulfilment of the norms they advocate through compromise, and risk unacceptable policy outcomes if they fail to participate in the political process. It is therefore reasonable to suggest that the philosophical problems in these three areas have political solutions, meaning that philosophical differences will not preclude the identification and pursuit of shared social goals. This situation does not hold for social consequences of technological change. Virtually every new technology redistributes the costs and benefits of producing and exchanging goods and services. Technologies can also undercut individual incentives for participating in social institutions that are essential to the formation of individual virtues such as citizenship and accountability, or group virtues associated with community and cultural heritage. Those who benefit from a new technology seek policies that promote it, while those who lose and those who fear the loss of less tangible social goods seek amelioration, at least, and may oppose the technology altogether. The allocation or distribution of costs and benefits throughout society is a classic element of *social policy*. Such economic issues are routinely debated by political theorists who propose alternative philosophical visions of community, the public good and progress. In actual politics these ethical norms are interwoven in a contest between the economic interests of winners and losers. Since animal biotechnology clearly produces winners and losers, social policy is the inevitable fourth category of public policy debate.

17.3 BIOTECHNOLOGY POLICY AND PHILOSOPHY

Perhaps the seminal philosophical article on the ethics of recombinant DNA controversies was published in 1978 by Stephen Stich. Stich's article was written in the wake of the 1976 conference at Asilomar where leading scientists debated the risks inherent in genetic engineering. Stich reviewed 'bad arguments' that surfaced in both scientific and lay debates. He defended an approach that took the ethical responsibilities of scientists seriously, but that interpreted those responsibilities largely in terms of anticipating and mitigating risks (Stich, 1978). Writing specifically on genetic engineering of animals, Bernard Rollin echoed Stich's message a few years later, stressing that the lesson to learn from Frankenstein metaphors was not that some things should never be done, but that scientists must avoid the fictional Dr Frankenstein's 'failure to foresee the dangerous consequences of his actions or even to consider the possibility of such consequences and take steps and precautions to limit them' (Rollin, 1986).

Stich and Rollin devote considerable attention to the argument establishing scientists' responsibility to consider risks or unwanted outcomes very seriously before pursuing their research. Stich and Rollin classify these risks and unwanted outcomes into categories that reflect a demarcation first of fact and value, and then of different kinds of value. Their approach purifies a vague and contentious thicket of issues by analysing how distinct burdens of proof might be applied to different components of the controversy. Both Stich and Rollin dismiss the possibility that genetic engineering could be intrinsically wrong. Movement of genetic materials is permissible, subject to consideration of the consequences. Moreover, the types of consequence that count are familiar: human health, animal well-being, environmental quality, and distributive justice. While genetic engineering allows humanity to do things that have never been done before, these papers define the ethical issues raised by molecular biology as familiar problems of technological risk.

Although biotechnology has progressed to the point that the conservatism of these early papers by Stich and Rollin might be questioned, it is still crucial to examine the ethical issues associated with the risk of unwanted consequences. Such issues are the exclusive focus of this chapter. Stich and Rollin do not agree in how to address the problem of unwanted consequences. Stich seems far more comfortable with consequentialist or optimizing solutions to the problem of technological risk. The classical characterization of this approach dictates that potential outcomes be assessed or subjected to a process of valuation. This value, whether positive or negative, is then discounted by the probability that the outcome will actually occur. When costs and benefits of an activity have been thus assessed, they may be summed. Each option available for choice can be analysed similarly. A decision maker must choose the option that is expected to produce the optimal ratio of benefit to cost. Stich is at most committed to the spirit rather than the letter of this approach, but he nonetheless seems comfortable with an assessment of biotechnology that compares its costs and benefits.

Rollin agrees that the ethics of animal biotechnology demand a prediction of its likely consequences. He differs from Stich in judging that at least some potential consequences should not be subjected to an evaluation of cost and benefit trade-offs. His article is primarily addressed to impacts upon animals. He describes the potential for creating dysfunctional animals, condemned to lives of physical pain or cognitive suffering. Rollin states that when such animals are inadvertently produced, scientists have an obligation to terminate the experiment, ending the animal's suffering. When there is knowledge that dysfunctional animals are likely to be produced, the experiment should not be done. However, Rollin goes on to describe the potential for using genetic engineering to change an animal's nature, so that, for example, pain cannot be sensed, or cognition does not occur. Such modifications would be permissible in Rollin's view. Indeed, they might be obligatory for scientists who wish to develop transgenic models for certain types of disease. In none of these discussions does Rollin endorse a cost–benefit type of accounting. Although biotechnology raises ethical issues in virtue of its unwanted consequences, Rollin shows that it is possible to bring non-consequentialist patterns of ethical reasoning to bear on the problem of unwanted consequences.

The international public controversy over the approval and adoption of rBST, the hormone that increases productivity of dairy cows, can serve as a model for analysing ethical issues related to animal biotechnology. With but few additions, the categories of unintended consequence in the Stich/Rollin theory of scientists' ethical responsibility are well represented. The next section explores how a Stich/Rollin assessment might be applied to rBST. The following sections elaborate each of the four areas of unwanted consequence. Two claims are argued in the balance of the chapter. First, although the philosophical dimensions of unwanted consequences are likely to remain controversial, the existence of a reasonably well-functioning political forum for human health, animal well-being and environmental impact constitutes a political solution. Second, the lack of a political forum for debating social consequences is a serious political deficiency in biotechnology policy.

17.4 rBST: ASSESSING UNWANTED CONSEQUENCES

Somatotrophin or growth hormone is produced naturally in mammals and regulates not only growth but also other functions, notably lactation. When somatotrophins are administered under carefully managed conditions, milk production can be increased, and the lactation cycle can be extended. Bovine somatotrophin can, therefore, be administered to cows under a herd management regime that results in significant increases in milk production. It is not economical, however, to use bovine somatotrophin harvested from cows because of the high production cost. Genetic modification of bacteria for production of somatotrophins was one of the first successful applications of recombinant DNA technology, and genetically engineered organisms are now used routinely to produce human growth hormone for medical applications. Several animal drug companies, including Monsanto, Eli Lilly and Upjohn have succeeded in developing a recombinantly produced bovine somatotrophin over the last decade, and the Monsanto version, trade-named Posilac, was approved for use in the United States in the autumn of 1993. The story in other countries with well-developed regulatory systems (and significant levels of dairy production) has been similar, though the final verdict is still unknown for Canada and Europe at time of writing.

The social history of rBST deserves a more extended treatment than is warranted in the present context. It must suffice to say that the technology has been opposed by a complex network of interested parties. Most specific objections to rBST can be classified into the categories of animal well-being, food safety and social consequences. These represent three of the four areas noted for our study, so a discussion of environmental quality is added to round out the issues. In the rBST case, food safety has become deeply contested. Concerns about the integrity of the food industry, the regulatory process, and agricultural research organizations are generally expressed as uncertainties about the safety of rBST milk, but scientists and regulators have been adamant about excluding these concerns from risk assessment. This difference of opinion, crucial to the ethical analysis of animal biotechnology in general, is represented below by introducing a distinction between safety and anxiety. However, social consequences associated with restructuring in the dairy industry precipitated the entire debate following a study by Robert

Kalter (1985), and are clearly the most politically contentious. In this respect, controversy over rBST is a particularly apt model for considering animal biotechnology.

17.4.1 Food safety

Food safety is perhaps the most obvious area of potential impact from genetic engineering as it affects agricultural animals. For present purposes, *food safety* will be defined as a function of the probability that consumption of a food will produce injury or debilitating disease, or that substitution of a food for reasonable alternative foods will adversely affect a person's health through nutritional deficiencies. Food safety policy represents a classic risk issue. Consequentialists treat risk in the manner described above: measure probabilities and expected values, then choose the course of action that optimizes the ratio of good to bad outcomes. The alternative view stresses rights. When dealing with risks to human beings, the approach generally stresses a right of informed consent on the part of the person exposed to risk, no matter how small. The consequentialist position translates into public policy where key decisions are made by experts who can assemble and interpret information on risk. Informed consent requires mechanisms where individuals are exposed to food-borne risk only under circumstances of their own choice.

Of all potential impacts from rBST, food safety has received the greatest technical specification. It is also the one on which there is the greatest unanimity (Munro and Hall, 1991). At present, the consensus standard is that foods produced using animal biotechnology must be at least as safe as conventional foods, and procedures for assessment of food products from biotechnology in all industrialized nations virtually assure that far more will be known about the probability of injury or disease from recombinantly produced foods than from foods of more conventional origin. There is, as a result, the possibility that ethics will weigh in on the side of *less* attention to food safety in virtue of disproportionate expenditure of resources on the assessment and mitigation of quantitatively minimal risks (Johnson and Thompson, 1991). Anyone even remotely inclined to take a consequentialist position on risk would deem the safety of rBST for human consumption a non-issue.

Despite this circumstance, food safety emerged as one of the most prominent public points of controversy in the rBST case. Samuel Epstein, a biomedical researcher at the University of Illinois, expressed early concerns about potential health impacts, but the overwhelming consensus of scientific opinion has been that use of rBST in the production of milk has no impact on the health risks associated with milk (Kroger, 1992). Critics of rBST then turned to the possibility that mastitis associated with elevated levels of milk production might create human health hazards (Hansen, 1991). The Pure Food Campaign under the leadership of Jeremy Rifkin organized chefs on both coasts to protest what they termed adulteration of milk by addition of rBST. However, with the exception of the few sources cited here, the vast majority of criticisms associated with food purity are addressed as factors that do not bear in any direct way on the probability of injury or other deleterious human health impacts.

Anecdotal evidence suggests that a significant number of people do not want milk from cows treated with rBST. There are several reasons why this might be the case. First, reasonable people may wish to dissociate themselves from foods produced using recombinant DNA technology on religious or aesthetic grounds. Nothing is more human than to adopt beliefs about the purity and authenticity of foods that would be difficult or impossible to support on scientific grounds. Is New York State Champagne an oxymoron? The French certainly think so. Avoiding impure or inauthentic foods may not be a safety issue in the narrow sense, but it can be extremely important to those who hold the relevant beliefs. Second, people routinely make consumer choices to express solidarity with other groups or political causes. This type of consideration overlaps with aesthetics to some extent, as the injunction to 'Buy American' echoes the French desire for authentic champagne. In the rBST case, however, solidarity may have more to with loyalty to small dairy producers or animal well-being concerns. In either case, it may be important for some consumers to choose so-called 'non-BST' milk.

Neither of these concerns relates to the probability of disease or injury that is associated with drinking rBST milk. They could be described as elements of *food anxiety* rather than safety in a narrow sense. Ironically, controversy itself creates anxiety. As questions are raised about the technology, people naturally wonder whom to believe. They may ultimately resolve this question by considering the costs of being fooled. If the critics of rBST are wrong, a consumer is losing several cents per gallon of milk purchased. Although this may add up to significant social costs, even a family purchasing a hundred or more gallons of milk every year may find the cost a reasonable price to pay for avoiding the anxiety of a new and unfamiliar form of milk. If the scientists are wrong, after all, the cost would be measured in ill health, especially to children, who drink more milk than adults. Even if one thinks it far more likely that the scientists are right, it may be rational to forego the marginal consumer price benefit in exchange for the familiarity of ordinary milk.

Critics of rBST have not produced reasons to ban the product. What they have produced are reasons why individual milk consumers might want an alternative. The issue, thus, is one of consent. Those who want the price savings, or who are confident in the product's safety, should have access to the product. Those who do not want it, for any reason, should have some mechanism for avoiding it. The mechanism is almost certainly a label that would allow those who want 'ordinary' milk to get it. This is not the place to analyse the ethics of labelling in detail, however. It must suffice to say that the most promising mechanism for milk and for other foods using animal biotechnology is a negative label, one that certifies the absence of any use of recombinant DNA technology in producing the food. Although negative labels are far from perfect in assuring consent, they represent a reasonable compromise between enabling consent for those who care about biotechnology in their food, and not stigmatizing a safe, beneficial technology for those who do not (Thompson, 1997).

17.4.2 Animal well-being

Although impact on animals may be a marginal category in some areas of science politics, it has always been prominent in discussions of animal biotechnology and for obvious reasons. Rollin's 1986 and 1992 papers on animal biotechnology stress the possibility that genetic engineering may produce situations that contribute to animal suffering. Certainly this potential has been one of the most controversial topics with respect to the ethics of rBST. Comstock raised the issue of animal welfare impacts associated with rBST in a 1988 paper, noting stress associated with the administration and with the pharmacological effects of rBST. Concerns over the linkage of rBST to enhanced milk production and in turn to increased incidence of mastitis have been the subject of considerable review and concern ever since. However, the concern for animal well-being noted by Rollin and Comstock is unlikely to be defined as a compromise to animal health, given current approaches that are standard in the animal sciences. Rollin introduces the concept of *telos* to describe the genetically encoded set of physical and psychological needs that determine 'the fundamental interests central to [animals'] existences, whose thwarting or infringement matters to them' (Rollin, 1990, p. 305). He suggests that any experimental or production practice which compromises an animal's *telos* is morally wrong, and specifically notes that a farmer's profitability (or a consumer's price reduction) does not provide a sufficient justification for practices that violate the package of rights an animal must be accorded in virtue of its *telos*.

For both Rollin and Comstock, these rights cash out in terms of practices that produce pain or suffering to individual animals, or that frustrate animals' ability to behave consistently with their genetic endowment or nature. Current regulatory approaches to animal well-being vary dramatically around the globe. In the United States, *ex ante* assessment of animal technology is limited to animal health. It is carried out by the Food and Drug Administration as a component of certifying the safety and efficacy of animal drugs. Anti-cruelty statutes provide an opportunity for animal advocates to bring charges on behalf of abused animals, and provide a basis for *ex post* regulation. In practice, however, anti-cruelty statutes are rarely successful in overturning an agricultural production practice, though they have been applied to generate reforms in transport of animals. It is nevertheless easy to see how the anticipation of impacts described by Rollin and Comstock fits under the general heading of responsibilities noted by Stich, once the pain and suffering of non-human animals is recognized as morally significant. Extensive physiological, behavioural and cognitive approaches to the assessment of impact on animals are in at least rudimentary stages of development; the task now is simply to apply them in the study of animal well-being.

Assessment of transgenic animals, animals whose genomes have been altered through manipulation of recombinant DNA, will be more difficult, however. It may be impossible to anticipate the impact of a genetic modification on an animal's needs. Rollin explicitly stipulates that it will not be wrong to change an animal's *telos*, even if doing so may result in chimerical beasts that cause aesthetic revulsion. What will matter is the *telos* of the new animal, and our ability to assess the vital interests and needs of animals whose genetic constitution departs significantly from that of animals whose genome is the result of evolutionary adaptation. Domestication and even conventional breeding rely on selection in a way that allows us to predict a rough fit between an animal's physiological, behavioural and psychological needs and the environment in which it will live and reproduce. It is less clear that the animals produced through recombinant techniques will have behaviours, interests and needs adapted to the environments in which they will live. Although this introduces uncertainty into our collective ability to anticipate impacts on animal well-being, it does not alter the conceptual basis of the scientists' responsibility to consider and assess such impacts.

17.4.3 Environmental impact

Agricultural technologies are routinely assessed with respect to environmental impact, though requirements to assess such impacts have arguably been less stringently applied to agriculture than to manufacturing and energy sectors of the economy (Thompson *et al.*, 1994). While the technical requirements of environmental assessment are becoming relatively well defined, the ethical significance of environmental assessment is extremely complex. There are, for example, environmental impacts that impinge on human health, but assessments also model technology's impact on broader ecosystem processes. Impacts on these processes may be considered adverse only when they affect human life, but they may also be considered significant simply because they challenge the stability or equilibrium of an ecological zone. A growing literature on environmental ethics in agriculture provides the basis for minimizing such challenges (see Aiken, 1984; Norton, 1991).

The rBST case is a relatively poor model for illustrating ethical issues associated with environmental impacts of animal biotechnology. The consensus of opinion on rBST was to regard environmental impact as one of the least serious of consequences or potential impacts associated with the technology. This consensus appears to have been based on the assumption that rBST would reduce the number of dairy cows, and since faecal wastes are regarded as the most serious environmental contaminant associated with dairying, the reduced number of cows was projected to produce a corresponding reduction in the total volume of waste. As such, the environmental impact of rBST was judged to be positive (Executive Office of the President, 1994). This conclusion is questionable on the ground that the social consequences of restructuring the dairy industry have secondary environmental impact (Lanyon and Beegle, 1989). When environmental risks are more direct or better understood, they are more likely to emerge as a category having ethical significance.

Other forms of animal biotechnology are likely to have more controversial environmental impact. For example, a recombinantly engineered rabies vaccine was tested in the wild in Belgium (Brochier *et al.*, 1991). Critics of this action represented both sides of a classic divide in environmental ethics. Although the issue was characterized as 'environmental,' the greatest volume of criticism stressed the potential for risks to human health. These critics feared the introduction of a potential pathogen into the environment. The ethical concern was 'human-centred', or anthropocentric. A quieter voice raised questions about the impact of releasing the virus on the wild populations themselves, and by implication, the ecosystem in general (Anonymous, 1991). The politics of the issue were not good for environmentalist complaints. It is difficult to argue that unconstrained spread of rabies should be permitted as part of natural ecosystem checks and balances. Nevertheless, the principle behind this concern reflects the type of ecocentric, holistic and non-anthropocentric thinking that characterizes the opponents of anthropocentrism in environmental ethics.

One can certainly imagine applications of animal biotechnology where the divide between human and ecocentric interests would create significant difference of opinion. A different type of recombinant vaccine may provide the first example. Researchers in Nairobi may be close to developing a recombinant vaccine for sleeping sickness. This disease has long been the bane of cattle herders in East and Central Africa. At the same time, however, the prevalence of the disease and its vector, the tsetse fly, has effectively protected large areas of habitat from human exploitation. The tsetse fly limits the success of both poor subsistence herders and large commercial operators in much of Africa. Where these human uses are excluded, African wildlife may thrive. Given the enormous pressure on habitat in Africa, ecocentric environmentalists will certainly regard the environmental consequences of this new vaccine with apprehension. Given the food needs of resource-poor African pastoralists, there will be compelling human-centred reasons to use it aggressively.

17.4.4 Social consequences

Social consequences are associated with all agricultural technologies. Some consequences, such as the elimination of hand labour jobs, may be intentional. Some technologies are too costly for poor producers, but can give large or wealthy farmers significant advantages over the poor. The economic structure of agriculture in both developed and developing countries means that aggressive early-adopting farmers derive short-term benefits from production-enhancing technology, but that the ultimate beneficiaries are food consumers. Although animal biotechnologies may be less susceptible to a farm size bias than are mechanical and chemical technologies, it is reasonable to think that many poor producers will be unable to compete with richer competitors as a direct result of biotechnology.

Robert Kalter's (1985) study predicted that relatively small-scale dairy producers might be disadvantaged when rBST became available. This prediction is itself somewhat complex, and a substantial literature on it is summarized by Tauer (1992). For the purposes of this discussion, the economic issues that arise in predicting a technology's effect on the size distribution of farms and the make-up of rural communities are less relevant than the general question of why alleged impacts on small versus large farms might be thought ethically significant. There are at least two strategies for approaching this issue. One begins with the assumption that those adversely affected by new technology are harmed in some way analogous to impacts described above. They may be deprived of income they would have received without the technology, and may also be harmed in more subtle psychological and social ways. These impacts must be weighed against benefits not only to other producers, but also to food consumers (Thompson *et al.*, 1994, pp.

242–5). A second strategy begins with the observation that those who make decisions about whether to develop and market a technology occupy a position of power over the small farmers who will be affected. On this more populist view, what is ethically significant is the distribution of power, not the distribution of risks and benefits. The remedies associated with this way of framing the issue enhance affected parties' ability to influence decisions that will have dramatic effect on their future livelihood and way of life. In this respect, it is crucial to note that in the United States no agency of government has the authority to monitor or regulate technology on the basis of its social consequences. Lacking an outlet for their frustrations, groups seeking remediation of social consequences will politicize the regulatory process for environmental, animal welfare and human health consequences (Thompson, 1992).

17.5 RESOLVING ETHICAL DISPUTES THROUGH POLITICS

Most of these problems are amenable to political solutions. This is not to say that the philosophical problems are solved politically. Anthropocentrists and ecocentrists will continue to be at odds over whether non-human species, or ecosystems themselves, are morally considerable. Furthermore, they will continue to have broad and deep political differences. However, close attention to the regulatory politics of animal biotechnology reveals that these continuing philosophical and political differences can often be blunted, if not finessed altogether. The rabies controversy in Belgium illustrates the point. Here, both anthropocentric and ecocentric critics of the vaccine tests were up in arms because procedures for anticipating environmental risk and for informing the public were ignored. Close adherence to fairly common procedures of environmental impact assessment and public notification would have likely mitigated the uproar. Thus, the ethical problem was less a problem of conflicting interpretations of environmental goals and responsibilities than it was a problem of failing to follow procedures that are philosophically non-controversial.

Returning to rBST, the problems of animal welfare and of food safety are similar. Animal rights activists clearly seek radical changes in society's practice, and these changes might well eliminate many practices in animal agriculture. On this point, animal activists might take a 'rights' view that challenges the consequentialist perspective of those who think that a compromise to animal health, well-being or needs fulfilment can be offset by benefits to human beings. Many technologies are being introduced in animal agriculture all the time, however. Focusing on technologies derived from recombinant DNA techniques will not necessarily fix the animal activist's attention on the most problematic technologies. The case of recombinant rennet illustrates this point. Rennet is an enzyme essential to cheese making. Traditionally it is harvested from the entrails of slaughtered calves. A bacterium has been engineered that produces a purified version of the enzyme under industrial conditions. Recombinant rennet has met with absolutely no resistance on either food safety or animal welfare grounds. In the latter case it is easy to see why. It is a technology that changes animal agriculture for the better from an animal welfare or animal rights perspective.

Evaluation of biotechnology's unwanted effects must be done on a product-by-product basis. If new biotechnologies do create dramatic risks to animal health or well-being, the radical activists' concern will certainly be matched by that of philosophically far more conservative animal protectionists. While there is, indeed, a philosophical difference of opinion on animal use (and presumably a corresponding political difference), there is little reason to think that this difference will surface predominantly or even especially in considering specific products of animal biotechnology. There is likely to be consensus on the extreme cases. Advocates of reform in human relationships with animals will have little to gain by raising concerns about relatively less-significant products and technologies, simply because they happen to be associated with biotechnology.

The issue is only slightly more complex for food safety. As already discussed, negative labels represent a compromise solution that allows those who wish to avoid rDNA products to do so. Developed nations have mobilized the scientific community to assess substantive risks associated with food additives and with residues of chemical technology. While these mechanisms are not perfect, it seems far more likely that problems will be associated with the continuing use of chemical technology than with biotechnology. Many of the new applications of plant biotechnology, for example, will reduce the application of pesticides and fertilizers, if they perform as promised by their developers. As such, consumer advocates should be generally supportive of biotechnology, especially if the problems of consent can be resolved. Lacking positive evidence that there are food safety problems, it is silly to object to a technology that will reduce food-borne risks solely on the ground that it involves biotechnology. The sheer novelty of recombinant DNA techniques and products presents reason for caution and for explicit assessment of risk, to be sure. As the experience base builds, however, the same principles that mandate caution over biotechnology may well shift toward favouring biotechnology over its chemical or mechanical alternatives. Whether this happens is not a philosophical question. Only time will tell.

There are, thus, political solutions to the unwanted health, animal well-being, and environmental impacts of animal biotechnology. To say that there is a political solution is not to say that philosophical problems are settled by a show of hands. Political solutions redirect philosophical disagreements away from regulatory decisions about animal biotechnology. In some cases they redirect those disputes to broader, more comprehensive debates over public policy. In other cases, the dispute is moved out of the political realm entirely. In purely philosophical terms, the dispute between anthropocentrists and ecocentrists may be the most fundamental of those discussed above. The tension between habitat preservation and wise use of resources creates seemingly irresolvable rifts in land use policy. The practical lesson may be that it will be prudent for developers of animal biotechnology to concentrate on products that are consistent with more environmentally benign forms of animal agriculture. Biotechnology may be useful in mitigating pollution from animal waste, for example. Technologies that would extend animal agriculture to new places, to new ecosystems, will be more problematic. The fact that scientists and technology planners can choose biotechnologies to avoid some of the most serious conflicts, not only in environment but in food safety and animal well-being as well, is further reason to regard these areas as amenable to a political solution.

17.6 SOCIAL CONSEQUENCES REDUX

Good policies represent political solutions to philosophical problems when they appeal to the overlapping consensus that exists in most industrialized democracies. In an uncharacteristically pragmatic moment, John Rawls argued that political philosophy must seek to identify the common principles and policies that would be endorsed even by persons having very different life philosophies (Rawls, 1987). Rawls was hopeful that the main elements of a just society could be identified by emphasizing these areas of consensus, rather than the dissent that often underlies them at the level of fundamental beliefs. Food safety, animal well-being and environmental impact represent three areas of policy where Rawls' hopes seem to have some chance of being fulfilled. The unwanted, unintended social consequences of animal biotechnology are less amenable to a hopeful solution. Ironically, it is in the area of distributive justice that Rawls' appeal to the overlapping consensus seems least promising.

Social consequences must be analysed at multiple levels (Berlan, 1991). New technologies routinely jeopardize some forms of employment, and ruin businesses that are wedded to obsolete technologies. The first level of analysis draws our attention to the individuals who are left without jobs and income during such transitions. Social welfare programmes in most developed countries moderate the effect of these transitions, offering temporary benefits and retraining to affected parties. The second level of analysis takes up the loss of these jobs to the communities in which the relevant industries are located. Job loss in one sector translates into failed businesses, schools and hospitals across the board. Public policies for coping with community transition are unevenly distributed across developed countries. The United States arguably does a poorer job of moderating transitions at this level than do most countries. Nevertheless, it is reasonable to claim that some policies and mechanisms exist for coping with the impoverishment and psychological harm that are associated with these aspects of technological change.

Another level of analysis is reached when we consider the impact of technological transitions on entire regions. With respect to transitions in mining and manufacturing, as well as agriculture, entire regions of the world have been effectively depopulated. When people leave the countryside in this way, it is not only individuals and community institutions that are lost. Entire ways of life, and networks of kinship and mutual affection, are dissolved. There is really no way to compensate the losers for these transitions, for the basic values that define what is a profit and what is a loss have been stripped away from them. They will, to be sure, wind up someplace else, but, mere survival aside, the systems of meaning that determine value will have to be reconstructed entirely. When they are, new values, new friendships, new senses of possibility will be in place, but to compare new with old is no more meaningful than comparing the life of a contemporary suburbanite with that of a 17th century aristocrat. Who is better off? Is it clear that one would trade places with the other? The texture of these lives is so different that such questions are ridiculous.

It is thus reasonable to say that when technological transitions have such systematic effects, a loss occurs that cannot be compensated. Whether such losses should be permitted, or whether they should be resisted, is a complex question. Clearly the dispute over the social consequences of rBST centres on just such a question. The dairy farmers who have opposed this technology are acutely aware that a delicate balance of subsidy and productivity keeps them in business. If productivity increases, there will be more milk at a lower price. Small-scale producers cannot recoup losses from a reduced margin of profit by increasing volume. Furthermore, increases in volume will put political pressure on policies that keep prices at current levels – a circumstance especially crucial for dairymen in Europe and Canada. The classic dairy, with 50 to 100 or 200 cows, is at risk. What will go with it are the businesses, schools and hospitals of a hundred counties, but what is worse is that a form of life that is thought particularly characteristic of agriculture will disappear from the landscape. It will be replaced by industrial plants servicing 2000, 4000, or even 10 000 cows in a single location, trucking the feed in and the milk and manure out (Lanyon, 1994).

Fairness demands that we recognize how this eventuality may be the result less of biotechnology and rBST than of transportation technology and the computerization of dairy management. Such a transition has been well under way in dairy farming long before the debate over rBST. This point illustrates that, even with regard to social consequences, other non-biological technologies will be as important as the applications of molecular biology. Nevertheless, what is relevant here is how with few exceptions, industrial democracies have failed to develop regulatory structures that would allow such questions even to be raised. The United States policy here was stated succinctly in the White House white paper on impacts of introducing rBST. 'At no time in the past has the Federal Government prevented a technology from being adopted on the basis of socio-economic factors' (US Government, 1994, pp. 35–6). The report goes on to describe relocation of cheese making and milk processing in the United States as large dairies in the West and southwest replace those of the traditional north-eastern crescent extending from Minnesota to Vermont. There is no suggestion, however, that the cited precedent should be reversed. No agency or office within the Federal Government of the United States save Congress itself has the authority to regulate technology on the basis of social consequences.

Why so much emphasis on the US? It is at least arguable that the US exercises hegemonic influence on the matter of social consequences for the rest of the world. Trade agreements and the sheer pressure of international competition make governments reluctant to place their own producers into an economically disadvantageous position relative to US producers. The lack (or weakness) of regulatory procedures for social consequences represents a form of international assurance problem. If every government would regulate on the basis of social impact, the rules of international competition would be fair. However, as long as one government does not, any government wishing to regulate technology on the basis of social impact risks losing its ability to compete in the relevant sector altogether. When the one actor is as large and dominant as the United States, the absence of political solutions to the social consequence issue is a foregone conclusion.

17.7 CONCLUSION

There are, therefore, four ethical problems associated with animal biotechnology, but only three political solutions. The significance of this fact is both political and philosophical. Politically, the lack of a policy framework for even raising, much less resolving, problems associated with social consequence introduces a high degree of uncertainty into the politics of animal biotechnology. Again, the rBST case illustrates this point. Why did rBST become an issue at the Food and Drug Administration (FDA) where questions of human and animal health were to be assessed? Why, especially when there was such unanimity among the science community that food safety risks were minimal, does it continue to raise public concern? Some of the answers have been discussed above, but the political contentiousness of rBST at FDA must have arisen partly because those interested in social consequences had nowhere else to go. The absence of a forum for debate and regulation of social consequences in either administrative or judicial branches of government leaves no alternative but the translation of these issues into trumped up and ultimately false concerns about human and animal health, or environmental consequences.

The political effects of the missing forum for social consequences are particularly significant for those interested in the animal welfare implications of biotechnology. On the one hand, those who oppose all forms of animal biotechnology (on animal rights or species integrity grounds, for example) will typically find willing allies among those who are concerned about agricultural technology's impact on small or resource-poor farmers, and on rural communities. Working together, these groups can slow the approval process for biotechnology products, and might even succeed in making regulatory approval for animal biotechnology so expensive that the entire industry becomes economically unattractive. On the other hand, if more radical animal activists are busy fighting the battle against *all* forms of animal biotechnology, it will be more difficult to act on the available consensus to regulate animal biotechnology on welfare grounds. As argued above, it should be possible for researchers and industry to agree with animal protectionists on many parameters for animal genetic engineering and for the approval of animal drugs. Policy based on this consensus is, I would argue, the course of action that is truly in the interests of non-human animals. Yet acting on *any* political consensus becomes difficult when social interests intervene.

The philosophical implication is that our notions of distributive justice need to be more carefully integrated with research ethics. Is it reasonable for scientists to think that if the responsibilities described by Stich and Rollin are addressed systematically, they have discharged their responsibilities as researchers? The Stich/Rollin approach leaves many philosophical questions unanswered, but it implies that if scientists are honest and diligent in conforming to the requirements of the regulatory process, they have done everything that we may reasonably require of them. Conformity with the regulatory process means that scientists are acting within the framework of an overlapping consensus on how human health, animal and environmental impact should be addressed. That consensus will change, of course, as well it should when subjected to ethical scrutiny, but the political apparatus for each of the three covered areas provides a procedure for dealing with unwanted consequences in a way that shares the burdens with the scientific conscience.

But scientists should not be as sanguine about social consequences as this view of research ethics suggests. Regulatory compliance is an empty moral imperative where social consequences are concerned. Partly this is just a case of getting social reform on a well-trodden issue, especially in the United States. However, social philosophy has done a poor job of raising questions about technological change. The pattern has been to see technological change as inevitable, and to see the problem as one of distributing its benefits and costs. This image treats technological change as if it were an act of God, like a hurricane or an earthquake, not as something for which human beings and human organizations could be held responsible. The model proposed by Stich and Rollin recognizes the importance of agency in producing technology, and implies that social consequences are as deserving of our attention as are risks to human health, animal well-being and environmental quality. The implication, however, is not followed out. Though molecular biologists are in a good position to discharge some of the responsibilities noted by Stich and Rollin, the responsibility regarding the social consequences or their work are not included among them.

Ironically, a key ethical problem associated with animal biotechnology is the neglected responsibilities that scientists and the developers of biotechnology have to human communities. Clearly other problems – not the least of which are those that relate to animals themselves – deserve contemplation and attention, as other chapters in this book attest. Yet considerations of distributive justice will influence our ability to discharge ethical responsibilities to animals, and it is important to consider the problems of justice and technical change among the ethical implications of animal biotechnology.

REFERENCES

Aiken, W. (1984) Ethical Issues in Agriculture, in *Earthbound: New Introductory Essays in Environmental Ethics* (ed. T. Regan), Random House, New York, pp. 257–88.

Anonymous (1991) Outfoxed: Preventing Rabies (Belgium). *The Economist* (December 21, 1991), **321**(7738), 106.

Berlan, J.P. (1991) The Historical Roots of Our Present Agricultural Crisis, in *Towards a New Political Economy of Agriculture* (eds W.H. Friedland, L. Busch, F.H. Buttel and A.P. Rudy), Westview Press, Boulder, CO, pp. 115–36.

Brochier, B., Kieny, M.P., Costy, F. *et al.* (1991) Large-Scale Eradication Of Rabies Using Recombinant Vaccine Rabies Vaccine. *Nature*, **354**(6354), 520–2.

Comstock, G. (1988) The Case Against BST. *Agriculture and Human Values*, **5**(3), 36–52.

Executive Office of the President (1994) *Use of Bovine Somatotropin (BST) in the United States: Its Potential Effects, A Study Conducted by the Executive Branch of the Federal Government.* Washington, DC, January.

Hansen, M. (1991) Consumer Concerns: Give us all the Data, in *Agricultural Biotechnology at the Crossroads: Biological, Social and Institutional Concerns* (ed. J.F. MacDonald), National Agricultural Biotechnology Council, Ithaca, NY, pp. 169–76.

Johnson, G.L. and Thompson, P.B. (1991) Ethics and Values Associated with Agricultural Biotechnology, in *Agricultural Biotechnology: Issues and Choices* (eds B. Baumgardt and M. Martin), Purdue Research Foundation, Lafayette, IN, pp. 121–37.

Kalter, R. (1985) The New Biotech Agriculture: Unforeseen Economic Consequences. *Issues in Science and Technology*, **2**, 125–33.

Kroger, M. (1992) Food Safety and Product Quality, in *Bovine Somatotropin and Emerging Issues* (ed. M. Hallberg), Westview, Boulder, CO, pp. 265–70.

Lanyon, L.E. (1994) Dairy Manure and Plant Nutrient Management Issues Affecting Water Quality and the Dairy Industry. *Journal of Dairy Science*, **77**, 1999–2007.

Lanyon, L.E. and Beegle, D.B. (1989) The Role of On-farm Nutrient Balance Assessments in an Integrated Approach to Nutrient Management. *Journal of Soil and Water Conservation*, **44**, 164–8.

Munro, I.C. and Hall, R.L. (1991) Food Safety and Quality: Assessing the Impact of Biotechnology, in *Agricultural Biotechnology, Food Safety, and Nutritional Quality for the Consumer* (ed. J.F. MacDonald), National Agricultural Biotechnology Council, Ithaca, NY, pp. 64–73.

NABC (1991) Summary, in *Agricultural Biotechnology, Food Safety, and Nutritional Quality for the Consumer* (ed. J.F. MacDonald), National Agricultural Biotechnology Council, Ithaca, NY, pp. 58–62.

Norton, B. (1991) *Toward Unity Among Environmentalists*, Oxford University Press, Oxford.

Rawls, J. (1987) The Idea of an Overlapping Consensus. *Oxford Journal of Legal Studies*, **7**, 3–24.

Rollin, B.E. (1986, republished 1990) The Frankenstein Thing, in *Agricultural Bioethics: Implications of Agricultural Biotechnology* (eds S.M. Gendel, A.D. Kline, D.M. Warren and F. Yates), Iowa State University Press, Ames, IA, pp. 292–308.

Rollin, B.E. (1992) The Creation of Transgenic Animal 'Models' for Human Genetic Disease, in *Animal Biotechnology: Challenges and Opportunities* (ed. J. F. MacDonald), National Agricultural Biotechnology Council, Ithaca, NY, pp. 85–94.

Stich, S. (1978, republished 1989) The Recombinant DNA Debate, reprinted in *Philosophy of Biology* (ed. M. Ruse), Macmillan Publishing Co, New York, pp. 229–43.

Tauer, L. (1992) Impact of BST on Small versus Large Dairy Farms, in *Bovine*

Somatotropin and Emerging Issues (ed. M. Hallberg), Westview, Boulder, CO, pp. 207–17.

Thompson, P.B. (1992) BST and Ethical Issues, in *Bovine Somatotropin and Emerging Issues* (ed. M. Hallberg), Westview, Boulder, CO, pp. 33–49.

Thompson, P.B. (1997) *Food Biotechnology in Ethical Perspective*, Chapman and Hall, London.

Thompson, P.B., Matthews, R. and Van Ravenswaay, E. (1994) *Ethics, Public Policy and Agriculture*, MacMillan, New York.

18

Advisory considerations on the scientific basis of the food safety evaluation of transgenic animals

Daniel D. Jones[1]

18.1 INTRODUCTION

Advances in animal biotechnology, and transgenic animals in particular, present many new challenges to society and governments in areas such as bioethics, patenting, food quality, food safety, and possible environmental effects. Our agency has co-sponsored conferences on a number of technical and societal issues in animal biotechnology (Burke, 1993; McGloughlin, 1994). It is evident from such conferences that scientific, environmental, and other interest groups often have divergent views on these issues, making it difficult to develop coherent public policies. One mechanism that has proven useful in the United States for organizing policy input from diverse interest groups and in building public confidence in new technologies is the use of Federal advisory committees.

The US Federal Advisory Committee Act provides for committees of mostly non-government experts to develop advice on government programmes and policies. Today, an average of 1000 advisory committees with more than 20 000 members advise the Federal government on a wide range of scientific, technical, economic, and social issues. Scientific advisory committees in the US usually consist of scientists and other

[1] The author is the National Program Leader for Biotechnology, Cooperative State Research, Education and Extension Service, US Department of Agriculture, Room 843, Aerospace Center, 14th and Independence Avenue S.W., Washington, DC, 20250, USA. The views expressed are those of the author and do not necessarily represent official policy or the interpretations of USDA.

experts who meet to discuss issues identified by the government and develop recommendations on how best to deal with them. With few exceptions, advisory committee meetings are open to the public and this provides an opportunity for the public and a variety of interest groups to make their views known as a committee develops its recommendations.

One of the highest profile advisory committees in the US has been the National Institutes of Health Recombinant DNA Advisory Committee (NIH-RAC). The NIH-RAC developed recommendations on emerging issues in recombinant DNA research including a research moratorium, laboratory containment, and agricultural field testing (Watson and Tooze, 1981). In the area of food and agriculture, other advisory committees have advised the US government on a variety of issues such as agricultural research priorities, the safety of food ingredients, standards for organically produced foods, genetic engineering, and the safety of whole foods from genetically modified organisms. This chapter focuses on recently completed scientific recommendations on the food safety evaluation of transgenic animals.

18.2 SCIENTIFIC ADVICE ON BIOTECHNOLOGY IN FOOD AND AGRICULTURE

The US Department of Agriculture established a scientific advisory committee in 1987 called the Agricultural Biotechnology Research Advisory Committee (ABRAC). The ABRAC was modelled in part after the NIH-RAC which had reviewed proposals in the biomedical and non-biomedical areas for a number of years. The establishment of the ABRAC arose in part from the interest of the NIH-RAC in limiting its future reviews to the health and biomedical areas. The ABRAC, until its termination in 1996, resembled the NIH-RAC, but differed in that it embodied expertise and experience specific to the food, agricultural and environmental research community.

The purpose of the ABRAC was to provide advice to the Secretary of Agriculture on policies, programmes, operations, and activities associated with the conduct of agricultural biotechnology research. This included administrative and procedural measures designed to promote the safety, effectiveness, and public acceptance of agricultural biotechnology research. The ABRAC could also, when it was asked, address the scientific aspects of regulatory questions and thereby help to strengthen the scientific basis and credibility of actions taken by regulatory agencies within USDA.

The ABRAC consisted of 15 members drawn from academia, industry and government with knowledge in such fields as animal/veterinary science, fisheries science, plant science, forestry, microbiology, bioprocessing, food science, environmental science and policy, laws and regulations and bioethics.

18.3 NON-TRANSGENIC ANIMALS FROM TRANSGENIC EXPERIMENTS

Current methods of producing most transgenic animals other than fish have a success rate in the range of 2% or less. One of the first issues to arise from experiments on transgenic animals is what to do with the animals that are not successfully transformed, sometimes referred to colloquially as 'non-takes' or 'no-takes'. Some US companies, in order to help defray their research costs for animals such as cattle with long reproductive cycles, expressed interest in entering non-transgenic animals from transgenic experiments into the food supply. In the US, this would require approval by the USDA Food Safety and Inspection Service (FSIS) which inspects cattle, sheep, swine, goats, equines and poultry intended for human food use.

Some consumers and interest groups may feel concern about the safety of food products from non-transgenic animals used in transgenic experiments simply because they have been exposed, albeit unsuccessfully, to attempts at genetic transformation. In June of 1990, the FSIS requested the ABRAC to review criteria that the FSIS had developed for the food safety evaluation of non-transgenic animals from transgenic animal research (ABRAC, 1990). The criteria proposed by the FSIS to determine if animals are non-transgenic were: (i) failure to detect the presence of the transgene by Southern hybridization or the polymerase chain reaction (PCR); (ii) absence of measurable transgene product; and (iii) a healthy phenotypic appearance. If an animal were found to be non-transgenic by these criteria, then it would be subject to inspection and slaughter under the regulatory procedures that the FSIS normally uses for livestock and poultry used in research (FSIS, 1990).

After review and discussion of the proposed FSIS criteria, the ABRAC expressed approval of this general approach, anticipating that some of the details might change as regulatory needs evolve and scientific progress is made. In December 1991, the FSIS published a notice affirming that animals involved in biotechnology experiments are research animals and that the FSIS would inspect for food use livestock and poultry which were involved in such research but which were not genetically modified products of biotechnology (FSIS, 1991). In its 1991 decision criteria for non-transgenic animals, the FSIS included an additional criterion, namely, the absence of transgene-associated traits, and it included other appropriate scientific methods in addition to Southern hybridization and PCR as suitable test methods.

18.4 TRANSGENIC ANIMALS

Non-transgenic animals from transgenic experiments are very similar in physiology and behaviour to conventional animals which breeders have not attempted to transform by artificial gene transfer. However, animals which have been successfully transformed by gene transfer may differ in ways that are small or large, intended or unintended, or significant or insignificant, from conventional animals. The task of evaluating the food safety of transgenic animals is thus generally a more challenging undertaking than for non-transgenic animals.

The FSIS, in considering several possible regulatory approaches to transgenic animals, requested the ABRAC to address the scientific aspects of the food safety evaluation of transgenic animals. For this purpose, the ABRAC formed a Transgenic Animal Working Group whose members are identified in Table 18.1.

The Transgenic Animal Working Group, with assistance from FSIS, developed a proposed approach for the definition, identification and food safety evaluation of transgenic animals, summarized in Table 18.1 (ABRAC, 1993a). This Working Group proposal was presented to the

Table 18.1 Summary of suggested points to consider for evaluating the food safety of transgenic animals*

For non-virally mediated genetic changes:

- Concentration and tissue distribution of gene product relative to non-transgenic animals
- Secondary effects of possible food safety concern:
  - Visual pathology
  - Detectable metabolic effects
  - Marker gene effects
  - Promoter–enhancer *cis* sequence effects
  - Traditional toxicological testing

For virally mediated genetic changes:
(in addition to the above)

- Spread or infectivity of virus/viroid/vector
- Effects of viral enhancers on nearby host genes
- Recombination of retroviral vector with endogenous viral sequences

*Developed by the USDA Agricultural Biotechnology Research Advisory Committee's Transgenic Animal Working Group. Members included Dr James Lauderdale (Chair), Upjohn Company; Dr Ann Boyd, Hood College; Dr Harold Hafs, Merck, Sharp & Dohme; Dr Susan Harlander, Land O'Lakes, Inc.; Dr Duane Kraemer, Texas A&M University; Dr Bennie Osburn, University of California, Davis; Dr Gary Weber, USDA Extension Service; Dr William Witt, US Food and Drug Administration; Dr Richard Witter, USDA Agricultural Research Service.

full ABRAC (ABRAC, 1993b) which, over a period of months, modified and refined the approach.

18.5 DEFINITION OF TRANSGENIC ANIMALS

The definition of transgenic animals developed by the Transgenic Animal Working Group and the ABRAC is as follows: Transgenic animals are animals and their progeny whose genetic composition has been changed by introducing specific genes (e.g. recombinant DNA). The ABRAC intended the scope of this definition to include animals developed by antisense, amplification and deletion technologies. The definition also appears to be sufficiently broad to include mosaic and chimeric animals. In its discussions, the ABRAC intended that the presence or detection of specific genes in an animal is the criterion for being transgenic rather than expression *per se*.

The ABRAC discussed extensively whether the definition of transgenic animals should distinguish between parental, intraspecific and exogenous sources of DNA. One view was that 'transgenic' implied the movement of genes from one animal to another and that the definition should capture both interspecific and intraspecific gene transfers. An opposing view contended that the definition of transgenic animals should not include intraspecific gene transfers, should not imply that the source of a gene determines its safety, and that any distinction between endogenous and exogenous genes is irrelevant to risk. However, as others pointed out, intraspecific gene transfers, with novel combinations of structural genes and regulatory elements, can lead to significant consequences such as gene amplification, increases in the level of gene expression, and expression in new tissues. In support of this view, endogenous genes generally have a history of safe expression in the host while exogenous genes may result in the expression of a new gene product with unknown consequences for safety.

The opposing view conceded that newness or novelty of gene expression may be important for safety, but reiterated that the endogenous/exogenous distinction was not. Under this view, an animal that is alive and in good health would be seen as its own test for safety. The ABRAC resolved the point by collapsing the endogenous/exogenous distinction into the single category of gene product on which the safety evaluation would focus (ABRAC, 1993b).

18.6 IDENTIFICATION OF TRANSGENIC ANIMALS

The ABRAC, with assistance from the FSIS, proposed criteria for detecting and identifying transgenic animals. The criteria for identifying transgenic animals were for the most part the opposite of the criteria for non-transgenic animals that the FSIS announced in 1991 (FSIS, 1991). The proposed basis for identifying transgenic animals was at least one of the following:

1. Detection of the transgene by Southern hybridization, polymerase chain reaction, or other appropriate method.
2. Presence of measurable gene product.
3. Presence of transgene-associated traits.

The minimum level of detection proposed was one copy of foreign DNA per haploid genomic equivalent as determined by DNA hybridization.

18.7 EVALUATING THE FOOD SAFETY OF TRANSGENIC ANIMALS

The ABRAC identified a number of different kinds of genetic changes that could be accomplished in animals, including gene additions, gene modifications (amplification, deletion, transcription/translation), mosaicism, marker genes and somatic cell therapy. The ABRAC divided the food safety evaluation of transgenic animals into two parts depending on whether the genetic change was accomplished using a viral vector or non-virally.

18.7.1 Non-virally mediated genetic changes

The first point to consider in the food safety evaluation of transgenic animals under the recommended ABRAC approach is determination of the concentration and tissue distribution of the transgene product relative to those of non-transgenic animals. This approach is similar to the general approach taken by other food safety authorities (International Food Biotechnology Council, 1990; Food and Drug Administration, 1992; OECD Group of National Experts, 1993). If the concentrations and tissue distributions of the gene product in transgenic animals are not significantly different from those in non-transgenic animals, the next step would be examination of secondary effects described below.

If the transgene product concentrations and tissue distributions are significantly different from those of non-transgenic animals, then the food safety implications would be assessed first on the basis of existing technical literature and regulatory history relating to that gene product. With sufficient technical information and regulatory history, a food safety assessment can be based on existing human food safety guidelines. If the technical literature and regulatory history are too limited to support a safety assessment, the ABRAC suggested convening an expert panel. The role of such an expert panel would be to define the questions pertaining to assays, animal models, and experimental sensitivity and specificity that must be addressed in order to determine the human food safety of the transgene product.

18.7.2 Secondary effects

The ABRAC addressed concerns about possible unintended secondary or pleiotropic effects of transgenic processes in animals on the safety of food products prepared from them. The presence and impact, if any, of such secondary effects on food safety could be determined by a variety of methods, including visual inspection, detectable metabolic changes, and traditional pathology and toxicological testing methods as appropriate. These methods would presumably also detect significant results of unintended genetic changes related to gene transfer such as insertional mutagenesis.

The question of the safety evaluation of marker genes transferred in tandem with the gene of primary interest also arose. The ABRAC recommended that if a marker gene is used to track the gene of primary interest, the marker gene and its expression product(s) should be subject to the same criteria of concentration and tissue distribution relative to non-transgenic animals as the gene of primary interest (Kline, 1993). Fuchs *et al.* have purified chemically analysable quantities of the gene product of *NPTII*, a commonly used marker gene, and shown chemical equivalence of the plant and microbial varieties, as well as digestibility and no observed adverse effects when fed to mice (Fuchs *et al.*, 1993a,b).

The ABRAC addressed the question of unintended effects of promoter–enhancer *cis* sequences on gene expression. ABRAC members noted that the orientation of the promoter to the inserted gene during *in vitro* cloning is subject to constraints of experimental design. Cloning vectors usually have specific relative positions in which the gene and its regulatory elements are placed in order to promote successful expression of the gene product. However, new enhancer elements can influence gene expression from distal sites in both the 3′ and 5′ directions from the insertion site and in positive and negative ways in different cell types.

Therefore, some testing strategy should be included in proposals concerning transgenic animals to assay or detect unwanted effects of inserted promoters/enhancers on host genes. The ABRAC recommended that secondary effects of enhancers on distant host genes should be evaluated either in the host or in scientifically defensible model systems in order to identify any deleterious side effects which may influence human food safety.

18.7.3 Virally mediated genetic changes

When a transgenic animal is produced with the aid of a viral vector, the ABRAC recommended three points to consider in addition to those described above.

1. A determination of the spread or infectivity of the virus, viroid, or other vector should be made. The FSIS normally refers questions involving infectious DNA or viruses in connection with food animals to the USDA Animal and Plant Health Inspection Service.
2. An evaluation of the effect of viral enhancers on host cell genes near the insertion site should be performed. The rationale for examining the effects of viral vectors is similar to that for expression vector enhancers described above.
3. If a retroviral vector is used, the possibility of recombination with endogenous viral sequences should be evaluated. Since this may require criteria beyond those used for transgenic animals, the ABRAC recommended the use of an expert panel for this purpose.

The ABRAC forwarded its recommendations for the food safety evaluation of transgenic animals to the US Department of Agriculture in September 1993 (Kline, 1993). In March 1994, the FSIS published a notice of availability of its 'Points to Consider in the Food Safety Evaluation of Transgenic Animals from Transgenic Animal Research' (FSIS, 1994). The FSIS points to consider document is very similar to the ABRAC recommendations with only minor changes.[2]

18.8 OTHER KINDS OF TRANSGENIC ANIMALS

Scientists in the US are developing other kinds of transgenic animals that may require environmental as well as food safety evaluations. For example, scientists at Auburn University in Alabama requested approval from USDA for studies of transgenic fish in outdoor research ponds. The agency that co-funded the research, the USDA Cooperative State Research Service (CSRS), prepared the required environmental assessment for transgenic carp and published it in 1990 (CSRS, 1990a).

[2] The FSIS 'points to consider' document may be obtained from the Director, Technology Transfer and Coordination Staff, Science and Technology, Food Safety and Inspection Service, US Department of Agriculture, Washington, DC, 20250, USA, Telephone: (202) 720-8623.

The agency later approved studies for transgenic carp (CSRS, 1990b) and transgenic catfish.[3] Environmental and other interest groups have often called for improved procedures and standards for research on genetically modified aquatic organisms used in aquaculture research. This could also reduce the need for labour- and resource-intensive, case-by-case evaluations of the safety of fish research proposals in the future.

The ABRAC has been involved in a broad-based effort to develop generic performance standards for assuring the environmental safety of aquatic research with genetically modified fish and shellfish. Toward this end, the ABRAC formed a Working Group on Aquatic Biotechnology and Environmental Safety which included aquaculture and aquatic research experts from outside the ABRAC. This ABRAC working group worked closely with the aquatic research community, the aquaculture industry, and Federal and state government agencies to develop appropriate and acceptable scientific standards for research with transgenic fish and shellfish.

The Working Group met in conjunction with a workshop for scientists, business people, government officials and other interest groups at the University of Minnesota in 1993. Workshop participants discussed scientific goals, means and criteria to help scientists with the genetic and ecological risk assessment of research involving genetically modified fish, crustaceans and molluscs and the design of adequate biosafety protocols. ABRAC recommendations on guidance to help assure the environmental safety of research on modified fish and shellfish were published in 1995 (ABRAC, 1995).[4]

18.9 BENEFITS FOR THE POLICY PROCESS

The openness of ABRAC meetings and the flexibility to form working groups with diverse expertise on specific issue areas have brought significant benefits to the scientific community, special interest groups, the Federal government, and the general public. The government benefits by having an outside source of scientific advice on biotechnology with a high level of scientific authority and public credibility. Agricultural researchers benefit by having a scientific source of public guidance for assuring the safety and environmental compatibility of

[3] In the interim between the carp and catfish approvals, the CSRS published agency procedures to implement the US National Environmental Policy Act (CSRS, 1991). The reason that a separate *Federal Register* notice was not published for the catfish approval is that the agency determined that the experimental protocol, as proposed, fell under a categorical exclusion in the published procedures.

[4] The performance standards are also available on the worldwide web at http://www.nbiap.vt.edu/.

agricultural research. Companies benefit by having a public and scientifically credible forum in which the safety and public acceptance of products of agricultural biotechnology can be discussed openly. Members of the general public, including interest groups, benefit by having an opportunity to make their views on biotechnology known and considered for incorporation into public policy on agricultural biotechnology.

18.10 CONCLUSION

The ABRAC has proved to be a useful mechanism for integrating input on emerging biotechnologies from diverse interest groups into the public policy process. The ABRAC has shown that it can develop science-based guidance for the safety evaluation of particular groups of genetically modified organisms, such as transgenic livestock and aquatic organisms.

Acknowledgements

The author acknowledges the original contributions of the members of the Transgenic Animal Working Group identified in Table 18.1. The author also gratefully acknowledges comments and suggestions on the manuscript made by Marylin Cordle, Alvin Young, Pat Basu, Anne Kapuscinski, and James Lauderdale.

REFERENCES

Agricultural Biotechnology Research Advisory Committee (1990) *Minutes, June 21–22, 1990, Document No. 90-02*, US Department of Agriculture, Washington, DC.

Agricultural Biotechnology Research Advisory Committee (1993a) *Minutes, Transgenic Animal Working Group, April 8, 1993, Document No. 93 WG-01*, US Department of Agriculture, Washington, DC.

Agricultural Biotechnology Research Advisory Committee (1993b) *Minutes, June 29–30, 1993, Document No. 93-01*, US Department of Agriculture, Washington, DC.

Agricultural Biotechnology Research Advisory Committee (1995) *Performance Standards for Safely Conducting Research with Genetically Modified Fish and Shellfish, July 31, 1995*, Document nos. 95-04, 95-05, US Department of Agriculture, Washington, DC.

Burke, W.S. (ed.) (1993) *Symbol, Substance, Science: The Societal Issues of Food Biotechnology, Conference Proceedings*, North Carolina Biotechnology Center, Research Triangle Park, NC.

Cooperative State Research Service (1990a) Research proposal on transgenic fish; publication of environmental assessment, in *Federal Register*, February 16, 1990, **55**(33), 5752–72.

Cooperative State Research Service (1990b) Availability of an environmental assessment and finding of no significant impact relative to USDA funding of research on transgenic carp, in *Federal Register*, November 21, 1990, **55**(225) 48661–2.

Cooperative State Research Service (1991) Agency procedures to implement the National Environmental Policy Act, in *Federal Register*, September 27, 1991, **56**(188), 49242–8.

Food and Drug Administration (1992) Statement of policy: foods derived from new plant varieties, in *Federal Register*, May 29, 1992, **57**(104), 22984–3005.

Food Safety and Inspection Service (1990) § 309.17, Livestock used for research, in *Code of Federal Regulations, Title 9, Animals and Animal Products*. US Government Printing Office, Washington, DC.

Food Safety and Inspection Service (1991) Livestock and poultry connected with biotechnology research, in *Federal Register*, December 27, 1991, **56**(249), 67054–5.

Food Safety and Inspection Service (1994) Update on livestock and poultry connected with biotechnology research, in *Federal Register*, March 17, 1994, **59**(52), 12582–3.

Fuchs, R.L., Heeren, R.A., Gustafson, M.E., Rogan, G.J., Bartnicki, D.E., Leimgruber, R.M., Finn, R.F., Hershman, A. and Berberich, S.A. (1993a) Purification and characterization of microbially expressed neomycin phosphotransferase II (NPTII) protein and its equivalence to the plant expressed protein. *Bio/Technology*, **11**, 1537–42.

Fuchs, R.L., Ream, J.E., Hammond, B.G., Naylor M.W., Leimgruber, R.M. and Berberich, S.A. (1993b) Safety assessment of the neomycin phosphotransferase II (NPTII) protein. *Bio/Technology*, **11**, 1543–7.

International Food Biotechnology Council (1990) Biotechnologies and food: assuring the safety of foods produced by genetic modification. *Regulatory Toxicology and Pharmacology*, **12**(3), part 2, i-S196.

Kline, A.D. (1993) Letter dated September 27, 1993 to R.D. Plowman, Acting Assistant Secretary, Science and Education, US Department of Agriculture, Washington, DC.

McGloughlin, M. (ed.) (1994) *Proceedings of the International Workshop on Animal Biotechnology Issues*, Biotechnology Program, University of California, Davis.

OECD Group of National Experts on Safety in Biotechnology (1993) *Safety evaluation of food derived by modern biotechnology: concepts and principles*, Organization for Economic Cooperation and Development, Paris.

Watson, J.D. and Tooze, J. (1981) *The DNA Story: A Documentary History of Gene Cloning*, W.H. Freeman and Company, San Francisco.

19

Legislation and regulation: a view from the UK

Michelle Paver

19.1 INTRODUCTION

The law relating to animal biotechnology falls broadly under two headings:

- *regulatory aspects*, that is, rules and regulations governing what research and industry are or are not permitted to do; and
- *monopolistic aspects*, that is, the law governing what research and industry may or may not monopolize by means of the patent system.

This review will touch briefly on the complex regulatory field, and concentrate on patent law, considering the extent to which biotechnological inventions can be monopolised under the patent system.

19.2 THE UK REGULATORY ENVIRONMENT

In the UK in recent years, two sets of regulations have been brought in which address particular concerns raised by biotechnology.

The so-called 'Contained Use Regulations' came into force on 1 February 1993, implementing an earlier EC Directive (90/219/EEC). Broadly, they aim to ensure the safe use and handling of genetically modified organisms ('GMOs'), in particular by limiting contact between GMOs and the environment. (The UK regulations are in fact broader than the Directive, since the Directive concerns only microorganisms, whereas the UK regulations also cover non-microorganisms such as plants and animals.)

The 'Deliberate Release Regulations' also came into force on 1 February 1993 and they too (along with Part VI of the Environmental

Protection Act 1990) implemented an EC Directive (90/220/EEC). In essence, the Deliberate Release Regulations require consent to be obtained from the Department of the Environment before the release or marketing of GMOs.

In the UK it has become increasingly apparent that both sets of regulations are highly complex and difficult to work in practice, as well as being hard to interpret. Further EC Directives (94/51/EC; 94/15/EC) were introduced in 1994, both of which have been implemented by regulations in the UK. The primary aim of these is to simplify the procedure for obtaining Government consents, particularly so as to avoid hampering long-term development projects. Yet further EC proposals to achieve this are currently under consideration.

19.3 UK PATENT LAW AND THE EUROPEAN PATENT SYSTEM

Before considering the extent to which animal biotechnology is capable of being monopolized, it is instructive to recall what a patent is, and how it may be used.

A patent is a monopoly granted by the State to an inventor in return for disclosing his or her invention. It operates as an exclusive right to stop others from practising the invention, rather than being *per se* a right to use the invention. In other words, a patent does not of itself give legal clearance to put an invention into effect: the invention must still comply with national and European standards of safety, ethics, and so on, before it can legitimately be used.

A patent may therefore be regarded as a kind of 'deal' between the State and the inventor. The State rewards the inventor by granting a 20-year monopoly, and in return the inventor tells the public how to work his invention. This he does in a specification, which describes the invention (generally giving examples of how to carry it out) and sets out claims which precisely define the scope of the monopoly.

The basis of European patent law is the European Patent Convention ('EPC'), which came into force in 1978. This provides the basis for the European Patent system – whereby an inventor files a single patent application at the European Patent Office ('EPO') and designates those European countries in which he wishes the patent to be effective. If and when the European patent is granted, it takes effect as a national patent in each of the designated countries.

The EPC also sets out the criteria for patentability, that is, the requirements which a patent application must satisfy if it is to mature into a granted patent. In respect of these criteria, Member States' national patent laws (for example, the UK Patents Act 1988) must harmonize with the EPC.

Thus, to be patentable in the UK (and in the EPO), an invention must be:

- Novel
- Inventive
- Capable of industrial application
- Sufficiently described
- Not inherently unpatentable – for example, not a discovery, immoral invention, essentially biological process, or plant or animal variety.

Even if a European patent application is found to satisfy the above criteria and is eventually granted, however, it is important to bear in mind that its validity may still be challenged. This may be done either in the EPO, by means of opposition proceedings commenced within 9 months of the grant of the patent, or in revocation proceedings brought by an interested party in a relevant national court, at any time during the life of the patent.

In considering biotechnological inventions, it is important to appreciate that the above criteria apply to *all* inventions – including biotechnological ones. Thus, the motto is: 'the normal patentability criteria apply' – irrespective of whether the subject matter of the invention is, or is not, living matter.[1] Clearly, however, biotechnology raises unique questions about the interpretation of these criteria.

The current approach in Europe is that the EPC is capable of accommodating the special problems that biotechnology raises, without the need for amendment. However, it has for some time been acknowledged that there may be a need for clarification in this area. Accordingly, a draft Directive on the patentability of biotechnological inventions ('the draft Directive') was promulgated in 1988.[2] The express aim of the draft Directive was to clarify the way in which the established patentability criteria should be applied to biotechnological inventions, both by the EPO, and by national courts and patent offices. There was no intention actually to change the law.

In subsequent years, the draft Directive went through a great many controversial revisions and met with considerable opposition, not least from the European Parliament. It seemed finally to have expired early in 1995, when it was torpedoed by the European Parliament. However,

[1] The same principle applies in the USA. In its oft-quoted decision Diamond v. Chakrabarty (206 USPQ 193 1980), the US Supreme Court held that patentable subject matter includes 'anything under the sun that is made by man'. The position was made even clearer in the US Patent and Trade Mark Office's statement of 1987, shortly after its first ruling on the patentability of a multicellular organism in *ex parte Allen* (USPQ 2d 1425 1987), when it said it considered 'non-naturally occurring non-human multicellular living organisms, including animals', to be patentable subject matter.

[2] OJ no. C 10, 13.1, 1989, p.3.

at the end of 1995, the Council and Commission proposed a further draft, in order to resolve the legislative log-jam and to prevent Europe being at a competitive disadvantage as compared with the rest of the world, particularly the United States, given the increasing importance of the biotechnology industry.

Whatever the eventual outcome, however, it should be borne in mind that any Directive which ultimately results will technically have no binding force in the EPO, although it must be assumed that EPO Examiners would still pay it some heed. It is also important to remember that the EPO has already, for the past decade or so, been deciding the validity of biotechnology patents, and granting them, without reference to the draft Directive. This has led some people to view the draft Directive as having been superseded.

The questions remain, however: do we need a Directive on biotechnology patents? Can the EPC hold good without amendment? This can best be seen by taking a look at how in the past the patentability requirements have been applied to biotechnological inventions.

19.3.1 Novelty

Even though a given substance already occurs in Nature, a claim to it will be novel if what is claimed is a preparation which is for some reason not available in Nature. For biotechnological inventions this requirement could therefore be met if, for example, a given naturally occurring substance were highly purified or otherwise in a form in which it does not occur in nature, such as a protein produced in an unglycosylated form.

19.3.2 Inventive

An invention must be inventive – in that it must not have been obvious to a person of ordinary skill in the relevant field at the date when the patent was applied for. So what does 'obvious' mean in this context? In the EPO, it means that a man of ordinary skill would have thought the idea worth trying with 'a reasonable expectation of success'. Similar principles have been applied in the UK courts, and, in certain decisions concerning biotechnological inventions, the vulnerability of such inventions to an attack of obviousness has become evident.

Thus, when the Court of Appeal considered the test for inventiveness in *Genentech Inc's Patent* [1989] RPC 147, it invalidated the patent. The Court applied a test to the effect that obviousness requires the skilled man to have thought the idea 'well worth trying out in order to see whether it would have beneficial results'.[3] In other words, a research group would have to 'have been directly led to try the idea in the expectation that it might well produce a useful result'.[4]

A differently constituted Court of Appeal took a similar line recently in *Biogen Inc. v. Medeva plc* ([1995] RPC 2), although for slightly different reasons. Here, the patent concerned the production by recombinant means of hepatitis B viral antigens. This was achieved by the inventors in the late 1970s, at a time when most scientists in the field regarded it as technically impossible. However, this was not enough, in the view of the Court of Appeal, to render the invention non-obvious – that is, inventive. The essential difference between Biogen and the others in the field at the time was (the Court held) that Biogen were:

> 'prepared to make the necessary investment of time and money. ... This is not a description of an inventive step. Nothing the plaintiffs did was not obvious if you took the business decision to do it. Others considered the odds against success too long to justify trying.'

Biogen's appeal to the House of Lords was heard in May 1996, and the decision is imminent. This will be the first occasion on which the House of Lords considers the question of obviousness (and indeed other aspects of patent validity) in relation to biotechnological patents.

The difficulty for patentees in making a case for inventiveness in biotechnology is clear. Much biotechnology is elegantly simple in theory, so it is only too easy to argue along *Genentech* and *Biogen* lines that a given invention is simply the application of known techniques to achieve a known desired end. But should this be enough to deprive the inventor of the reward for his endeavour? It will be interesting to see what the House of Lords has to say on the question.

19.3.3 Capable of industrial application

To be patentable, an invention must be capable of industrial application. This tends not to be a problem for most biotechnological inventions – although it might have proved a bar to the controversial attempt by the US National Institute of Health (NIH) to obtain patents covering cDNA fragments [known as Expressed Sequence Tags (ESTs)]. These ESTs correspond to partial sequences of genes expressed in human brain tissue, and for most of the ESTs claimed, the genes to

[3] See Johns-Manville Corporation's Patent ([1967] RPC 479).
[4] See Olin Mathieson Chemical Corporation v Biorex Laboratories [1970] RPC 157.

which they correspond and the proteins which they encode are unidentified. Thus, any alleged industrial applicability remains speculative. As the NIH withdrew its patent applications in view of the public protest they elicited, the question remains unresolved. However, the indications from the commentary to the draft Directive are that in Europe such applications could be regarded as patentable, provided all other patentability requirements are met.

19.3.4 Sufficiency

An invention must be clearly and completely described in the patent specification, to the extent that someone skilled in the art could perform it.

In practice, patent claims tend to be drafted as broadly as possible – so as to cover as many variants of the invention as possible, and thereby obtain as valuable a monopoly as possible. However, with most inventions, the patent specification itself only discloses how to work *some* of these embodiments – and indeed, not all of them may in fact be workable.

The question then arises, does a broad claim satisfy the requirement for sufficiency if the specification actually discloses how to make only a few such variants? Another question: is such a claim sufficient if it covers variants that do not work? Or if some of the examples in the specification are not repeatable with 100% certainty? The answer to the above questions in the EPO and the UK seemed recently to be 'yes', but that position is now in doubt.

Taking the EPO first, the basic approach was set out in *Decision T 292/85 Genentech I/Polypeptide Expression*. The patent claimed recombinant plasmids coding for functional polypeptides. It covered variants which were as yet unknown or inaccessible to the public, and the patent specification only disclosed certain specific examples. The patent was nevertheless held to be sufficient. The EPO said that:

> 'An invention is sufficiently disclosed *if at least one way is clearly indicated enabling the skilled person to carry out the invention.*' (emphasis added)

Similarly, in *Decision T301/87 Biogen/Alpha-interferons* the EPO held that:

> '... it is in the nature of processes starting from natural sources and aiming at genes coding for polypeptides that individual variations inevitably occur ...'

But then in *Exxon/Fuel Oils [1994] EPOR 149*, the EPO Board of Appeals set a limit to the above test (which had become widely known as the 'one-way rule'). The EPO held that:

> '...the disclosure of one way of performing the invention is only sufficient within the meaning of [the term] if it allows the person skilled in the art to perform the invention in the whole range that is claimed...'

It was this decision that provided the English Court of Appeal in *Biogen* (see above) with the basis for their finding that the Biogen invention in question was insufficient. The Biogen patent covered recombinant DNA molecules expressed in suitable host cells to produce polypeptides having the immunological characteristics of hepatitis B viral antigens. The main claim covered the expression of all types of hepatitis B viral antigens, in *all* suitable hosts – whereas, at the relevant time, the Court held that Biogen had only succeeded in expressing *surface* antigen in a *prokaryotic* host. In October 1994 the Court of Appeal held the patent invalid for insufficiency, expressly approving the EPO *Exxon* decision in its judgement:

> 'In our judgement, what the [EPO] Board have said correctly represents the law. The disclosure must be sufficient to enable the whole width of the claimed invention to be performed. ... It is not the law that the disclosure of a single embodiment will always satisfy the requirement regardless of the width of the claim. ... The disclosure must be wide enough to enable the man skilled in the art to perform the claimed invention across its full width not just by reference to one type of antigen or one type of host. The plaintiffs [in the present case] had a choice as to how widely they would draw their claims. If they choose to draw it widely they must accept the co-relative obligation to make a correspondingly wide disclosure. If they are unable to make that disclosure, that shows that they are seeking to claim an invention to which they are not entitled. That is the position here.'

What this decision seems to say is that the 'one-way rule' cannot be relied upon to justify a patent's sufficiency, and that patentees must be able to satisfy the requirement across 'the whole width of the claimed invention'. If allowed to stand, this decision could have major implications for biotechnological patents in the UK – although as mentioned above, Biogen has appealed to the House of Lords. It is of note that a differently constituted Court of Appeal, in *Chiron Corporation* v. *Murex Diagnostics* and *Chiron Corporation* v. *Organism Telenita* (*no. 12*), although bound by the decision in *Biogen*, was nevertheless able to find some flexibility in the *Biogen* test. The Court held that a claim covering a class of products covers a single invention, if all the products within that class all have in common the same essential discovery.

19.4 INHERENTLY UNPATENTABLE: DISCOVERIES

Under European patent law, a discovery, scientific theory, or mathematical method is not patentable. At first glance, it would seem arguable that that which occurs in Nature (such as, for example, a DNA sequence) could amount to a discovery, and so would not be patentable. However, it has been the approach of European patent law that in general, an invention which is the practical application of a discovery will be patentable. Thus, the mere fact that an invention embraces a discovery does not render it unpatentable. In other words, an invention may embrace a discovery (provided it claims a practical application of it) – but it may not consist of one. The distinction is not always easy to draw.

19.4.1 Inherently unpatentable: immorality

European and UK patent law prohibits the patenting of an invention which would be contrary to morality.[5] This is perhaps the area of the greatest public interest as regards biotechnological inventions, and in particular those concerning animals, and those derived from human tissue.

So, what is an immoral 'invention'? The EPO Guidelines for Examination state that an invention is 'immoral' if it is probable that the general public would regard it as so abhorrent that patenting would be inconceivable. Apart from that general statement (and certain proposals in the draft Directive), there are at present no other guidelines on how to apply this morality exclusion. It is up to the individual EPO Examiner to determine the question for himself, on the facts of each case.

The most celebrated application of the morality test to date is in *Decision T19/90 Harvard Mouse/OncoMouse*. Here, the Examining Division at first declined to consider the question at all, on the grounds that patent law is the wrong forum for doing so. However, the Technical Board of Appeal (also declining to do the job itself) remitted the case back to the Examining Division for consideration of the question, and told it to weigh the possible suffering to the animal against the invention's usefulness to mankind. This the Examining Division duly did, finding the invention to be moral, and hence patentable. This is not the end of the story for *OncoMouse*, however, as the patent is being vigorously opposed by a coalition of animal welfare groups.

[5] Article 53(a) EPC (and s.1(3) Patents Act 1977). A similar provision in the US denies patentability to inventions 'injurious to public health or morals' (US 35 USC 101), although interestingly, this does not seem to have been considered by the US Patent and Trademark Office in the prosecution of the US OncoMouse patent.

By contrast with *OncoMouse*, in the case of an Upjohn application, the Examining Division upheld an objection under the morality requirement. The patent claimed transgenic animals for use in screening agents for wool production and hair growth. The EPO decided that the potential suffering to the mouse was not outweighed by the invention's benefit to mankind, even though the invention was commercially important and useful.

The interesting thing about the two mouse cases is that they are relatively polarized in terms of morality – cancer research on the one hand, and hair promotion on the other – but that even so, the EPO did not seem to find it easy to perform its moral balancing act, as well it might not. One can imagine many hard questions arising in the future. For example, which factors should be considered? What weight should be placed on an invention's predicted benefit to mankind, when that benefit is usually only speculative?

Moreover, even knottier questions arise in the context of inventions derived from human tissue. For example, European patent application no. 0,112,149 (in the name of the Howard Florey Institute of Experimental Physiology and Medicine, Australia) covers a human gene obtained from a human ovary coding for relaxin, a hormone which relaxes the womb during childbirth. In 1992 the Green Party in the European Parliament opposed the application on the basis, among other things, of immorality. It argued that it is immoral to take advantage of pregnant women by using ovarian tissue for a technological process with a profit motive, and that it is inherently immoral to patent human genes, on the basis that it is wrong to monopolize Man's genetic material for profit. In January 1995, the EPO rejected the opposition. In its decision the EPO acknowledged that it is not the right institution to decide such fundamental ethical questions, and went on to say that it would apply the immorality exclusion strictly, and only in the clearest of cases. The relaxin case is under appeal.

The draft Directive attempts to clarify the morality question, and a consensus at last seems to be emerging. The main points concerning morality are as follows:

(a) *'The human body and its elements in their natural state shall not be considered patentable inventions.'* However, products obtained from the human body, including genetic material, *may* be patentable, where the element is not in its natural state (see above).
(b) *'Methods of human treatment involving germline gene therapy'* are not patentable. This is much clearer than in previous drafts. (It is interesting to note that there are currently several hundred patent applications pending in the EPO that relate to gene therapy, some of which cover germline as well as somatic gene therapy.)

(c) Also unpatentable will be *'processes for modifying the genetic identity of animals which are likely to cause them suffering or physical handicaps without any substantial benefit to man or animal, and also animals resulting from such processes, whenever the suffering or physical handicaps inflicted on the animals concerned are disproportionate to the objective pursued.'*

(d) Generally, *'inventions shall be considered unpatentable where exploitation would be contrary to public policy or morality.'*

Both (c) and (d) above endorse the tests already used by the EPO. They leave difficult decisions of judgement to the relevant patent Examiner – which could lead to inconsistencies, and does not take account of the fact that the EPO does not seem to consider itself qualified to determine issues of immorality. The aims are laudable, but the practice may well be difficult.

19.5 PLANT AND ANIMAL VARIETIES

To understand this exception one needs to go back a few years. In the 1950s, the idea of plant variety rights became popular, and in 1961 the first UPOV convention[6] specifically prohibited patent protection for plant varieties, which were already protected by plant variety rights. Subsequently, the EPC appears to have borrowed that idea, by prohibiting the patenting of 'plant and animal varieties'. The concept of an 'animal variety' was born. However, no-one seems to know what the term actually means. As the Examining Division acknowledged in 1989 in the first round of *OncoMouse,* it is not a 'criterion sufficient to delineate patentable from non-patentable subject matter'.

This difficulty is made worse by the fact that the French, German and English texts of the EPC are supposed to be equally authentic. Thus, 'animal variety', 'race animale' and 'Tierart' are all presumed to have equal force. The problem is that they mean different things. 'Tierart' is apparently akin to a species, whereas the English and French terms are more vague, connoting some kind of variation, apparently below the species level.

In *OncoMouse,* the Examining Division eventually decided that the patent in question fell outside this exclusion from patentability, because its claims were drafted in terms of non-human 'mammals' and 'rodents' – both of which are higher taxonomic classification units than either Tierarten, races animales or animal varieties. Thus, at least we know what an animal variety is *not*. However, when we go below the species level, we still do not know where the lower taxonomic line of patent-

[6] UPOV: The International Union for the Protection of Plant Varieties.

ability is drawn. Nor do we really know why there is such a lower limit of patentability, since (unlike plants) there are no animal variety rights which could make double protection a real danger.

The draft Directive maintains the non-patentability of plant and animal varieties, without shedding further light on what the terms mean.

19.6 ESSENTIALLY BIOLOGICAL PROCESSES

Under the EPC (and the UK Patents Act), 'essentially biological processes for the production of plants and animals' are *not* patentable – with the important proviso that 'micro-biological processes or the products thereof' *are* patentable. The questions therefore arise: what is an 'essentially biological process', and what is a 'microbiological process'?

There is well-established guidance on the meaning of 'essentially biological process' from the EPO. It is 'the routine manipulation of a known and naturally occurring biological event'. For an invention to fall outside this exception (and so be patentable), there needs to be 'significant technical intervention', which goes beyond the above-mentioned routine manipulation of biological processes.[7] Thus, traditional methods of selective breeding would be classed as essentially biological processes, and be unpatentable. By contrast, the production of, for example, recombinant antigens would be regarded as something more than just the routine manipulation of a natural biological event, and would therefore fall outside this exclusion.

The meaning of 'microbiological processes' is not as clear – although this does not seem to have caused too many problems in the past. This is probably because the EPO has construed it so broadly as to include, among other things, human cell lines – which have been patented for the past decade or so. The draft Directive defines microbiological process as any process 'involving or performed upon or resulting in microbiological materials'. Microbiological material is not actually defined, but the commentary to the draft Directive refers to it as being 'any biological material made up of micro-organisms or cellular or subcellular biological material derived from plants, animals or the human body'.

19.7 SOME UNANSWERED QUESTIONS

As a general principle, it seems sensible that the existing patentability criteria should continue to be applied to biotechnological inventions, as

[7] See *Decision T320/87 Lubrizol Genetics Inc.*

they are applied to all other inventions. However, the controversies that have emerged over the past few years, for example concerning the NIH applications, transgenic animals, and the patenting of human genes – to name but a few – have highlighted the need for a clear and consistent approach to biotechnological inventions. This is perhaps most acute as regards questions of morality, which have not traditionally been the province of patent law. There is therefore a need to satisfy public concerns over ethics and animal rights, while ensuring that the political and economic climate (of which the patent system is an important element) is not made so restrictive as to inhibit invention and investment in the European Community.

Many people feel that, even if the draft Directive eventually becomes law, it may not provide the assistance required. Even if it were much more clearly worded than previous drafts, since it will not actually be binding on the EPO, it would not in itself provide a complete answer.

Of the number of options that have been suggested, perhaps the most sensible would involve arranging for the signatory states to the EPC to adopt a protocol to the EPC, which would set out in some detail the approach to be taken by patent examiners and national courts when applying the immorality exclusion to biotechnological inventions. This would seem to have the advantage of providing the required guidance – provided, that is, that the protocol is itself clearly worded, and avoids the ambiguities and obscurities that have so bedevilled the draft Directive. Such a protocol would also, it is hoped, go some way to ensuring at least a degree of consistency in the approach of patent examiners and national courts to this most difficult of areas.

REFERENCES

Commission Directive 94/51/EC.

Council Directive 90/219/EEC Genetically Modified Organisms (Deliberate Release) Regulations 1992 (S1/1992/3280).

Council Directive 90/220/EEC.

Decision T320/87 Lubrizol Genetics Inc.

Genetically Modified Organisms (Contained Use) regulations 1992 (S1 1992/3217).

Genetically Modified Organisms (GMOs) Proposed New Regulations, A Consultation Paper, Department of Environment.

Genetically Modified Organisms (Contained Use) (Amendments) Regulations 1996, and Genetically Modified Organisms (Deliberate Release) Regulations 1995.

Johns-Manville Corporation's Patent ([1967] RPC 479).

Olin Mathieson Chemical Corporation v. Biorex Laboratories [1970] RPC 157.

Proposals for the amendment of the Genetically Modified Organisms (Contained Use) Regulations 1992, Health and Safety Commission.

20

Animal patenting: European law and the ethical implications

Peter Stevenson

20.1 INTRODUCTION

The patenting of animals has become a major issue in Europe as a result of two separate developments. Firstly, from 1988 to 1995 a proposed EU Directive, which would have permitted the patenting of genetically engineered animals, wound its way through the protracted process of law-making in the European Union. Finally in March 1995, the European Parliament voted to reject the Directive, in part due to the opposition of some MEPs to animal patenting. In 1996, however, the European Commission re-issued the proposed Directive, thereby giving fresh impetus to the debate on this issue.

Secondly, in 1992 the European Patent Office (EPO) in Munich granted the first European patent for an animal, the oncomouse patent. The EPO operates under the European Patent Convention. This Convention is not a piece of European Union law. However, all the EU Member States are members of the Convention. The following non-EU countries are also members of the Convention: Switzerland, Monaco and Liechtenstein.

20.2 ETHICAL CONCERNS

Both the public and politicians have of late been expressing increasing concern about the ethics of patenting transgenic animals. Patenting is often opposed for two separate, though interconnected, reasons. Firstly, the wider availability of patents will give a massive commercial boost to genetic engineering, a process which all too often poses very considerable threats to the health and welfare of the animals involved.

Secondly, there is a growing belief that the patenting of animals is in itself unethical.

We are living at a time when people are increasingly concerned about certain existing uses of animals – such as the export of live animals and factory farming – and it is a cause of dismay that at this very time a new source of animal abuse – genetic engineering and cloning – is being developed.

20.2.1 Enhanced productivity

Some of the dangers of transgenesis can be illustrated by the genetic engineering of farm animals for greater productivity. Through selective breeding to make them grow more quickly, or bigger, many modern farm animals have already been pushed to, or in some cases beyond, their physiological limits. Genetic engineering is likely to exacerbate this process. Patented farm animals are likely to be the victims of severe stress, their bodies designed to grow ever more quickly, or bigger, or to produce yet more milk or eggs.

In 1990, an application (since withdrawn) was made to the EPO for a patent for a transgenic chicken with the bovine growth hormone gene. The purpose: faster growth, leaner meat and earlier sperm production in males.

Traditional selective breeding has already had a highly detrimental impact on broiler chickens (broilers are the birds reared for their meat). The modern broiler has been bred to reach its slaughter weight in just 6 weeks. This is twice as fast as 30 years ago. What grows quickly is the muscle, which is what is eaten as meat. The bones, however, fail to keep pace and cannot properly carry the overdeveloped body. As a result, each year millions of chickens suffer from painful, sometimes crippling, leg disorders. The very rapid growth of modern chickens also puts an enormous strain on their hearts and lungs. As a result, many die of heart disease before reaching their slaughter age of just 6 weeks.

All this is just the result of selective breeding (and rich diets). A very high degree of stress could await a transgenic chicken spurred on to yet faster growth by the bovine growth hormone gene. Such an animal is, moreover, unnecessary. If more poultry meat is needed, the simple answer is to breed more ordinary chickens. There is no need for a transgenic superchick.

The dangers of genetic engineering can also be seen in the case of the Beltsville (USA) pigs, into which bovine growth hormone genes were inserted. The Beltsville pigs suffered from a range of health problems, including lameness, severe synovitis, degenerative joint disease, ulcers, certain heart diseases and inflammation of the kidneys.

Selective breeding has already imposed considerable stress on pigs as they have been pushed to faster growth rates. Some have painful joint problems and others die of heart disease. It is arguably unethical to risk imposing yet more pain and suffering on pigs by genetically engineering them to grow at an even faster rate.

In Australia and the UK a number of animal patents have already been granted. These include a patent for sheep genetically engineered for increased wool growth. Other patents are for methods of producing transgenic farm animals, designed, for example, to have increased weight gain or feed efficiency. A patent has been granted by the EPO for a method for the creation of a pig with extra porcine growth hormone gene.

20.2.2 Disease resistance

One branch of genetic engineering aims to produce farm animals which are resistant to certain diseases. At first sight this may appear to be a benign development. However, much of this work concentrates on diseases of the factory farm. These are diseases which inevitably develop when large numbers of animals are kept crammed in close proximity to each other.

By genetically engineering animals to be resistant to the endemic diseases of the factory farm, we are in real danger of condemning them to a continued existence in those restrictive, barren conditions. Surely the real answer does not lie in genetically engineering disease resistant animals; instead we should rear animals in humane conditions, conditions which by their nature are less likely to be a breeding ground for infectious disease.

20.2.3 Xenotransplantation

The threats posed to animal welfare by advanced technologies have been highlighted by two recent developments: cloning and the incorporation of human genes into pigs to make their organs more suitable for transplantation into people.

In the excitement about these 'scientific triumphs', the animal suffering inherent in these developments has largely been ignored.

The production of genetically engineered animals – be it for xenotransplantation or other purposes – involves subjecting a number of animals to invasive surgical procedures. Firstly, the donor animal is superovulated with hormone injections; then in the case of pigs and sheep, after artificial insemination, the donor undergoes surgery to remove her fertilized egg cells. After they have been genetically modified, the embryos are implanted – surgically – into surrogate mothers. In the writer's view, it is ethically unacceptable for surgical procedures to be carried out on animals for the benefit not of the animals but of humans.

Pigs destined to be organ donors are likely to be kept in unnatural, ultra-sterile conditions with no straw or bedding material. Indeed, the Advisory Group, which in 1996 prepared a Report on the Ethics of Xenotransplantation for the UK Department of Health, concluded that for pigs involved in xenotransplantation, husbandry methods 'are likely to be more restrictive and closely controlled than is usual in, for example, good agricultural practice' (Kennedy, 1996).

20.2.4 Cloning

The production of one cloned sheep can involve killing several sheep and subjecting a number of others to surgical interventions. Once the cloned embryo is formed, it is generally placed – surgically – in a temporary recipient ewe. After a few days, this 'foster mother' ewe is killed and the embryo extracted. If developing well, the embryo is then placed – again surgically – into another ewe who carries it to term.

Cloned animals can suffer from serious abnormalities. Three of the five lambs whose cloning in Scotland was reported in 1996 had malformed internal organs; all three died shortly after birth. Moreover, cloned fetuses often grow abnormally large, which can lead to painful or caesarean births.

Researchers suggest that cloning will be a useful tool in multiplying the number of genetically engineered animals. Previous work on the genetic engineering of sheep has, however, resulted in problems. When bovine transgenes were incorporated into sheep, they resulted in pneumonia and a diabetes-like condition prior to death.

20.2.5 Ethical issues

For thousands of years people have made use of animals. Genetic engineering, however, represents a major departure in that 'making-use'. It involves saying: the way God/Nature has created this animal and provided for its procreation is not good enough; we will manipulate its genetic makeup to make it of more use to us. In contrast to this approach, an increasing number of people feel the need to rediscover a sense of awe and respect for the world in which we live rather than seeking to manipulate it to our own ends. And there can be no manipulation more profound than that of another being's genetic structure.

Having manipulated animals, some people then wish to patent them. The thinking behind a patent is that if someone has spent a great deal of skill and time in inventing something, they should be able to prevent others from marketing the invention without their permission.

It will be clear from this that at the heart of the patenting system is the need for there to be an invention. If you have not invented something, you cannot get a patent. Now, it is the notion that a person has invented an animal, or a process for creating an animal, which underscores much of the opposition to patenting.

The patenting of animals is seen by many as being out of step with both religious and modern secular beliefs. Most religions regard animals as part of God's creation. From a religious viewpoint it is offensive, even blasphemous, for people to claim to have invented an animal; it is God, not humanity, who creates life. The Reverend Professor Andrew Linzey of Mansfield College, Oxford, has argued that:

> '...the granting of patents involving sentient creatures is wrong because it violates fundamental Christian principles ... animal patents should not be given, not now, not ever'.
>
> Linzey (1992)

He goes on to argue that the chief end of creation is not the service of human beings. Dominion, as given by Genesis, must be exercised in conformity with God's moral will. Professor Linzey then questions whether human beings have misunderstood:

> '...their authority and responsibility in creation – their essential diaconical, servicing role – so that they now exercise a power which threatens to morally obliterate all legitimate claims to proper treatment except their own?'
>
> Linzey (1992)

From a secular viewpoint, many people accept the need to reassess our relationship with the world in which we live. There is a new awareness of our responsibility to the environment. This is allied to a growing belief that animals are to be seen not as something placed in the world for our convenience, for us to use as we wish, but as our fellow creatures capable, like us, of feeling pain and stress. Patenting, involving as it does regarding animals as inventions, as things, is out of step with this contemporary approach to animals.

In 1995, the Banner Committee prepared a report for the British Ministry of Agriculture on the ethical implications of emerging technologies in the breeding of farm animals. They stated that in their view 'harms of a certain degree or kind ought under no circumstances to be inflicted on an animal'. They supported the provision on animals in the rejected EU Directive 'insofar as it goes some way towards protecting animals from unwarranted genetic modification'.

The Committee stressed, however, that they would prefer EU law to include 'a stronger moral criterion – specifically one that ensured patents would not be granted where the resultant animal is such that its natural good or integrity has not been respected by the modification' (Banner, 1995).

20.3 THE PROPOSED EU PATENTING DIRECTIVE: DECEASED 1995; REBORN 1996

The continuing debate about the ethical implications of animal patenting has for many years focused on the EU's proposed Directive on the Legal Protection of Biotechnological Inventions (commonly called the Patenting Directive).

The European Commission's original proposal, back in 1988, was so anxious to lay out the red carpet for the biotechnology industry, that it gave no protection to animals at all. Happily, as so often on animal welfare matters, the European Parliament came to the rescue. In October 1992, the Parliament adopted an amendment which excluded from patentability inventions which 'would offend against public order or common decency'. The Parliament also adopted two further amendments which gave guidance as to the kinds of inventions involving animals which are deemed contrary to public order. One amendment excluded from patentability animals which cannot be kept without adverse effects on their health or which are unnaturally interspecific. The other excluded inventions which involve unnatural processes for the production or modification of animals or which cause unnecessary suffering or physical harm.

Towards the end of 1992 the Commission produced a revised draft. This largely ignored the Parliament's wishes. It contained wording which at first sight seemed helpful, but which a careful second reading revealed to be hollow. The Commission's revised draft provided that the following shall be unpatentable:

> '... processes for modifying the genetic identity of animals which are likely to inflict suffering or physical handicaps upon them without any benefit to man or animals'.

The effect of the revised draft would have been that an applicant would simply have to show that some benefit, however small, however trivial, would result and a patent would be granted even if massive suffering or physical handicaps were to be inflicted on the animals involved.

On 8 February 1994, the Council adopted its Common Position. This was a marked improvement on the Commission's dismal offering. Under the Council's wording, if animals are likely to have suffering or physical handicaps inflicted on them, the patent applicant must show that there would be a *substantial* benefit to man or animals. Moreover, unlike the Commission, the Council excluded from patentability not just certain processes for modifying the genetic identity of animals but also the animals resulting from such processes.

The proposed Directive then returned to the Parliament for a Second Reading. In Spring 1994, the Parliament's influential Legal Affairs Committee adopted an amendment to the Council's Common Position which insisted that no patent could be granted where suffering or physical handicaps were likely to be imposed on animals *even if* a substantial benefit to humans or animals could be shown by the applicant for the patent. This was a significant improvement while still falling short of the position of animal welfare groups throughout Europe who wanted to see animals excluded altogether from patentability.

After the Parliament's Second Reading in May 1994 there remained a number of outstanding differences between the Council and the Parliament. Accordingly, the Directive then went to the Conciliation Committee, a new procedure established under the Maastricht Treaty (properly called the Treaty on European Union). In early 1995 this Committee came up with wording which, as regards animals, was largely similar to the Council's Common Position.

To the surprise of all – and the delight of many – in March 1995 the European Parliament voted not to accept the Conciliation Committee's proposal and rejected the Patenting Directive, thereby killing this insensitive proposal. For Compassion in World Farming and other animal welfare groups this was a major triumph as the passing of the Directive would have given the green light to an upsurge in the patenting of transgenic animals throughout the EU. In fact the Parliament's move had been foreshadowed when in January 1994 it adopted the Amendola Report which, amongst other things, called for a prohibition on the granting of patents for animals.

During the prolonged debate on the Directive an Opinion on patenting was delivered by the Commission's Group of Advisers on Ethical Aspects of Biotechnology. This was a truly dreadful document. It failed at any point to engage with the ethical objections voiced by many bodies and individuals. The Group acknowledged that such objections had been made, but did not go on to give them any detailed consideration. Indeed, at times the Group appeared more concerned with not discouraging genetic engineering than in living up to its name as Advisers on Ethical Aspects.

This critical view was echoed by the Intellectual Property Committee of the Law Society of Scotland, not a body noted for being a radical bunch of trouble-makers. They wrote:

> '... the unease which is, from time to time, expressed about "patents on life" derives from a concern that the availability of a property right in such a context indicates an inappropriate attitude to the Natural World, and to the place of humankind in it. ... There is nothing in the Opinion [of the Group of Ethical Advisers] to indicate that this unease has been comprehended and no serious effort is expended on dispelling it.'
>
> Law Society of Scotland (1993)

Indeed, it is possible that the Commission established the Group of Ethical Advisers not because they wanted an in-depth consideration of the issue, but so that they could project themselves as being sensitive to the public's concerns. And, needless to say, the Group duly provided a reassurance that animal patenting presented no significant problems.

Of course there never was any real danger they would conclude otherwise. Ethical advisory bodies are arguably often set up simply to lend respectability to decisions by government that animal welfare concerns must not be allowed to impede 'progress'. They are not meant to rock the boat.

No more had the Parliament been meant to rock the boat when in 1995 it rejected the Patenting Directive. The Commission refused to accept the position and early in 1996 it issued a new proposed Directive. As regards animals, this contains wording similar to that included in the earlier version of the Directive. The new proposal again makes it clear that animals can be patented, but it excludes from patentability:

> '...processes for modifying the genetic identity of animals which are likely to cause them suffering or physical handicaps without any substantial benefit to man or animal, and also animals resulting from such processes, whenever the suffering or physical handicaps inflicted on the animals concerned are disproportionate to the objective pursued'.
>
> European Commission (1995)

Compassion in World Farming reacted swiftly, again making it clear that they are wholly opposed to the granting of patents on transgenic animals and processes for creating such animals. It should be noted that the TRIPS (Trade-Related Aspects of Intellectual Property Rights) section of GATT does allow Member States to exclude animals from patentability. Another long battle over the EU Patenting Directive is under way.

20.4 THE ONCOMOUSE CASE

The oncomouse is a mouse genetically engineered to be highly susceptible to developing cancer. Its advocates argue that it is a helpful 'laboratory tool'.

Initially (July 1989), a patent for the oncomouse was refused by the Examining Division of the European Patent Office (EPO). The Examining Division took the view that Article 53 (b) of the European Patent Convention (EPC), which provides that patents shall not be granted in respect of 'animal varieties', excluded animals as such from patentability.

The Examining Division also considered Article 53 (a) which excludes from patentability 'inventions the publication or exploitation of which would be contrary to "ordre publique" or morality' (the morality Article). The Division tried to side-step this issue, taking the view that patent law was not the right tool for resolving ethical issues.

The Applicants appealed, and in October 1990 the EPO's Technical Board of Appeal remitted the case to the Examining Division for further consideration. In particular, they instructed the Examining Division to consider whether the morality Article precluded the granting of a patent in this case. The Board of Appeal ruled that Article 53 (b) (prohibition on patenting animal varieties) did not exclude animals as a whole from patentability; the Board did not, however, determine what was meant by the term 'animal varieties'.

Article 53 (a) is at the heart of this case, but the article gives no guidance as to how it is to be determined whether the use of a transgenic animal is 'contrary to morality'. In their Decision T19/90, when remitting the oncomouse case to the Examining Division, the EPO Board of Appeal stated that this 'would seem to depend mainly on a careful weighing up of the suffering of animals and possible risks to the environment on the one hand, and the invention's usefulness to mankind on the other'.

On 13 May 1992 the Examining Division granted a patent for the oncomouse, having concluded that its usefulness outweighed the animal suffering and environmental risks. On 12 January 1993, Compassion in World Farming (CIWF) and the British Union for the Abolition of Vivisection (BUAV) filed a formal legal Opposition under the European Patent Convention to the oncomouse patent; this Opposition is supported by some 40 European animal welfare societies.

20.4.1 Principal arguments of the Opposition

The patent is commonly referred to as the oncomouse patent. It should be emphasized, however, that in fact it extends to any onco-mammal, i.e. any mammal with an inserted oncogene sequence. Likewise the patent extends to a method for producing any onco-mammal.

The Opposition argues that the use of

- the oncomouse and indeed all onco-animals covered by the patent, and
- the methods for producing such animals covered by the patent,

are contrary to morality and thus excluded from patentability by virtue of Article 53 (a).

The Opposition begins by taking something of a 'Judicial Review' approach. The Opponents argue that the Examining Division failed properly to carry out the 'careful weighing up' of competing considerations required by the Board of Appeal. They contend that if such a careful weighing up had been performed, the Examining Division ought reasonably to have concluded that the suffering of the animals outweighs the usefulness of the invention to humanity.

20.4.2 Suffering of the animals

The Opponents believe that the Examining Division gave insufficient consideration to the very considerable suffering of animals inherent in the 'invention'.

Moreover, inasmuch as the Division considered the issue of suffering, they only looked at the oncomouse, whereas the patent extends to all onco-mammals. Bearing in mind the mammals regularly used in research, it is not fanciful to believe that onco-primate, onco-dog and onco-rabbit could be developed and, if and when this happens, they too will be covered by the patent. There is no evidence in the statement published by the Examining Division when they announced their intention to grant the patent that they gave any consideration to the suffering of onco-animals other than the oncomouse.

The Opponents believe that the majority of the public would condemn as morally unacceptable the creation of dogs, monkeys and rabbits genetically engineered to develop cancer.

20.4.3 Usefulness to humanity

The Opponents contend that the Examining Division has seriously overestimated the oncomouse's usefulness. They believe that it is of limited value in the development of anti-cancer drugs and in carcinogenicity testing (these being the two claims made by the Applicants for the oncomouse's usefulness). Indeed, the Patentee has conceded that oncomice make only a modest contribution to the science of cancer treatment. The Opponents argue that much of the contribution made by oncomice can be made as effectively by the use of non-animal alternatives together with the use of non-transgenic mice.

Support for the Opponents' arguments can indeed be found within the scientific community. Many scientists now agree that animals are of limited value in developing new anti-cancer drugs. In particular, the use of oncomice (or other onco-animals) does not overcome the serious problem of species difference in tumour development and responsiveness to anti-cancer agents; this limits the extrapolation of results from rodents to humans. There is, moreover, considerable scientific evidence that the oncomouse is unlikely to be of value in testing whether a particular substance is a carcinogen, i.e. whether it causes cancer. This view has been vividly expressed by Salsburg (1983), who has written:

> '...the lifetime feeding study of mice and rats appears to have less than a 50% probability of finding known human carcinogens. On the basis of probability theory, we would have been better off to toss a coin'.

20.4.4 Risks to the environment

It is hard to believe that over a period of years there will be no escapes from laboratories. If oncoanimals did escape, they could quite probably breed with wild animals or, in the case of dogs, with domestic dogs, either with strays, for example, or with dogs in parks. Their offspring could well inherit the oncogene. A large number of animals could rapidly become 'polluted' in this way.

The first limb of the Opponents' case concludes that the Examining Division failed properly to carry out the necessary weighing up and that if they had done so, they should reasonably have formed the view that the suffering of the animals and the environmental risks outweigh the usefulness of the 'invention'.

20.4.5 Board of Appeal's test not appropriate for determining morality

The second limb of the Opponents' case is that a weighing up of competing considerations is not the right way to determine whether a particular development is contrary to morality.

The Opponents believe that it is inherently contrary to morality to alter an animal's genetic structure with the clear purpose that it should develop a painful, lethal disease. If this is immoral, it does not become less immoral to do so simply because in a particular case it is believed (misguidedly in the Opponents' view) that humanity may derive a significant benefit. If we take such a pragmatic approach our morality ends up taking second place to our sense of what is expedient. In fact, morality frequently requires us to refrain from acting in the way we wish or in a way that may serve our interests. Surely, certain principles are worthy of our respect, even when such respect proves *not* to be to humanity's benefit.

In their Opposition, the Opponents also present the arguments referred to earlier in this chapter that the patenting of animals is out of step with both religious and modern secular beliefs.

Sixteen other Oppositions to the oncomouse patent were filed in addition to that of CIWF and BUAV. As well as the formal Oppositions, it should be noted that in 1993 the European Parliament passed a resolution by 178 votes to 19 condemning the oncomouse patent.

Oral proceedings in the oncomouse case were conducted at the EPO in Munich in November 1995. As the 4-day hearing wore on, the EPO's three-person Opposition Division Tribunal began to appreciate the strength of the Opponents' case. To everyone's surprise on the third day they eventually suggested that the patent holder might be well advised to re-draft the patent in a more limited form. This they did (though clearly they would prefer the patent to be upheld in its original form). One of the patent holder's re-drafts would restrict the patent to onco-rodents rather than applying it to all onco-mammals. Then, in another totally unexpected twist, the hearing broke up in anger and disarray, with the Tribunal saying that the hearing was at an end and that the case must continue in writing.

It is not possible to predict when the oncomouse case will finally be settled. Clearly, its outcome could well determine the future of animal patenting in Europe.

Already a number of other animal patents have been granted by the EPO. In May 1995, for example, a patent was granted for 'a method for introducing a gene of interest into a gallinaceous bird' such as a chicken. The description of the invention refers to the introduction of genes to induce immunity to disease. Disturbingly, however, it also refers to the possibility of increasing 'fowl growth rate'. The dangers of increasing animals' growth rate have been described earlier in this chapter.

If the oncomouse patent is upheld, the floodgates for more animal patents could be opened, whereas revocation of the oncomouse patent could severely restrict the granting of such patents.

20.5 ANIMAL VARIETIES

Article 53 (b) of the EPC provides that 'plant and animal varieties' shall not be patentable. The meaning of the term 'animal varieties' remains undecided. However, in the Plant Genetic Systems (PGS) case (where the Opponent was Greenpeace), the EPO's Technical Board of Appeal in Decision 356/93 defined 'plant variety' as 'any plant grouping within a single botanical taxon of the lowest-known rank which is characterised by at least one single transmissible characteristic distinguishing it from other plant groupings and which is sufficiently homogeneous and stable in its relevant characteristics'.

This decision reflected Decision T49/83 which defined 'plant varieties' as 'a multiplicity of plants which are largely the same in their characteristics and remain the same within specific tolerances after every propagation or every propagation cycle'.

In short, a plant variety is a group of plants which is distinct from other groups and which share a characteristic which is transmitted in a stable manner throughout succeeding generations.

In the oncomouse case CIWF and BUAV are arguing that 'animal varieties' should be given a corresponding definition and that, accordingly, the patent should be revoked as the oncomouse constitutes an animal variety.

Crucially in the PGS case, the Board of Appeal ruled that a claim which in effect embraces plant varieties is excluded from patentability even if specific plant varieties are not individually claimed. In other words, a patent application which as a matter of fact includes plant varieties will fail even if it is directed to plants generally and not to specific plant varieties individually. It is reasonable to assume that the EPO will apply the same thinking to claims which embrace animal varieties.

Unfortunately, early in 1997, the European Parliament's Rapporteur introduced an amendment which would overturn the key aspect of the EPO's decision in the PGS case, i.e. the aspect referred to in the previous paragraph. If his amendment is incorporated into the Directive, it will mean that the rule that animal varieties cannot be patented will become meaningless in practice and will be robbed of any power to exclude animals from the patenting system.

Article 53 (b) of the EPC provides that the exclusion from patentability of plant and animal varieties does 'not apply to microbiological processes or the products thereof'. In the PGS case the Board of Appeal ruled that a multi-step process is not a microbiological process if only one step, however decisive, in that process is a microbiological process.

This contrasts with the position in the new proposed EU Directive. This too excludes plant and animal varieties from patentability while providing that this exclusion does not apply to microbiological processes and the products thereof. There is, however, a crucial difference in that it stipulates that 'a process consisting of a succession of steps shall be treated as a microbiological process if at least one essential step of the process is microbiological'. In proposing this wording, the Commission is clearly determined to make sure that the exclusion of animal varieties from patentability becomes meaningless in practice. Nothing must be allowed to impede the onward march of biotechnology, certainly not mere concerns about animal welfare.

20.6 CONCLUSION

An increasing number of people are coming to believe that animals are not to be seen as existing simply to satisfy human needs, they are not our tools, and nor are they 'agricultural products' (the term used by the Treaty of Rome, the cornerstone of EU law). If we are to use them, we must do so with respect and with awareness that they too are living creatures. And altering their genetic makeup does not show much respect. Patenting shows them even less.

The rejection in 1995 of the Patenting Directive by the European Parliament was an important blow against the patenting of transgenic animals. The Parliament's decision was a sensitive response to increasing public concern about animal welfare. There is a widespread fear, however, that the new version of the Directive will be adopted by the EU. Moreover, the law of many individual European countries allows animal patenting. Many believe that those laws and the European Patent Convention should be amended to clearly exclude such patenting. We are at a crucial watershed. If we allow the genetic engineering and patenting of animals to go ahead we will in effect be giving our blessing to a whole new era in humanity's exploitation of animals.

REFERENCES

Banner, M.C. (1995) *Report of the Committee to Consider the Ethical Implications of Emerging Technologies in the Breeding of Farm Animals*. HMSO, London.

European Commission (1995) Proposal for a European Parliament and Council Directive on the legal protection of biotechnological inventions. COM(95)661 final.

Kennedy, I. (1996) Animal tissue into humans. Report by the Advisory group on the Ethics of Xenotransplantation. Department of Health, HMSO, London.

Law Society of Scotland (1993) Letter dated 15th October 1993 to the UK Patent Office.

Linzey, A. (1992) A theological critique of patenting animals. Paper submitted with the Opposition under the European Patent Convention to the oncomouse patent by Compassion in World Farming and the British Union for the Abolition of Vivisection.

Salsburg, D. (1983) The lifetime feeding study in mice and rats: an examination of its validity as a bioassay for human carcinogens. *Fundamental and Applied Toxicology*, **3**, 63–7.

21

Controls on the care and use of experimental animals

Donald W. Straughan and Michael Balls

21.1 INTRODUCTION

During the last century, public concern has increased over the actual or potential adverse effects experienced by animals as a consequence of scientific activities. There has been a corresponding increase in special provisions for controlling the use of laboratory animals, since general provisions for animal protection are unlikely to cope with complex scientific activities involving the deliberate use of procedures likely to be painful. Such controls on the use of animals for scientific purposes attempt to balance the legitimate needs of society, science and industry (to have information and products derived from the use of experimental animals) against the equally legitimate ethical concerns of society that animals should not be used except when necessary and for worthy ends, and that animal use and suffering should be minimized as far as possible, through optimal deployment of the Three Rs (*replacement*, *reduction* and *refinement*), a concept pioneered by Russell and Burch (1959).

The purpose of this chapter is to review the general principles that might underlie such control systems, and to comment on the ways in which these principles are implemented through statutory and voluntary controls in some major communities. It is intended to complement key texts (OTA, 1988; Smith and Boyd, 1991; Orlans, 1993). The term *procedure* is used to mean any technique or series of techniques applied to a vertebrate animal to meet a particular scientific objective.

21.2 STATUTORY AND VOLUNTARY CONTROLS

The effectiveness of any control system depends very heavily on the attitude and commitment of those involved both in using animals and in supervising and monitoring such use. Whether the controls are statutory or voluntary is important. Most nations have some form of statutory control. This ensures a consistent framework for compliance and may allow legal sanctions against non-compliance. Detailed comparative studies are still required to establish the effectiveness of individual national systems. As a general principle, we support a statutory framework, since it is ultimately governments, not voluntary groups, who should act as a 'guardian of the public interest' (Zegers, 1989).

In Australia, there is no federal legislation and only some states (e.g. New South Wales, South Australia and Victoria) have comprehensive legislation regulating animal research. Exceptionally, Canada has adopted a comprehensive national system of voluntary self-regulation, which is administered by the Canadian Council on Animal Care (CCAC; Anon, 1980). Arguments for this system include the view that 'legislation has always tended to enforce only minimal acceptable standards' and that 'no government inspectorate could expect to enjoy the benefit of the scientific expertise presently provided, without honorarium, to the CCAC assessment program'. The general success of this system confirms the importance of individual attitudes, but ultimately, in our view, it seems unlikely that voluntary self-regulation of animal experimentation can provide the same degree of protection against abuse as statutory controls with enforcement powers.

21.3 GENERAL PRINCIPLES OF THE IDEAL CONTROL SYSTEM

To be effective, the controls should be comprehensive and should be applied in a consistent, competent and reasonable fashion. They need to have regard to what we will call the 'Seven Ps':

- the **p**rotected species;
- the **p**urposes for which animals are used;
- the scientific **p**rocedure(s) applied to them;
- the **p**roject, i.e. the programme of work to be undertaken;
- the **p**laces where animals are used (and those where animals are bred and supplied from);
- the **p**eople using and caring for the animals; and
- the need for **p**ublic information and accountability.

21.3.1 Protected species

Ideally, as is specified in the Council of Europe Convention (the Convention) on the use of animals for experimental and other scientific purposes (Anon, 1986a), the controls should apply to the use of *any* living vertebrate animal for *any* scientific purposes which may involve pain, suffering, distress or lasting harm. The aim is to ensure that any possible adverse effects are kept to a minimum. There appears to be no rational justification for having controls which are not totally species-comprehensive. However, in the USA, although the statutory federal controls cover all warm-blooded animals, only some species were named specifically, and this has allowed the US Department of Agriculture (USDA) to exclude from their controls the most commonly used laboratory animals, i.e. rodents such as rats and mice, as well as birds. Cold-blooded vertebrates such as amphibians or fish are not covered by the USA Animal Welfare Act (AWA). However, species-comprehensive controls are required in institutions receiving funding from the US Public Health Service (PHS)/National Institutes of Health, and by some individual states in the USA, as well as being applied by many individual Institutional Animal Care and Use Committees (IACUCs).

Currently, the UK also applies formal legal controls on the use of an invertebrate cephalopod (*Octopus vulgaris*). In Canada, the CCAC include the cephalopods (octopus and squid) under categories of invasiveness (Davis, 1990) and also require enumeration of their use if assigned to a research protocol.

Article 7.3 of Directive 86/609/EEC (the Directive: Anon, 1986b), which applies to the 15 Member States of the European Union, requires the selection of 'animals with the lowest degree of neurophysiological sensitivity', where there is a choice. This phrase is not included in the Convention. There is support, but no conclusive evidence, for the proposition that 'lower' animals suffer less than 'higher' animals. Nevertheless, in our view, at a minimum, there should be support for additional controls and special protection for higher non-human primates such as Old World monkeys and chimpanzees. However, it is not clear that there is justification for giving all non-human primates (including those which are smaller and less developed neurologically, such as marmosets) more protection than rodents! Such speciesism may have regard to public concern, but it is hard to justify in terms of level of neurological development and presumed sentience. The Convention (Articles 21 and 17) applies additional controls to the acquisition and marking of companion animals (cats and dogs), and the Directive also applies them to non-human primates (Articles 21 and 18). As well as providing special protection for cats, dogs and primates, the UK also gives special protection to equidae (Anon, 1986c).

Both the Convention and the Directive protect free-living larval forms such as tadpoles, but exclude fetal and embryonic forms on the basis that procedures on fetuses of viviparous species are carried out via the mother, which is covered. Exceptionally, the UK protects mammals from halfway through their gestation period and embryonated avian and reptile eggs from halfway through their incubation period (Anon, 1986c). The Australian Code of Practice also gives consideration to fetuses and embryonated eggs (Anon, 1990).

21.3.2 Purposes

A general principle is to restrict the purposes for which scientific procedures can be performed, based on what is considered broadly acceptable by society. Thus, Article 2 of the Convention specifically states that procedures may be performed only for certain approved purposes, which include basic scientific research, education and training, as well as certain types of applied research. The quality, efficacy and the safety testing of drugs, substances or products are specifically mentioned in the context of avoiding or preventing disease. The Convention implicitly covers procedures for the production of blood products, including antibodies and vaccines, as these are used for diagnosis, scientific research or treatment purposes. Article 3 of the Directive also includes the development, manufacture, quality assurance and safety testing of drugs, foodstuffs and other substances or products. Nevertheless, within Europe, it is not clear whether there are consistent policies on the use of animals for blood products, including antisera, and on their enumeration and reporting to the relevant authorities. Indeed, the German Animal Welfare Act presently covers experiments rather than scientific uses, and is considered to exclude the use of animals for the manufacture of vaccines and sera, including the production of blood products such as antibodies. Unlike the Convention, Article 3 of the Directive does not specifically mention scientific research, education and training, or forensic inquiries, though these types of purpose were accepted by the governments of the EU Member States in a Council resolution in November 1986.

Education and training

The commentary on Article 25 of the Convention makes it clear that animal procedures for education and training should only be carried out in preparing for professional activities involving the performance of procedures or the care of animals, i.e. in some areas of tertiary education or in training for certain professions or other occupations.

Where potentially painful animal procedures for education and training are not specifically authorized by legislation, they may be in breach of national laws against animal cruelty.

The control of procedures for tertiary/professional education and training varies within and among countries. For example, the use of animals for purely educational purposes even varies between different establishments teaching similar courses. Since alternative replacement methods are already used widely for tertiary education and training, the need for *in vivo* procedures and their educational effectiveness should be regularly reconsidered. In some countries where the quality of professional competence is not disputed, as in the UK, human and veterinary surgeons gain the necessary experience through properly supervised training on genuine clinical material, without the need to practice on healthy animals. Thus, it is difficult to understand why practice on laboratory animals should still be allowed in medical and veterinary training in many other countries. In our view, all animal procedures within university/college education courses and in professional training should always be fully justified and subject to the same controls and close scrutiny as any research proposal – as in The Netherlands and the UK. Only procedures producing minimal adverse effects should be allowed, which include those under terminal anaesthesia. In the UK, microsurgery is considered to be a special case and a permissible purpose, but the justification for this could be challenged, since some UK and German institutions manage to train microsurgeons and maintain their skills without using animals.

Some of the other difficulties involved are illustrated by the situation in Germany, where animal experiments for education are currently excluded from the controls, since freedom in teaching is guaranteed by the German Federal Constitution. Attitudes to such experiments vary among the different Bundesländer.

Schools

In line with Article 25 of the Convention, there is consensus within Europe that potentially painful procedures should not be performed on any live vertebrate animal in elementary or secondary schools. In Canada, the CCAC Guide also specifically forbids, in pre-university classrooms, 'any procedure in live vertebrates which is liable to cause pain, distinct discomfort or prejudice health; the performance of surgery; or the use of toxic substances'. However, in the USA, though experiments on animals in high schools have decreased considerably in recent years, they are only prohibited at present in a few States, e.g. Maine and Massachusetts.

21.3.3 Procedures

Procedures can be classified according to their propensity to cause adverse effects.

Major procedures

Ideally, as in Article 8 of the Convention, additional controls are applied where the use of anaesthetics, analgesics or physical methods to eliminate pain throughout the procedure is incompatible with its aim. Also, in these situations, 'appropriate legislative and/or administrative measures should be taken to ensure that such procedures are necessary'. Additional controls should also be applied for procedures causing (or likely to cause) severe pain or severe distress which is likely to endure. The Convention covers this in Article 9, by stating that such a *procedure* must be specifically declared and *justified* to the responsible authority, to ensure that it is *necessary*. Similarly, both the USA IACUC Guidebook and the Australian Code of Practice specify that justification must be required for any procedure which may cause pain or distress, but in which alleviation of the pain or distress cannot be reasonably assured. Another key issue is whether there is a fundamental requirement to immediately kill humanely any animals in severe pain which cannot be alleviated (as in the UK, the CCAC Guidelines and the Australian Code of Practice), or whether there is an option to wait until the end of the procedure (USA Animal Welfare Act). Moreover, where this requirement exists, how and when is the policy implemented? There are considerable differences in practice both within and among countries. Some countries (e.g. UK, Australia, Canada) have lists of severe procedures and/or those which cause special concern.

Article 8 of the Convention also encourages the use of anaesthetics, analgesics or physical methods to eliminate pain, etc. throughout the procedure, as far as is practicable. Another issue is whether experiments wholly under terminal anaesthesia or after decerebration should still count as procedures. Animals will suffer pain if drugs are administered inadequately or if brain transection is incomplete, so the prevailing view is that such procedures should be identified and controlled, as is stated in Article 1 of the Convention.

Minor procedures

Procedures which do not cause pain or significant adverse effects, or which are very minor, can be disregarded. The Convention does not attempt to define them, but in 1992 the parties to the Convention agreed that the amount of pain caused by the introduction of a needle into the body of an animal illustrated the level at which the use of an animal becomes a procedure. This provides a rich potential for inconsistency. On grounds of practicality, most control systems do not restrict the killing or marking of animals 'by the least painful method accepted in modern practice (that is humane methods)', as in Article 1 of the Convention. This usually permits the humane killing of animals to provide cells or tissues for *in vitro* studies, without specific authority and enumeration (though such animals will have been subject to the general provisions for care in the scientific establishment concerned). However, the killing of untreated animals to obtain tissues for scientific purposes is controlled in Belgium, The Netherlands and Sweden.

Killing and marking

What constitutes 'acceptance' of the many methods of killing or marking also provides a rich potential for diversity and inconsistency. There is an expectation that all methods of killing or marking will be applied competently, but this can only be assured by appropriate training, supervision and monitoring of the competence of individuals (which also requires controls). Within the EU, a working party has recently produced a report on the euthanasia of experimental animals, which may ultimately become the basis for Commission guidelines. The killing of rodents by decapitation or by microwave irradiation requires specific approval in the UK, and decapitation of small animals, including rodents under 450 g, without prior anaesthesia, requires specific justification in the USA. In our view, the manipulation of accepted methods of rapid and humane killing to allow perfusion or tissue sampling in still-living animals should be subject to formal legal controls.

Breeding of animals with genetic abnormalities

It is reasonable to protect genetically abnormal animals where special husbandry or medical attention are needed to maintain their well-being, and/or where they are likely to show harmful defects, e.g. *oncomouse*. However, the Convention (Article 1) only appears to control the (initial/experimental) course of action which is intended or liable to result in the birth of animals which may show adverse effects, rather than their continued breeding thereafter. Thus, an explanatory note states that 'the breeding of animals with genetic abnormalities is not a procedure as long as the aim is their propagation'. Though this note indicates that animals with genetic abnormalities may have 'special requirements for their well-being', it does not require proper controls to be applied to the routine breeding of animals with genetic abnormalities. This is an unsatisfactory state of affairs, and it is not surprising that there is considerable variation among countries, including the EU Member States, as to the controls to be applied to the continued breeding of transgenic animals for subsequent experimental use.

21.3.4 Project review

Every system of controls should be based on two fundamental principles, namely, the optimal deployment of the Three Rs, and the need to consider the likely benefits of the proposed work and to weigh them against the likely caused harm to the animals ('cost'), before determining whether a proposal is justified and really necessary (Anon, 1986c; Smith and Boyd, 1991; Balls, 1995). This requires optimized protocols, and adequate review by properly informed persons trained in making such decisions and independent of the applicant scientist.

The Three Rs

There is little or no dispute about this general objective. For example, Article 6 of the Convention requires that a procedure should not be performed, if another scientifically satisfactory non-sentient replacement method is reasonably and scientifically available (*replacement*). Article 7 requires that, in a choice between procedures, those selected should require the minimum number of animals (*reduction*), cause the least pain or suffering (*refinement*) and be those most likely to provide satisfactory results. The Directive makes the same requirements.

Project content

Clearly, before starting animal experiments, all scientists should have protocols which meet basic scientific and ethical principles and which can provide a basis for an informed review. These protocols should be adequate in terms of:

- detailing and justifying the design of the procedures to be performed and their subsequent statistical analysis;
- showing that animal suffering and use are minimal in terms of refinement and replacement alternatives;
- noting how these procedures conform to any national guidelines and justifying any deviations from them;
- describing likely technical failure rates;
- describing the nature, severity and likely incidence of adverse effects;
- describing the humane end-points and the anaesthetic/analgesic regimes to be used; and

- clarifying the actions to be taken by the scientist or the establishment welfare officer should the limits on severity be exceeded.

In addition, to permit effective review, additional information needs to be supplied to justify the necessity for the procedure and its importance, so that informed decisions can be made on 'benefit' and the weighing of benefit against 'cost' to the animals to be used.

Justification

The need for justification provides basic problems both in interpretation and consistency of application. In many countries, the review process presupposes that the scientific objectives are satisfactory and justified and is limited mainly to considering the extent to which the Three Rs have been implemented. Thus, for example, the controls in Belgium, Canada, France and Ireland do not weigh benefit against severity. In Canada, the CCAC document on the role and responsibilities of ACC (Anon, 1992a) notes that, 'Generally, the ACC does not have the time, expertise or proper resources required for peer review of scientific merit.' The ACC's responsibility is to be satisfied that a reliable review has been undertaken elsewhere, research council funding being accepted as sufficient evidence of merit! In the USA, IACUCs should attempt an evaluation of the research, but in practice, it is likely that they rarely consider whether the research is justified, probably because such committees, 'Often lack the scientific expertise to perform the relevant evaluation' (Anon, 1992b). Also, it is argued that since the bulk of research in the USA is externally funded, peer review will have taken place. However, such a review is unlikely to involve an assessment and weighing of likely benefit and suffering, so IACUCs should attempt some evaluation. An excellent account of the general factors to be considered has been provided by Orlans (1993).

Protocol review

How and by whom protocols are reviewed and approved is crucial. Most countries now devolve advice and/or approval of protocols to specific review committees. In countries with a review system, experiments should not start before approval, usually written, is received. The review committees may be supra-institutional, or more usually, institutional. In Germany, as in Sweden, protocol review is by independent regional local committees and not institutional committees. In The Netherlands, the role of the previous regional review committees has just been devolved to institutional review committees, which continue to pass their recommendations to local state officials. In Sweden, the seven local review committees are regionally based (Anon, 1994). Uniquely in the UK, all projects and protocols are reviewed nationally by individual Home Office inspectors and not by committees, though in a few instances, projects will also be reviewed by other inspectors and perhaps also by a national advisory committee. The occasional institutional committees in the UK have no statutory recognition (though they may advise the institutional management on whether or not particular types of research should be conducted in their establishment and also on what persons should be given facilities for research involving animals).

It is important to know the status and power of such committees (or individual inspectors in the UK) to make advisory or binding decisions. A system which allows advice to be regularly disregarded to any significant extent is thereby flawed. Where such committees exist in Europe, they have advisory rather than decision-making powers, but their recommendations are usually accepted and implemented by officials. In Australia, the national Code of Practice, though not mandatory, is generally applied and requires institutional Animal Experimentation Ethics Committees (AEEC) to examine and decide on research protocols – irreconcilable differences between the AEEC and an investigator are referred to the local governing body. In the USA, the IACUCs have federally mandated authority to 'review and approve, require modifications in (to secure approval), or to withhold approval'.

Institutional review committees have much to commend them in terms of raising local awareness of intended animal experiments and of the ethical and scientific issues involved. They probably have a key role in maintaining standards of technical competence and animal care. However, a matter of concern must be the extent to which institutional committees can be independent when considering applications from possibly prestigious and well-funded colleagues. This is an argument for close national monitoring of their work. The alternative system of formal project review by regional or national experts overcomes some of these difficulties and appears to spare the use of academic time and resources, but it may tend to diminish (or even undermine) the critical role of institutions and increase bureaucracy.

A key issue must be whether the need for national or local approval applies to all procedures (or only the most severe), to all species, to all places, and to all purposes. In Germany, the local review committees have a limited scope, being allowed only to review and approve academic or basic industrial research (which accounts for only 14% of total animal use). They do not consider the remaining animal use [including use for new drug development (57%), for vaccine testing or in the production of blood products (Spielmann, 1995)]. Regulatory experiments need no licence, but have to be notified. While it is true that such regulatory experiments follow standard protocols, these protocols often allow some flexibility in design and also vary among major trading blocks. Thus, it is not clear that the German system ensures that regulatory protocols are minimal in terms of animal numbers and animal suffering.

In the USA, AWA policies (including those on IACUCs, inspection and data collection) do apply to individual industrial companies, when they are registered as using AWA-regulated species.

Retrospective reporting

Reporting by experimenters on the outcome of project work involving animals is not routinely required in most countries, except for Sweden. Detailed reports on an *ad hoc* basis for projects and procedures which cause concern are often required in the UK and may be sought elsewhere. Regular reporting should be encouraged, to provide valuable information on how outcome matched expectation and on how the level of adverse effects matched what was predicted.

21.3.5 Places

The controls should apply to all scientific establishments where animals are held for use in procedures, whether or not the animals are to be subject to scientific procedures, as is required by the Convention. This is rational and promotes consistency.

To achieve effective control, all scientific user establishments should have to be registered with, or otherwise approved by, the responsible authority and controlled, as specified in Article 20 of the Convention. Most countries do this. There appears to be no clear justification for distinguishing between establishments on the basis of whether or not they are supported by grants from national agencies, as in the USA and Canada. In the USA, formal controls do not apply to establishments which do not use AWA-protected species or move them across State lines, or to establishments not receiving PHS grants. Thus, USA laboratories *only* using rats, mice or birds, e.g. for antibody or vaccine production, would be exempt from registration.

There are practical reasons for extending control, as in Article 14 of the Convention, to establishments which breed or supply animals, including surgically prepared animals or transgenic animals, for use in subsequent scientific procedures.

It also makes sense to limit the supply of animals from uncontrolled sources, as is required by the Convention. Unless exemptions are given, certain species (mice, rats, guinea-pigs, golden hamsters, rabbits and quail) must be obtained from registered breeders and suppliers. In view of public concern, the ideal system places additional controls on the source and use of companion animals such as cats and dogs. Thus, both Article 21 of the Convention and Article 19.4 of the Directive prohibit the use of stray cats and dogs in procedures and require that they be acquired directly from, or originate from, registered breeding establishments. A general exemption is not allowed for the use of stray cats and dogs, but a special exemption for such use is apparently not prohibited. The Directive includes non-human primates in the list of animals which must be specially bred (unless an exemption has been obtained), while Article 21.1 of the Convention requires the use of purpose-bred non-human primates when there is a reasonable prospect of a sufficient supply. A great deal of attention is likely to be focused on sources of non-human primates in the near future.

21.3.6 People

Whether the controls are statutory or voluntary, a variety of people are involved at many levels, and these people are all vital components of any controls.

Controls can be applied, particularly through licences or certificates, to individuals performing procedures, to individuals ordering and planning projects, and to the heads of establishments where work on animals is conducted. The UK covers all three, through personal and project licences and certificates for user establishments. In France, Germany and Sweden, only principal investigators are licensed. In contrast, in The Netherlands and Norway, only the heads of establishments are licensed to perform experiments, although they do have the power to delegate. In The Netherlands, heads of establishments have the legal responsibility to ensure that certain requirements are met, including an imaginative system of education and training for those planning and performing experiments and caring for animals (van Zutphen and van der Valk, 1995).

Ideally, the person finally in charge of the administration of the establishment, facilities and staff should be identified and specifically authorized in all user establishments. Unfortunately, both Article 20a of the Convention and Article 19.2a of the Directive adopt a more limited approach, by only requiring identification of the person or persons administratively responsible for the care of the animals and the functioning of the equipment. This appears to allow some ambiguity in the personal responsibility for, and management control of, the staff actually performing scientific procedures. The person in charge of administration or arranging for animal care in breeding and supply establishments also needs to be specified, as in Article 15 of the Convention.

The prompt detection and treatment of defects in meeting animal needs and/or of animal suffering is very important, as is specified in Article 5 of the Convention. Provision also needs to be made for a veterinarian or other competent person to advise on the well-being of the animals, as in Article 20d of the Convention and Article 19.2d of the Directive. However, such persons (and key care-staff deputizing for them) should also have the clear duty and power to order the immediate destruction of animals which they consider to be suffering excessively, or to stop experiments which exceed the severity or technical failure rate previously authorized, or to institute treatment when necessary, ideally after consultation with the experimenter. This is broadly allowed for in UK and Swedish law, and in the Australian Code. IACUCs in the USA are federally mandated to 'suspend an activity involving animals when necessary, take corrective action, and report to the funding agency and USDA'. In any country, the interpretation and application of these powers seems likely to vary among institutions.

Health monitoring is also necessary, as is the provision of care, accommodation, an environment, and food and water 'appropriate to its health and well-being' for any animal used or intended for use in a procedure. A daily check on environmental conditions should be required, as in Article 5 of the Convention. Ideally, any restriction on the extent to which an animal can satisfy its physiological and ethological needs (including facilities for exercise) should be limited as far as is practicable, as is also stated in Article 5. To define standards and ensure consistency and accountability, a national and enforceable Code of Practice (rather than guidelines) should be published in each country. Both the Convention and the Directive set out detailed guidelines in an Appendix, and most European countries apply them. The guidelines for housing cats, dogs and non-human primates, albeit setting minimum standards, provide a useful marker for comparing different national attitudes and practices. Some European countries have not yet fully implemented the standards in the Convention. In the USA, the federally prescribed minimum floor areas for dogs and monkeys are significantly less than those laid down in the Convention.

Most systems involve national or regional inspectors to monitor the effectiveness of the controls. Such inspectors must be competent and their number should be adequate. To make properly informed judgements, we believe that they need to have a broad, but expert, background in the biological sciences and in research and its applications, and experience in the assessment of pain. Where inspectors are concerned only with compliance, animal welfare and animal facilities, a veterinary qualification is appropriate. Where inspectors have an additional and significant role in protocol review, as in the UK, then additional numbers will be needed and medical qualifications are useful.

Whether or not they have a role in protocol review, all inspectors should have the clear duty and power to order the immediate destruction of animals which they consider to be suffering excessively, or to stop procedures which exceed the severity or the technical failure rate authorized, even without the agreement of the experimenter. On occasions, this duty may also need to prevail over the advice of establishment animal welfare officers. In the USA, the federal inspectors [from the Regulatory Enforcement & Animal Care Unit (REAC) within the USDA Animal & Plant Health Inspection Service] can inspect animals undergoing research, but are prohibited from interfering with or confiscating animals whilst under experiment, though they can challenge practices and procedures. However, IACUCs can stop ongoing research, and REAC inspectors will advise IACUCs of their concerns and, where appropriate, can initiate regulatory actions (including fines or prosecutions) to bring about correction. REAC inspectors are also prohibited from inspecting facilities run by other federal agencies, though these facilities are expected to adhere to federal laws.

Training

Animal experiments should be both designed and analysed, as well as performed, in a competent fashion, and only competent staff should carry out animal care and husbandry.

Festing (1992) has shown that many papers published in scientific journals involved studies which are unsatisfactory in design and analysis. This must represent at least a partial failure to meet Article 26 of the Convention, which requires appropriate education and training for those concerned. Poor science is unacceptable in its own right and cannot justify *any* use of animals or the causation of *any* suffering. Key issues to ensure the technical competence of experimenters include:

- the quality and quantity of training required by individuals in different professional groups;
- who determines whether the training is satisfactory (and how); and
- the provision, if any, for continued assessment and training in new techniques and methodologies.

Any notion that people with medical qualifications should be generally exempt from training in the use (and care) of animals and in experimental design is insupportable. Further consideration of this aspect is not possible here, except to note that most countries are now tackling the problem, albeit in different ways.

Records

User establishments should be required to keep detailed records, as stipulated in Article 28 of the Convention, to show the number and species of all animals acquired, their source and arrival date, and the numbers and species of animals used for procedures, both overall and in selected categories. Supplying establishments should be required to record the number, species and date of animals acquired or leaving, together with details of sources and recipients, as in Article 16 of the Convention. Such record keeping is now standard in most countries, but there is no way of knowing at present whether these data are of an acceptable quality.

21.3.7 Public accountability

Members of the general public in many countries are concerned about the treatment and use of laboratory animals, and this concern is expressed through animal welfare organizations, political activity and in many other ways, not all of them either peaceful or legal. One way of serving the legitimate public interest is by the collection and publishing of statistical information showing the numbers and kinds of animals used in particular procedures and for what purposes. The publication of such information is required by both the Convention and the Directive, but the current situation in Europe is not satisfactory (Straughan, 1994) and the position is worse elsewhere. Steps have now been taken to remedy the problems in Europe.

When such data are published, they should be accompanied by estimates of their completeness and likely accuracy. This rarely happens, though the data collected in The Netherlands and in the UK are considered to be as accurate as is possible. As noted earlier, there are substantial gaps in US data, because federal law does not cover the most commonly used species.

An important aspect of accountability is for the general public to be able to find out in some detail, not only what scientific procedures were authorized on their behalf, but the reasons for such authorization. The USA and Sweden have a strong commitment to Freedom of Information, but this is sadly lacking in many countries, including the UK. Confidentiality (commercial and academic) is a genuine difficulty on occasion, but in these instances public access after a modest delay, say of 5 years, might solve the problem.

21.4 THE EFFECTIVENESS OF CONTROLS

It is not easy to measure the effectiveness of any system of controls. The usual approach is to look at the failures, rather than the successes (which are presumed to be the majority). Some figures on individual breaches of the controls are available, but these reveal very little. There are no measures of the procedures which fail to give a result for predictable and avoidable reasons. There are no measures for other 'failures', such as the use of excessive numbers of animals in badly designed procedures or projects, or the causation of avoidable suffering. On-site monitoring and detailed inspection of records may reveal some of the failures and are important qualitative approaches. An alternative approach, used by some animal welfare groups, is to look at the work which is published and therefore in the public domain, and then to list and discuss those publications which cause concern. A good example of this approach is the RSPCA/FRAME report on the use of non-human primates in Great Britain (Hampson *et al.*, 1990). This is an acceptable approach to the effectiveness of controls, but it is imperfect, because much industrial work is never published, scientists do not publish failed experiments, and work submitted to journals but rejected may never be known. When 'moles' reveal unsatisfactory occurrences within scientific establishments, one is tempted to ask whether and when these would have been discovered and corrected by the normal monitoring processes and inspections.

21.5 CONCLUSIONS

There are a variety of systems for controlling animal experiments in different countries. Whether the broad framework is statutory or voluntary, and whatever the formal balance between local and national controls, the actual implementation of any system of controls ultimately depends on responsible attitudes and effective cooperation from the scientists concerned and from their institutions. Some nations and institutions work to the highest standards and embody best practices – ideally, there would be consistency, so that all worked to the best standards. How this ideal could best be achieved and monitored is a matter for debate.

It seems very likely that the controls and/or their implementation are not satisfactory in some countries, but a firm conclusion on this must wait for a detailed assessment of the effectiveness of national controls in terms of an agreed set of parameters of best practice. Several nations have devised lists of experimental procedures which are unacceptable or which cause concern and require careful evaluation by ethical review (and other) committees. These should form the basis of international debate and, hopefully, agreement.

Most, if not all, countries assert a commitment to the Three Rs of Russell and Burch (1959). However, the optimal implementation of the Three Rs seems unlikely, where countries do not require any formal weighing of the benefits of animal use against the cost to the animals. In all countries, it should be a fundamental principle to insist that 'the more extensive the likely adverse effects and the greater the numbers of animals used, then the greater the justification needed'. There are also differences in the nature of the monitoring systems within Europe, particularly in the role and powers of independent inspectors. When compared with Europe, controls in the USA appear less than ideal, in that they do not consistently cover the most widely used species of experimental animals, federal establishments are immune from inspection, though not from the law, and REAC inspectors are prohibited from 'interfering' with ongoing experiments.

In our view, animal procedures within university (college) education courses and in professional training should always be subject to the same controls and close scrutiny as are applied to any research proposal.

Further, we suggest that proposals to use animals for commercial and alleged legislative reasons, such as toxicology/safety studies, should also be reviewed and subject to proper scrutiny. Such scrutiny should include the nature of the requirement, and should ensure that details supplied in any standard operating procedure/protocol are satisfactory with respect to: (i) adequate experimental design; (ii) the use of the most ethically acceptable validated methods and humane end-points; and (iii) adequate working practices. Some of the tests currently performed for legislative purposes still do not require the use of the improved and validated tests accepted by the most progressive international and national guidelines to minimize suffering and/or to use the smallest number of animals. Such improved tests include: the fixed dose procedure instead of the LD_{50} test; the use of non-lethal end-point potency tests for diphtheria and pertussis toxin; and the use of minimal Draize eye tests as per the current OECD Guideline rather than the larger and more severe tests expected by the USA EPA.

It would be desirable to achieve a general commitment to provide comprehensive and reliable national statistics and greater freedom of information on scientific projects involving animals and on the work of ethical review committees and other project review systems.

Acknowledgements

We are very grateful to David Morton and Andrew Rowan for discussion and comment and to regulators, scientists and welfarists in many countries, particularly The Netherlands, Germany and the USA, for background material. Any mistakes or important omissions are our own, not theirs!

REFERENCES

Anon (1980) *Guide to the Care and Use of Experimental Animals*, Vol. I, Canadian Council on Animal Care, Ontario.

Anon (1986a) *European Convention for the Protection of Vertebrate Animals used for Experimental and other Scientific Purposes* (ETS123), Council of Europe, Strasbourg.

Anon (1986b) Council Directive 86/609/EEC of 24 November 1986 on the approximation of laws, regulations and administrative provisions of the Member States regarding the protection of animals used for experimental and other scientific purposes. *Official Journal of the European Communities*, **L358**, 1–29.

Anon (1986c) *Animals (Scientific Procedures) Act 1986*, HMSO, London.

Anon (1990) *Australian code of practice for the care and use of animals for scientific purposes*, Australian Government Publishing Service, Canberra.

Anon (1992a) *Animal Care Committees: Role and Responsibilities*, Canadian Council on Animal Care, Ottawa.

Anon (1992b) *Institutional Animal Care and Use Committee Guidebook*, National Institutes of Health Publication No. 92-3415, Bethesda, MD.

Anon (1994) *Provisions and general recommendations relating to the use of animals for scientific purposes*, CFN:S Skriftserie Nr. 25, CFN Publications, Stockholm.

Balls, M. (1995) On the ethical and scientific need for assessing the potential benefit of research likely to cause suffering to protected animals, in *Proceedings of the World Congress on Alternatives and Animal Use in the Life Sciences* (eds A.M. Goldberg and L.F.M. Van Zutphen), Mary Ann Liebert, Inc.. New York.

Davis, C.C. (1990) Invertebrate experimentation: CCAC restrictions (letter). *Canadian Zoological Society Bulletin*, **21**, 18.

Festing, M.F.W. (1992) The scope for improving the design of laboratory animal experiments. *Laboratory Animals*, **26**, 256–67.

Hampson, J., Southee, J., Howell, D. and Balls, M. (1990) An RSPCA/FRAME survey on the use of non-human primates as laboratory animals in Great Britain, 1984–1988. *ATLA*, **17**, 335–400.

Orlans, F.B. (1993) *In the Name of Science; Issues in Responsible Animal Experimentation*, Oxford University Press, New York.

OTA (Office of Technology Assessment, Congress of the United States) (1988) *Alternatives to Animal Use in Research, Testing, and Education*, Marcel Dekker, Inc., New York.

Russell, W.M.S. and Burch, R.L. (1959) *The Principles of Humane Experimental Technique*, Methuen, London [Facsimile edition (1992) UFAW, Potters Bar, Herts.]

Smith, J.A. and Boyd, K.M. (1991) *Lives in the Balance: The Ethics of using Animals in Biomedical Research*, Oxford University Press, Oxford.

Spielmann, H. (1995) Legal aspects in Germany, in *Alternatives to Animal Experimentation* (eds M. Cervinka and M. Balls), Nuclear HK, Prague, pp. 19–22.

Straughan, D.W. (1994) First European Commission report on statistics of animal use. *ATLA*, **22**, 289–92.

van Zutphen, L.F.M. and van der Valk, J.B.F. (1995) Education and training: a basis for the introduction of the Three Rs alternatives into animal research. *ATLA*, **23**, 123–7.

Zegers, L. (1989) Animal experimentation: a government's responsibility? In *Animal Experimentation: Legislation and Education, Proceedings of the EC Workshop* (eds. L.F.M. van Zutphen, H. Rozemond and A.C. Beynen), Department of Laboratory Animal Science, Utrecht, pp. 13–19.

Part Five

Concluding essay

22

Ethics, society and policy: a way forward

Michael Banner

22.1 INTRODUCTION

The recent *Report of the Committee to Consider the Ethical Implications of Emerging Technologies in the Breeding of Farm Animals* (hereafter 'the Banner Report') seeks to address the questions raised by the application to animals of the broad range of techniques of biotechnology.

Such a sentence, for all its seeming innocence as a factual description of the Report, will doubtless trouble the philosophically minded readers of this volume of essays. Those who have been brought up *After Virtue*,[1] so to speak, can hardly fail to be aware that amidst the moral fragments of Western thought no questions about biotechnology are straightforwardly 'the' questions about biotechnology. MacIntyre's sequel to *After Virtue* asks *Whose Justice? Which Rationality?*, and here we might echo the point by asking 'whose questions?' The heirs of Aristotle, Thomas Aquinas and Descartes will, for example, with their different views of animals, ask different sets of questions of the new technology.

The Banner Report can be distinguished from certain related reports, so I shall argue, in that it shows itself to be aware of what we might term the embeddedness of questions in metaphysical and moral theories. Naturally, it takes a view as to what questions are raised by biotechnology as regards, for example, the status and protection of

[1] A. MacIntyre's *After Virtue* has been an immensely influential text in moral philosophy, maintaining that the intractability of moral argument in our society is to be explained by the post-Enlightenment rejection of traditions of thought in which moral disagreements might have been capable of resolution.

animals, but it does so quite consciously and with a degree of argument appropriate if not for philosophers, then at least for Government Ministers to whom the Report is addressed. To be specific, the Report challenges the tendency to assess the new technologies solely in terms of questions of risk and benefit, and contends that this tendency uncritically privileges a particular philosophical position. Instead, it proposes a policy and system of moral evaluation which allows and requires questions of a different sort. This system of moral evaluation, so I shall conclude, calls into doubt the current pattern of regulation of the use of animals, and provides a basis for its reform.

In this chapter I shall focus on genetic manipulation as the technology which has been at the centre of controversy. In the first place I shall illustrate the rather unselfconscious tendency in certain recent reports dealing with genetic engineering to presuppose philosophical theories which suppress particular questions. The reports I shall mention are not concerned, or not concerned exclusively, with genetic modification of animals, but this does not matter for my purpose, which is to demonstrate how a certain narrowing of debate in bioethics occurs as a result of often undeclared, and certainly undefended, philosophical presuppositions. I shall then set out the principles which the Banner Report enunciates and on which it is based. I shall contend that these principles provide a more satisfactory framework for the regulation of the new technology than the framework used in these other reports, and shall briefly consider the significance of these principles, in the context of current UK legislation, for the practice of genetic engineering in relation to animals.

22.2 RISKS, BENEFITS AND PRUDENCE

The knowledge which is yielded by the recent advances in genetics renders possible essentially three projects of indirect or direct genetic engineering. In the first place, it allows for the more effective selection of progeny according to genotype. This can be thought of as *indirect* genetic engineering, since it does not involve the direct alteration of a gene but does affect the shaping of genetic inheritance. The possibility of effective selection may, then, depend merely on a breeding programme which makes use of knowledge of the genetic character of the 'parents', where this knowledge allows choices to be made for and against the occurrence of certain genotypes; with the growing understanding of the links between particular genes and traits, this knowledge will yield an increasingly powerful means of predicting and determining the characteristics of progeny in plant and animal breeding. In the second place, *selection* for specific genotypes may occur after, rather than before, breeding has taken place. Again, this does not involve direct genetic engineering, but rather the use of knowledge of the genotype of progeny to allow choices to be made for or against particular types through cloning, through the implantation only of certain embryos where there has been *in vitro* fertilization, or through the abortion of unwanted offspring after *in utero* testing. In the third place, *direct* genetic engineering (i.e. that which involves altering genes as opposed merely to selecting them on the basis of knowledge of actual or likely genotype) may occur either through manipulation of somatic cells with a view to changing the characteristic expression of an organism's genes or, more radically, through the manipulation of the germline itself.

If the science which has put into our hands these various possibilities is dazzling in its dogged rigour and sophistication, the same cannot be said of much of the moral reflection which has followed the scientific developments, declaring the newly feasible advancing of certain ends, or the means by which they are pursued, to be either acceptable or unacceptable. The moral reflection found, for example, in three recent reports is, for the most part, *ad hoc*, unsystematic and superficial, to such an extent that one is forced to say of these reports not simply that the answers they give are problematic, but that their very framing of the questions to be treated is itself questionable. Specifically – and they have this in common – they demonstrate a tendency to suppose that an ethical analysis of the new technology is exhausted by a prudential consideration of its potential risks and benefits, when this supposition in fact serves to conceal from view many ethical concerns.[2]

The *Report of the Committee on the Ethics of Gene Therapy* (hereafter 'the Clothier Report') is remarkable for its very limited engagement with the questions raised by the practice of genetic manipulation in humans. This limitation is two-fold, having to do with both the scope and the depth of the discussion.

[2] One suspects that this view of the matter would be more widely and vigorously expressed – or at least, more sympathetically entertained – were it not for the fact that the consensus which seems to emerge from the supposed moral discussion of these issues in these three reports is not one to which the pragmatically minded research scientist or medical practitioner is likely to object. As a matter of fact, this consensus proposes what might well be regarded, from this point of view, as a satisfactory division of the spoils – that is to say, it makes some concessions to those who would restrain or limit the practice of genetic engineering, but not such concessions as might seriously hamper either the advance or application of research. This consensus could be broadly characterized by its countenancing widespread genetic screening and abortions within certain limits, somatic cell therapy subject to the usual principles governing experimental procedures, and, for non-human organisms, genetic manipulation even of the germline where any harms associated with their production or release of such organisms can be regarded as warranted by the expected gains. There are things which, at least for the present, this consensus does not countenance – most obviously the manipulation of the germline in humans. But since the technicalities of germline manipulation are considerable, the constraint can hardly be thought unduly onerous.

It might be thought that a perfectly satisfactory explanation of the scope of the Clothier Report can be given by noting its terms of reference. These were:

> 'To draw up ethical guidelines for the medical profession on treatment of genetic disorders in adults and children by genetic modification of human body cells; to invite and consider proposals from doctors wishing to use such treatment on individual patients; and to provide advice to United Kingdom Health Ministers on scientific and medical developments which bear on the safety and efficacy of human gene modification.'
>
> Clothier (1992: 1.3)

Since the Committee was asked for advice on the 'treatment of genetic disorders. . . by genetic modification of body cells', the fact that the Report has next to nothing to say on the use of genetic manipulation with a view to enhancement of certain traits or on germline therapy is perhaps unsurprising. But as the Report itself makes clear (Clothier, 1992: 1.7), the Committee did not judge itself to be strictly bound by these terms of reference, but free, in fact, to go beyond them. However, having declared itself free to consider either of these topics, the Committee discusses neither, and limits its discussion to somatic cell therapy, saying that, given the novelty of the techniques involved, this therapy should be regarded as a form of research involving human subjects, and as such bound by the regulations which govern experimentation. On the other matters it merely declares, first, that 'In the current state of knowledge it would not be acceptable to use gene modification to attempt to change human traits not associated with disease,' and second, in relation to germline manipulation, that 'there is at present insufficient knowledge to evaluate the risks to future generations' (Clothier, 1992: 4.22 and 5.1).

As well as being limited in scope, the Report is limited in regard to the depth and seriousness of its discussion. Central to its recommendations is that whereas somatic cell therapy to alleviate a disease should be permitted on the same basis as other medical research, such research should not be allowed into the enhancement of human traits. But the distinction between the two is not without difficulties. In the first place, the distinction is just not that easy to make since the concept of disease, which is crucial to the distinction, needs considerable analysis. Without this analysis, one is left in doubt as to what exactly is permitted and what is not. Suppose, however, that a distinction between enhancement and treatment is satisfactorily drawn. There is, in the second place, a problem with the moral weight it is supposed to bear. Why should research into the enhancement of human traits be forbidden, notwithstanding that it could be subject to the very same conditions as regards consent, feasibility, and so on, which govern research into the alleviation of disease? At one point in the Report it is declared, as we have seen, that the attempt to modify human traits by genetic manipulation would not be acceptable 'in the current state of knowledge'. This suggests, although it does not explain, a pragmatic and rather limited objection having to do with the risks which might be involved in such a project, and seems to have nothing to do with 'the profound ethical issues that would arise were the aim of gene modification ever to be directed to the enhancement of normal human traits', to which issues the Committee declares itself to be 'alert' (Clothier, 1992: 2.16). What are 'these profound ethical issues' and would they arise where the normal human trait to be enhanced was, let us say, immunity to viral infection? The Report has nothing to say in answer to such manifestly pertinent questions.

If we ignore for the moment the seemingly stray reference to 'profound ethical issues', the Clothier Report illustrates very well the tendency to which I have referred, to consider the new technology almost solely in terms of categories of prudence, benefit and risk. It is prudent that somatic cell therapy, given the risks which may be involved in any novel treatment, should be governed by the protocols which relate to experimentation. Somatic cell enhancement or manipulation of the germline, for whatever end, are, however, held not to be prudent 'in the current state of knowledge'. The dominance of the vocabulary of prudence and risk is, however, even more evident in two other reports I shall mention. And here the consequences of that dominance become absolutely clear – the utter exclusion of the possibility of certain moral questions from a debate which has been framed (and one uses the word in its pejorative sense) in this way.

According to the report from a working party set up by the BMA, *Our Genetic Future: The Science and Ethics of Genetic Engineering*, 'biotechnology and genetic modification are in themselves morally neutral. It is the uses to which they are put which create dilemmas. The challenge which faces us is to try to achieve an optimal future: one which maximizes the benefits of genetic modification and minimizes the harms' (BMA, 1992, p. 4). This conception of the 'challenge which faces us' in virtue of our ever-expanding understanding of genetics is essentially the same as that which is found in a book written by J.R.S. Fincham and J.R. Ravetz in collaboration with a working party of the Council for Science and Society, entitled *Genetically Engineered Organisms: Benefits and Risks*. They too think the challenge, as the title indicates, is that of maximizing benefits and minimizing risks. Given this conception, it is hardly surprising that, like the BMA working party, Fincham and Ravetz find the chief difficulty for the application of genetic understanding to lie in our lack of a complete knowledge of the consequences of our interventions and manipulations. It follows then that, in principle, the dilemmas in this field – to which the BMA Report refers – are really the familiar difficulties which arise when we must act with imperfect empirical information. As the BMA Report puts it:

> 'Our predicament would be a great deal simpler if we could simply ask, in respect of any proposed development, questions such as "Is it safe?", "Will it enable us to find cures for genetic diseases?", and "How much will it cost?", and receive unequivocal answers. Unfortunately life is not so straightforward. The totality of scientific knowledge which we should like to have when making judgements about the future is rarely available. Consequently, the judgements which need to be made, and the decisions which need to be taken, are complex, contestable, and often incomplete. Until we have answers to these questions it is not possible to form settled views about the acceptability of some developments.'
>
> BMA (1992, p. 5)

The characterization of the 'challenge' which faces us in coping with the new biotechnology (as that of maximizing the benefits and of minimizing the harms) and of our 'predicament' (as that of doing so in a situation of imperfect knowledge), shared by these two reports, has an air of common sense about it. Nonetheless, we should reject it as begging important moral questions. Specifically, it will be argued that the tendency of both reports to locate the contestability of decisions about the application of genetic knowledge in the empirical realm, and in particular in the realm of uncertain futures, betrays an unconscious (or at least undeclared) commitment to a highly questionable moral framework.

Two things must, however, be said by way of clarification, for it is not being contended that the considerations which dominate these reports have no place in the discussion. First of all, there can be no doubt that the determination of the likely outcomes of alternative courses of action represents an element, and an important element, in many or most moral decisions. It follows, of course, that any uncertainty about these outcomes may be at the root of controversy and disagreement. Thus, to take an obvious example, a dispute as to whether capital punishment really serves as a deterrent may underlie a difference of opinion as to whether or not it should be permitted. Similarly, a dispute as to whether a genetically engineered organism will or will not affect the ecology of the environment into which it is proposed to release it may be the source of a disagreement between those who advocate its release and those who do not. But such disagreements would be essentially non-moral and trivial in comparison with the deep ethical disagreements which may occur in practical reasoning – a disagreement here, perhaps, over whether the state has a right to take life, or whether the natural ecology of an area is as such deserving of protection from human interference or change.

In the second place it may also be that a consideration of likely outcomes is the only issue at stake in certain decisions as to the application of biotechnology. It might be agreed, for example, that the release of a genetically engineered plant should be permitted only if it is prudent. Now there may be disagreement over whether that release is indeed prudent – a disagreement which may in fact be complicated by a dispute as to what constitutes a risk, or a risk worth taking (these last two being quite plainly ethical, not empirical issues). But supposing that the disagreement is not so complicated, the question of prudence may well be the only one which arises – as it may well be the only question which arises in connection with many instances of the introduction of a new technology. But if, for the sake of argument, we suppose that a straightforwardly empirical disagreement between those with a common prudential concern for genetic diversity is the only disagreement likely to arise in relation to genetic engineering in plants, it can hardly be supposed that this circumstance is characteristic of all the questions which the new genetics will cause to arise.

Take, for example, the controversy which might be expected to surround the application of genetic knowledge, by whatever means, to broadly eugenic ends. According to the BMA report:

> 'Using the science of genetic modification to produce a "master race", or to select children with particular attributes, is unacceptable. Even if parents are entirely free to reproduce as they choose, considerable social and ethical problems could arise if we eventually reach the currently remote possibility of being able to choose not just the gender but also some of the physical, emotional, and intellectual attributes of our children. If it became commonplace, for example, for parents to choose a boy as their first child, then this might well make it even harder to diminish sexual discrimination in our society.'
>
> BMA (1992, p. 209)

The reasoning here is uncertain, but the opposition seems to focus on the practice's undesirable consequences. Similarly, Fincham and Ravetz are inclined to portray possible controversies in these terms. Thus, having concluded that for those who do not object to antenatal diagnosis on the basis of 'religious principles', the practice 'would seem to offer only benefit', they take note of the fact that 'even some who are not absolutely opposed to abortion may still be worried by the possibility that, given too much information about the unborn child, some people might resort to abortion for reasons unconnected with predicted handicap', and instead having to do only with a preference for male over female children. They comment that:

> 'Most people would probably consider aborting a fetus or discarding an embryo because of its gender far less defensible than doing so because it was destined to suffer from a crippling handicap. Apart from anything else, to give free rein to parental choice in this matter might jeopardize the approximate numerical equality of the sexes, which it is clearly desirable to maintain.'
>
> Fincham and Ravetz (1991, p. 117)

In fairness, we should note the 'apart from anything else' by which Fincham and Ravetz qualify the giving of this particular reason for opposing abortion on the basis of sex, since it at least leaves open the possibility that this 'anything else' would reveal the authors to have greater insight than they here display. For in entertaining an objection to this practice based solely on a prediction of future consequences, Fincham and Ravetz, along with the BMA Report in its consideration of selection in general, betray the distinctly odd standpoint from which they approach these matters. For suppose we were persuaded, contra the BMA, that, as a matter of fact, sex selection would cause the diminution of sexual discrimination. Or suppose, contra Fincham and Ravetz, we took the view that numerical inequality between the sexes was desirable – it would, after all, give a certain power to the sex about whom the BMA is so anxious. Are we to suppose further that all but 'religious' objections to sex selection would melt away in the light of the desirability of the envisaged outcome and thus that any controversy about this practice must reside in differences over the likely consequences of this intervention? Could it not be that someone might hold the selection of children (by sex or any other characteristic) to be wrong in and of itself as a simple refusal of the fact that children are gifts and not acquisitions?

The inclination of these two reports to think of our predicament in making use of genetic knowledge as to do with the ensuring of good outcomes in circumstances of empirical uncertainty is a way of thinking which seems to neglect or exclude the possibility of genuine and deep moral disagreements over and above the prudential consideration of risk; that is to say, the conceptualization of the challenge presented by the new technology as being that of balancing benefits and harms converts what were thought to be ethical, into essentially empirical, disputes. It thus rules out the very sort of disagreement one would anticipate in the matter, for example, of sex selection (regardless of the method of selection), and in relation to genetic engineering too – a disagreement which centres on the question as to whether these practices are objectionable not in virtue of the balance of consequences, but in virtue of particular effects or even simply as such. Is genetic engineering in relation to animals, for example, wrong as fundamentally disrespectful, or as causing them harms to which they ought not to be subject no matter what the benefit?

But from where does the way of thought and conceptualization which dominates these reports come? If its only warrant were from common sense one might happily pass by with a warning against following such guides in these matters. But one suspects a certain indebtedness in these reports, albeit indirect, to consequentialism, which continues to exert an undue influence in the field of bioethics.

Consequentialism holds that actions are good or bad, right or wrong, solely in virtue of their consequences and more specifically that a good action is one which, in the given circumstances, will maximize overall benefits. Specifying or defining a benefit (and, of course, a harm, burden or cost) becomes all-important for such a moral theory, and spawns a variety of consequentialist approaches to the resolution of ethical questions. (One might note, then, how surprising it is that both the BMA Report and Fincham and Ravetz treat the notions of burden and benefit as if they did not require careful analysis but are intuitively plain and in order.) But no matter what the variety, consequentialist theories have this in common, namely that they think of all moral problems as in reality problems about the maximization of a good. This way of thought thus transforms moral dilemmas into problems of a more familiar sort, for the difficulties involved in moral decisions become the difficulties of calculation which are involved in making prudent choices of all sorts, moral or not: Shall I buy a dishwasher or a freezer? Would it be better to go to York by train or car? This transformation is not, however, without its own problems, for it seems to overlook the distinctive quality and character of genuine moral dilemmas. When we are confronted by such dilemmas, we have a sense that the difficulty is not simply one of calculating the probabilities of certain outcomes and comparing their desirability. It is rather the difficulty which occurs, for example, when a particular and worthy end can be secured only by a seemingly prohibited means, or when one recognizes the claim made by two (or more) obligations where the fulfilment of one necessarily entails reneging on the other or others.

Of course, consequentialism contests the existence of such dilemmas – a good action just is that action which maximizes beneficial consequences, so that a sense of difficulty which does not derive from empirical uncertainty can stem only from confusion, supposing, that is, that the notion of benefit is not itself contested. But consequentialism is itself a controversial moral theory, and one ought not to accept, without a good deal of argument, its understanding of moral decisions. Certainly one ought not to allow both the BMA Report and Fincham and Ravetz to foist upon us the seemingly innocent picture of our challenging predicament as that of balancing benefits and burdens in a state of imperfect knowledge, without an awareness of the underlying commitments from which it stems. As it is, we can say of these reports, before we even consider their answers, that they are unsatisfactory in the very questions they do (or more to the point, do not) pose. Specifically, for no reason which can be found in either report, they decline to ask questions about these practices which society at large is asking – such as whether they are intrinsically objectionable and impermissible.

22.3 INFLICTING HARM: A MATTER OF PRINCIPLE

The underlying and undeclared philosophical commitment of these recent reports prevents them from posing the questions about genetic engineering which are raised by those who do not share these very particular commitments. But if, as we said at the outset, there is no neutral account to be given of the questions to be put to biotechnology, how is society to proceed in forming public policy? More immediately, how is a committee charged with advising Ministers to proceed in framing its advice?

The Banner Report notes, on the basis of the many submissions it received, that:

> '...though they may not use this language, many people have intrinsic objections to the use of the emerging technologies. They may well be concerned about the effect of these technologies on animal welfare, genetic diversity, the environment, the pattern of farming and rural life, etc., but their concerns would not be exhausted by a consideration of these matters. For as well as worrying about the effects of the new technology, they feel a distinct unease about its very use.'
>
> Banner (1995: 3.3)

Furthermore, even some of those who did not express what could be described as an intrinsic objection expressed objections to various, particular perceived consequences of the application of biotechnology which were clearly not ones which they would reckon could be outweighed by associated benefits – thus many of the objections having to do with animal welfare. Now even had the Committee been persuaded of the philosophical merits of consequentialism (which it was not), it would be difficult to see how a workable and acceptable public policy could be established which, on the basis of such a philosophy, would simply refuse to entertain intrinsic objections and, in addition, would maintain that any harm could, in principle, be outweighed by sufficient goods. Thus, without feeling the need for sustained engagement with the philosophical problems of consequentialism, the Banner Report sets out a policy based on a quite different set of assumptions.

These assumptions are contained, in effect, in the three principles the Report enunciates, and which provide the basis for its proposals for the direction of future practice. The three principles are as follows:

1. Harms of a certain degree and kind ought under no circumstances to be inflicted on an animal – a principle which provides the rationale for various current prohibitions relating, for example, to non-therapeutic operations on farm animals such as tongue amputation in calves, tail-docking in cattle, and so on.
2. Any harm to an animal, even if not absolutely impermissible, nonetheless requires justification and must be outweighed by the good which is realistically sought in so treating it. This is the principle which underlies, in effect, the Animals (Scientific Procedures) Act 1986, which sets out to ensure that animals are used in experimental work only where the end result of the experiment can reasonably be expected to be commensurate with the harm which the animal is likely to suffer.
3. Any harm which is not absolutely prohibited by the first principle, and is in the particular circumstances considered justified in the light of the second, ought, however, to be minimized as far as is reasonably possible. This principle, which says that all reasonable steps should be taken to minimize the harm caused even by procedures which are justified when tested against the first two principles, gives rise to regulations such as the Bovine Embryo Collection and Transfer Regulations 1993, which seek to ensure good practice in regard to those particular techniques.

Well, it might be said, if these principles already underlie the existing animal legislation what is the point in enunciating them and, indeed, proposing them as a framework in which to evaluate the use of animals? The point is, however, that though these principles can be found to underlie aspects of the contemporary regulatory regime they do so in a rather haphazard fashion. That is to say, though they surface from time to time in various pieces of legislation, codes and so on, it cannot be said that they are applied systematically and rigorously across the board to the whole range of issues which ought to be governed by such principles. Indeed, very broadly speaking we could characterize the present state of affairs by saying that the first principle (the straightforwardly prohibitive one) has some (but limited) application in relation to farm animals, whereas the second principle (the cost–benefit one) is (though, not exclusively) significant in relation to laboratory animals. The importance of this point is that whereas the cost–benefit analysis tends to allow a good deal of research which might be questionable if the prohibitive principle applied, the prohibitive principle tends to allow certain practices which might be questionable in relation to the cost–benefit analysis. Put it another way – on the farm, animals lack the protection of a form of cost–benefit test; in the lab, they lack the protection of something like the test contained in the first principle.

The present situation has an unmistakable hint, therefore, of a strategy of divide and rule – the different principles give rise, within the domains in which they operate, to various provisions which doubtless do much to prevent harm to animals, but by being divided, they do far less than they otherwise would. It is not, of course, that one would wish away either of these principles, or the provisions which they warrant. What the framework recommended by the Report requires, however, is that these principles, together with the third, should apply consistently in all cases.

What is the significance of this proposed framework for the regulation of genetic engineering? Are the harms it threatens to do to animals ones which ought to be ruled out straightaway under the first principle; or, even if they are not of that kind, are they such that the goods which are sought in practising genetic engineering are insufficient to outweigh them?

If things were black and white it would, of course, be a lot easier. If the effects of genetic modification were invariably deleterious for an animal's welfare, or if the very use of genetic modification expressed a contempt for animals and a disregard for their natural characteristics (as some objections to genetic engineering maintain), a system of regulation might be devised which would aim to prevent all genetic modification. (And if one believed that about genetic engineering, one would not be deterred in advocating its prohibition by the argument which is so often trotted out, that we ought not to take unilateral action on this or any other welfare question, since to do so would simply be to export our welfare problems, since farmers on the continent will now produce for the British market what British farmers are prohibited from doing. Doubtless the same arguments could have been used 150 years ago by those who opposed legislation to prohibit child labour, and they would have been as unpersuasive then as they are now.) If we were persuaded that genetic engineering was seriously and invariably deleterious, there would be no difficulty in principle – we could prohibit it. If, on the other hand, genetic modification were invariably neutral in relation to an animal's welfare, or conducive to it, or could never be used in ways which are fundamentally objectionable, then no regulations would be needed.

As it is, however, genetic modification cannot be regarded as a single moral entity – some genetic modification may be intrinsically objectionable as manipulative of an animal's good, some not; some may be neutral in relation to an animal's welfare, some may actually result in an improvement, and some may do severe harm. Transgenesis may give us, that is to say, the notoriously deformed Beltsville pigs (modified with a human growth hormone), or it may give us the *seemingly* (and I stress the word seemingly) unaffected Edinburgh sheep (producing a human protein in their milk as a result of the incorporation of a human gene). The trick in any system of regulation, then, is to discriminate between the acceptable and unacceptable uses of genetic engineering. Does the present regulatory regime do that?

The key element in the present regulation of genetic engineering in relation to animals is the Animals (Scientific Procedures) Act 1986 (hereafter 'ASPA'). This, as already mentioned, has at its heart the requirement that experimentation should only be undertaken if 'the likely adverse effects on the animal concerned' are outweighed by 'the benefit likely to accrue as a result of the programme' (ASPA, 5(4)). It is obvious that, in any case, this balancing is a highly intuitive matter. We do not have a scale which tells us how to compare unmeasurable quantities of animal suffering against unmeasurable quantities of human well-being. But here there is a more mundane difficulty, of the type which the BMA Report and Fincham and Ravetz are inclined to think exclusively significant in this field. There is, that is to say, an empirical problem in making the required judgement – the effect of modifications is highly resistant to prediction, and may in any case emerge not immediately, but only as homozygous offspring are produced, or as animals are subjected, outside laboratory conditions, to the varied conditions and pressures of contemporary systems of production. Thus, though the Secretary of State must weigh the likely adverse effects on the animal against any likely benefit, the present inability of researchers to predict or control the expression of a gene makes such a task more than problematic, and so renders uncertain the effectiveness of this provision in preventing harm to animals.

It is clear, however, that in the light of the principles it adopts, the Banner Report must judge the cost–benefit test to be insufficient, on its own, whether or not such a test has these further problems of application. The Report maintains that there are certain things which might be done to animals which are impermissible no matter the good which may result. But judgement on the ASPA must be reserved, since there is a further provision which, in effect, introduces a more obviously deontological element into the picture, at least as regards genetic modification. Section 15(1) requires that any protected animal which, at the end of the regulated procedures, 'is suffering or likely to suffer adverse effects' must be killed and cannot be released from the Act's control. There is no qualification on this requirement – the Secretary of State cannot consider, in other words, whether the benefit of goods over harms would warrant the continued suffering of an animal. This means that even were the Secretary of State to grant a licence for an experiment which produced transgenic animals whose welfare was seriously impaired, the harm done to the modified animals would prevent their being released from the control of the Act and passing into commercial conditions. And since the purpose of any modification is to produce a transgenic line, this provision effectively blocks the achievement of such a purpose where there are 'adverse effects'.

As the Banner Report notes, the interpretation of the phrase 'adverse effects' is a crucial matter in judging the significance of these provisions. The Report recommends, therefore, that the Animal Procedures Committee (the body charged with advising the Secretary of State on the operation of ASPA) be invited to give an account of this phrase, and of the methods by which the existence of adverse effects are established. In making this recommendation it had in mind two principal concerns. In the first place, that 'adverse effects' may not be construed sufficiently widely to include within its scope what might be deemed intrinsically objectionable modifications, and in the second that the matter of the existence of such effects may not be the subject of an appropriately rigorous and wide-ranging review of an animal's condition. Until there is clarity in relation to these issues, the adequacy of the Act in relation to the problems posed by genetic modification is not finally settled.

22.4 CONCLUSION

Prior to the requested clarifications being obtained, it is impossible to judge the adequacy of the current regulatory regime. In principle, however, ASPA provides the basis for quite stringent control of genetic engineering of animals. Indeed, the effect of section 15(1), depending on its interpretation and application, is to introduce what I have referred to as the prohibitive principle into a piece of legislation which has at its heart a principle which derives from a quite different approach to the problem of the treatment and use of animals – the approach favoured by those reports which I have criticized as implicitly begging some important moral questions.

If the Animal Procedures Committee can give the assurances the Banner Report seeks, then it could be said that as things stand, animals produced by transgenic modification are more thoroughly protected than animals produced by conventional means. Conventional breeding, and other techniques such as embryo transfer and artificial insemination (unless they are specifically regulated), could, intentionally or unintentionally, adversely affect the welfare of farm animals and would not be subject to the so-called 'cost–benefit' analysis which is required before the licensing of a proposed programme of genetic modification – thus, breeding programmes which have produced oversized turkeys with considerable welfare problems are not subject to regulation of this kind. While, then, an appropriate structure of regulation is in place in relation to genetic engineering, much remains to be done to ensure that anything like such adequate provisions apply to other, less newsworthy procedures.

It is in the light of these and other considerations that the Banner Report recommends the establishment by the Government of a standing committee with responsibilities which would include advising Ministers on the ethical questions relating to current and future developments in the use of animals. Such a committee would have a number of advantages, but in relation to the concerns of this chapter it recommends itself in these terms – that it would provide a forum in which the different approaches to the human use of animals, with their different questions to biotechnology, can, instead of being ignored, be acknowledged and elucidated, and on the basis of such elucidation, grounds sought on which to establish a workable policy for the future.

REFERENCES

Banner, M.C. (1995) *Report of the Committee to Consider the Ethical Implications of Emerging Technologies in the Breeding of Farm Animals*, HMSO, London.

BMA (1992) *Our Genetic Future: The Science and Ethics of Genetic Technology*, Oxford University Press, Oxford.

Clothier, C. (1992) *Report of the Committee on the Ethics of Gene Therapy*, HMSO, London.

Fincham, J.R.S. and Ravetz, J.R. (1991) *Genetically Engineered Organisms: Benefits and Risks*, Open University Press, Milton Keynes.

MacIntyre, A. (1981) *After Virtue*, Duckworth, London.

MacIntyre, A. (1988) *Whose Justice? Which Rationality?* Duckworth, London.

Index

Page numbers appearing in **bold** refer to figures and page numbers appearing in *italic* refer to tables